Study Guide to Accompany

MICROBIOLOGY

Fifth Edition

Study Guide to Accompany

MICROBIOLOGY

Fifth Edition

George A. Wistreich
East Los Angeles College

David Smith
University of Delaware

MACMILLAN PUBLISHING COMPANY
New York
COLLIER MACMILLAN PUBLISHERS
London

Printed in the United States of America

Macmillan Publishing Company
866 Third Avenue, New York, New York 10022

Collier Macmillan Canada, Inc.

ISBN: 0-02-428970-1

Printing: 1 2 3 4 5 6 7 8 Year: 8 9 0 1 2 3 4 5 6 7

How to Use the Study Guide

This study guide has been prepared to help you to *identify, learn,* and *remember* basic concepts, principles, and terminology of microbiology. Think of it as an extension of your textbook. Diligent use of the guide will provide considerable benefits to you. Undoubtedly, you will develop your own method of using the study guide; however, here are some suggestions that may help you to get the most from your efforts and time in studying.

1. First, read the appropriate chapter in the text, using the stated objectives as indications of what you should know to achieve mastery of the subject matter. Underlining, or highlighting with a suitable colored pen, all important information can help to emphasize basic facts, concepts, and terms.
2. Next, read the *Introduction* and *Preparation* sections of the corresponding study guide chapter. This will help you to visualize the overall organization and content of the chapter. The *Preparation* section for each chapter also contains phonetic pronunciations for selected microbiological and related terms. Taking the time to sound out new and familiar terms and to say them aloud several times will help you to learn and to use a specialized vocabulary.
3. Now go to the *Pretest* section of the guide and test your knowledge. Write the answers for each question in the spaces provided. Compare your responses with the answers given at the end of the section. Add up the number of correct answers. The purpose of the *Pretest* is to enable you to evaluate your understanding of the subject matter and to reinforce your mastery of concepts, principles, and terminology. If your score of correct answers is 75 percent or better, proceed to the *Chapter Self-test* and *Enrichment* sections. If your score is below 75 percent, proceed to the *Concepts and Terminology* section.
4. It could be helpful at this point to scan the *Preparation* section before continuing. If, however, you are short of time, read and study the *Concepts and Terminology* component. If certain terms or concepts prove difficult for you, underline them. Review underlined material after you have completed the section. The phonetic pronunciations of most microorganisms are given the first time they are mentioned. This aid should make learning the names of microorganisms easier.
5. Now you are ready to find out how well you have mastered the subject matter presented. If you have gone through the various sections of the study guide, you should not find the *Chapter Self-test* to be difficult. Each of these sections contains objective types of questions and short essay questions. This last component has been included to provide you with the experience necessary to function well in a variety of testing situations. Separate and distinct *Disease Challenges,* another means for you to apply the knowledge gained, also are included in several chapters of Part VIII. Answers to all questions appear at the end of the section.
6. The last section for most chapters—*Enrichment*—will emphasize or extend the discussions of some material presented in the chapter. The suggested readings, discussions, and challenge exercises presented here not only offer a diversion from the usual pattern of study, but should provide an opportunity to extend or apply the knowledge you have gained.
7. Each Part of the Study Guide ends with a *"Microbiology Trivia Pursuit."* This section challenges your attention to detail and gathering of facts.

Contents

PART I INTRODUCTION TO MICROBIOLOGY

1

The Scope of the Microbial World

I. INTRODUCTION

Microorganisms are essentially the same as other forms of life. Many are single celled, whereas others are loose combinations or long filaments of independent cells. These forms of life perform the same fundamental activities within their individual cells as do higher organisms within their many-celled structures. This chapter provides general descriptions of the various types of microorganisms and branches of microbiology. The significance and uses of microorganisms are outlined to give you an overview of the text.

Following the discovery of microorganisms, studies indicated that these microscopic forms of life figure in the production and spoilage of various foods and alcoholic beverages. Further discoveries and experiments showed that microbes are also responsible for various diseases. Findings such as these suggested that certain processes affecting human welfare have a biological explanation.

After reading this chapter, you will have some insight into the problems faced by early scientists in establishing the nature and functions of microorganisms and their relationship to fermentation, putrefaction, and infectious diseases. The bases for microbial control and the applications of microbes as research tools also are presented to broaden your perspective.

II. PREPARATION

Chapter 1 should be read before continuing. The following terms are important for you to know. Refer to the glossary, the index, and the appropriate chapters if you are uncertain of any of them. A pronunciation guide for selected terms is provided as a learning aid.

1. acidity
2. acquired immune deficiency syndrome (AIDS)
3. aerobe (AIR-robe)
4. agar (AG-ar)
5. algology (al-GOL-oh-gee)
6. alkalinity
7. anaerobe (an-AIR-robe)
8. antibiotic (an-tie-by-OT-ik)
9. antibody (AN-tie-bod-ee)
10. antiseptic (an-tie-SEP-tik)
11. arthropod (AR-thro-pod)
12. autoclave (AW-toe-klave)
13. bacteria (back-TE-ree-ah)
14. biogeochemical
15. bubonic plague
16. catalyst (KAT-ah-list)

17. chemotherapy (key-mow-THER-ah-pee)
18. colony
19. contagion (kon-TAY-jun)
20. culture
21. deoxyribonucleic acid
22. diatom (DYE-a-tome)
23. ecology (ee-KOL-oh-jee)
24. enzyme
25. experimentation
26. fermentation (fur-men-TAY-shun)
27. filtration
28. fungus (FUN-gus)
29. genetics (je-NET-icks)
30. helminth
31. host
32. hypothesis (hi-POTH-ee-sis)
33. immunology (im-u-NOL-oh-jee)
34. infection (in-FEK-shun)
35. infusion
36. inoculation (in-nok-u-LAY-shun)
37. interferon (in-ter-FEER-on)
38. maggots (MAG-ots)
39. medium (ME-dee-um)
40. metabolism
41. microbiology (my-kro-bi-OL-oh-gee)
42. microorganism (my-kro-OR-gan-ism)
43. microscope
44. mold
45. molecular biology
46. morphology (more-FOL-oh-gee)
47. mycology (my-KOL-oh-gee)
48. parasite (PAIR-ah-site)
49. parasitology
50. pasteurization
51. pathogen (PATH-oh-jen)
52. *Penicillium* (pen-ee-sil-EE-um)
53. penicillin (pen-ee-SIL-in)
54. Petri dish (PET-ree)
55. postulates
56. protozoa (pro-toe-ZOH-ah)
57. protozoology (pro-toe-zoh-OL-oh-gee)
58. putrefaction (pu-tree-FAK-shun)
59. resolving power
60. simple microscope
61. smallpox
62. spontaneous generation
63. spore
64. sterile (STER-il)
65. streaking
66. taxonomy (taks-ON-oh-mee)
67. thermolabile (ther-moh-LAY-bil)
68. thermostable (ther-moh-STAY-bil)
69. turbid (TUR-bid)
70. tyndallization
71. ultrastructure
72. vaccine (VAK-seen)
73. virus (VYE-rus)
74. virology (VYE-rol-oh-gee)
75. yeast (yeest)

III. PRETEST

Correct answers to all questions can be found at the end of this section. Write your responses in the appropriate space provided.

Completion Questions

Provide the correct term or phrase for the following. Spelling counts.

1. Organisms capable of reproducing, metabolizing, and performing other activities in the absence of free oxygen are called ______________________.
2. ______________________ are completely dependent on the cells of higher forms of life.
3. ______________________ are organisms capable of growing in the absence of oxygen.
4. Naming branches or subdivisions of a scientific area of study according to the topics on which they focus is known as a(n) ______________________ approach.
5. Viruses are obligate ______________________ parasites.
6. One of the first scientists to recognize the true biological functions of microbes was ______________.
7. ______________________ is the study of algae.
8. ______________________ involves the study of molds and yeasts.
9. Diagnostic procedures, disease preventive methods, and the study of a host's defense mechanisms are all aspects of ______________________.
10. ______________________ are important to the production of several kinds of foods and commercially available antibiotics.
11. African sleeping sickness and malaria are examples of two diseases caused by ______________.
12. ______________________ are obligate intracellular parasites.
13. ______________________ is noted for the recognition of a submicroscopic nonbacterial infective agent.
14. ______________________ are disease-causing microorganisms.
15. Providing the validity of a hypothesis is done by ______________________.
16. A hypothesis that has been supported by various types of observations and experiments is called a(an) ______________________.
17. ______________________ provided the first descriptions of microorganisms.
18. The ______________________ microscope was the type used in these early observations.
19. The souring of milk is an example of ______________________.
20. ______________________ can be used to destroy microbes that cause wine spoilage without altering the quality of the wine.
21. The theory of ______________________ holds that life could and did appear spontaneously from nonliving or decomposing matter.

22. By a series of experiments, ______________________ showed that maggots and flies did not emerge spontaneously from putrefied meat.
23. The filtration of air or the exposure of air to chemicals such as sulfuric acid and sodium hydroxide will render it ______________________.
24. Pasteur's various experiments to disprove spontaneous generation showed that life in organic infusion could not occur without the existence of ______________________.
25. The heat-resistant bacterial structures found by Tyndall are known as ______________________.
26. The ______________________ incorporates steam under pressure for sterilization.
27. The process by which wound infections are prevented through the use of sterilized instruments and the application of carbolic acid to wounds is known as ______________________.
28. ______________________ were the first microorganisms shown to be pathogenic.
29. ______________________ can be used to show the direct role of a specific bacterium as the cause of a specific disease.
30. Preparations containing various nutrient combinations for the cultivation of microbes are known as ______________________.
31. The accumulations of bacteria on solid nutrient surfaces are called ______________________.
32. ______________________ is the solidifying agent commonly used for laboratory media.
33. ______________________ is noted for the isolation of penicillin.
34. ______________________ is concerned with explaining life processes in molecular terms.
35. The chemical nature of heredity material in bacteria was discovered to be (a) ______________________ by (b) ______________________, (c) ______________________, and (d) ______________________.

Matching

Select the answer from the right-hand side that corresponds to the term or phrase on the left-hand side of the question sheet. An answer may be used more than once. In some cases, more than one answer may be required.

_____ 36. Interactions between microbes and their environments

_____ 37. Cellular properties of microorganisms

_____ 38. Naming and cataloging microorganisms

_____ 39. Metabolic activities of microbes

_____ 40. The manufacture of fermented milk products

a. microbial morphology and ultrastructure

b. biophysics

c. microbial taxonomy

d. microbial genetics

e. medical microbiology

f. dairy microbiology

g. microbial physiology

h. none of these

Answers

1. anaerobes
2. viruses
3. asexual reproduction
4. taxonomic
5. intracellular
6. L. Pasteur
7. phycology or algology
8. mycology
9. immunology
10. fungi
11. protozoa
12. viruses
13. Iwanowski
14. pathogens
15. experimentation
16. theory
17. Anton van Leeuwenhoek
18. simple
19. fermentation
20. pasteurization
21. spontaneous generation
22. F. Redi
23. sterile
24. microbes
25. spores
26. autoclave
27. antiseptic surgery
28. fungi
29. Koch's postulates
30. media
31. colonies
32. agar
33. A. Fleming
34. molecular biology

35a. deoxyribonucleic acid (DNA)
35b. O. Avery
35c. C. McLeod
35d. M. McCarthy

36. h
37. a
38. c
39. g
40. f

IV. CONCEPTS AND TERMINOLOGY

The Microbial World

Microscopic forms of life are present in large numbers in nearly every environment, including soil, water, food, air, and surfaces of animals and plants. The majority of such microbes are not harmful to humans or to other forms of life. For the most part, microorganisms exhibit characteristic features common to biological systems. Several of these features, together with brief explanations, are listed in Table 1-1.

TABLE 1-1 Characteristics of Microorganisms

Characteristic	*Description*
Reproduction	The ability to duplicate or multiply. Many microorganisms are capable of reproducing both asexually (single cell division) and sexually (forming a genetically new individual from the union of nuclear material from two different cells).
Metabolism	The sum total of chemical reactions through which the energy needed for cellular activities is produced. Two categories of metabolism are recognized: *anabolism* (includes reactions involved with the formation of cell parts needed for growth, reproduction, and repair) and *catabolism* (includes reactions of digestion).
Growth	Most microorganisms increase in size as building materials are produced inside the cell—a process of growth from within.
Irritability	The ability to respond to environmental stimuli, such as acidity, intense light, temperature, and poisonous (toxic) substances.
Adaptability	Ability to adjust to environmental stimuli. Several microorganisms can survive unfavorable environments by altering certain of their activities.
Mutation	A permanent change in the genetic information of the microorganism; can be brought about either naturally or experimentally and is passed on to future generations.
Organization	Needed to perform the various activities essential to existence and survival. It is appropriate to refer to microbes as small, organized units of life, or *microorganisms*.

Microbiology and Its Subdivisions

Microbiology includes the study of bacteria, fungi (molds and yeasts), protozoa, certain microscopic forms of algae, and viruses. Table 1-2 lists and describes the major branches and specialty areas.

TABLE 1-2 Microbiology and Its Subdivisions

Subdivision	*Description*
Major Branch	
Bacteriology	The study of bacteria and their activities, which include causation of disease, decomposition of decaying or dead organic matter, and the production of various chemicals, foods, and other useful substances.
Immunology	A study of an individual's defense mechanisms against disease and materials foreign to the body. Involves determining the contributions of different body parts to this defense; also deals with diagnosis and development of new methods of disease detection and prevention.
Mycology	The study of fungi, which include molds, mushrooms, and yeasts. Activities include production of various types of food and antibiotics, and causation of disease.
Phycology (algology)	The study of algae, which include microscopic unicellular forms. Studies the activities of algae in organic matter decomposition and food production as well as harmful activities.
Protozoology	The study of protozoa, which can be found in various environments such as sewage, bodies of water, and damp soil, and can cause diseases such as malaria and African sleeping sickness.
Virology	The study of viruses, submicroscopic intracellular parasites that need living cells for their survival and activities.
Microbial morphology and ultrastructure	The study of microscopic and submicroscopic structural details of microorganisms.
Microbial ecology	The study of relationships between microorganisms and their environments (e.g., how microbes respond to unfavorable situations).
Microbial genetics	The study of the activities and functions of the nuclear elements of microorganisms. Genetic engineering recombinant DNA techniques, and investigations of how to regulate changes in growth and development of nuclear material brought on by mutation-causing agents are included in this specialty.
Microbial physiology	The study of microbial functioning, which includes metabolism, nutritional needs, and effects of environmental factors on essential microbial activities.
Specialty Areas	
Molecular biology	The principal aim of molecular biology is to determine the relationship between the chemical structure and genetic makeup of microbial and higher forms. While this specialty is not limited to microbiology, microorganisms have served as important tools in uncovering basic knowledge of all life forms (e.g., how genetic information flows from deoxyribonucleic acid to ribonucleic acid during the formation of proteins).
Microbial taxonomy	The naming and classification of microbes; involves determining similarities and differences among organisms, which serve as bases for classification and the demonstration of relationships.
Biochemistry	This specialty is concerned with the chemical basis of living matter and associated reactions. Determining the chemical composition of cells and how chemicals are formed and interact are just a few of the areas of investigation.
Biophysics	The study of the principles of physics as they apply to all living matter. Explores the basis of movement—how chemicals are combined and held together.

Applied Microbiology

The principles, basic information, and techniques of the different branches and specialties of microbiology can be applied to many areas. Examples are listed in Table 1-3.

TABLE 1-3 Examples of Applied Microbiology

Applied Area	*Description*
Food and dairy microbiology	The microbial conversion of raw materials into desirable end products. This process, known as fermentation, is responsible for the characteristic aroma, flavor, and general quality of foods such as yogurt, pickles, sauerkraut, and some cheeses.
Industrial microbiology and biotechnology	These subdisciplines, wherever applicable, are concerned with the use of microorganisms to produce economically important products and with the development of techniques used to prevent microbial destruction of economically important products formed by other means.
Medical microbiology	Includes studies dealing with the properties of microbial disease agents, developing methods for diagnosis and prevention of diseases, and incorporating both chemical and physical methods for the management and control of infectious diseases.
Veterinary microbiology	Concerns studies of disease agents affecting pets and livestock and the prevention, detection, and control of such agents. Worms and related parasites also are considered, as is the use of laboratory animals in studies concerned with disease processes.

Early Development of Microbiology

From 1673 to 1723, Anton van Leeuwenhoek designed and constructed one-lens (simple) microscopes to observe a variety of biological specimens, including single-celled microorganisms such as algae, bacteria, protozoa, and yeasts. Even though compound microscopes were in use at this time, Leeuwenhoek found that the lens of his instruments provided greater detail of *resolving power*. The study of microorganisms ceased after Leeuwenhoek's death, and was not resumed until the mid-1800s when Louis Pasteur demonstrated the biological functions of microorganisms.

The Germ Theory of Fermentation

Fermentation is recognized today as a natural process in which alcohols and organic acids such as ethanol and acetic and lactic acids result from the enzymatic action of microbes on substances containing sugar. Despite the discoveries of Leeuwenhoek, the biological basis of fermentation was not established until late into the nineteenth century. Two viewpoints were offered to explain the process: the nonvital (nonbiological) theory and the vital (biological) theory. According to the nonvital theory of fermentation, yeasts are by-products of fermentation. The vital theory of fermentation, which eventually proved to be true, held that yeasts such as *Saccharomyces cerevisiae* (sak-uh-raw-MY-seez sair-a-VIS-ee-eye) were responsible for the reaction.

Louis Pasteur made several contributions to fermentation research. He proved experimentally the microbial nature of fermentation and the specificity of fermentation reactions; he developed the heating process, called pasteurization, that kills most disease- and spoilage-causing organisms; and he discovered *anaerobes*, microorganisms that can live only in the absence of free oxygen.

The Spontaneous Generation, or Abiogenesis, Controversy

In the fourth century B.C., Aristotle proposed that lower forms of animal life arose spontaneously from nonliving or decomposing organic matter. This view became known as the doctrine of spontaneous generation. Disproving spontaneous generation at both the macroscopic and microscopic levels involved numerous individuals. Several of these scientists and their contributions are listed in Table 1-4.

TABLE 1-4 Contributors to Disproving Abiogenesis

Scientist(s)	*Contribution*	*Approximate Year*
F. Redi	Demonstrated that flies do not develop spontaneously from putrefied meat.	1665
L. Joblot L. Spallanzani	Showed independently that heating of infusions under controlled conditions prevents microbial growth.	1771 1765
T. Schwann	Showed that bacterial growth does not occur in nutrient-containing flasks when such materials are exposed to heated air.	1836
F. Schulze	Similar experiments to those of Schwann, but used air exposed to strong chemicals such as sulfuric acid and sodium hydroxide.	1836
H. Schröeder T. von Dusch	Showed that cotton serves as a filter when air is passed through it; introduced cotton plugs for bacteriological culture containers.	1854
L. Pasteur	Demonstrated that air free of microbes cannot create life in organic infusions; showed, with the aid of open swan-neck flasks, that microbial growth occurs only when nutrient solutions come in direct contact with microbes.	1862
J. Tyndall	Showed that dust particles carry microorganisms and that microbial growth occurs in sterile broth only after it has been introduced from an outside source. Demonstrated the existence of heat-resistant bacterial spores; developed a sterilization procedure known as *Tyndallization.*	1877

Antiseptic Surgery

Joseph Lister devised procedures to prevent microorganisms from entering wounds. His approach, which became known as aseptic surgery, includes sterilization of instruments and the application of chemicals to wounds. The concept of antiseptic surgery provided indirect support for the germ theory of disease.

The Germ Theory of Disease

The concept of infectious diseases preceded the proof of the existence of pathogens by several centuries. Fungi and protozoans were among the first microorganisms associated with disease. Robert Koch, from experiments with anthrax in 1876, was the first to demonstrate the role of bacteria as disease agents. Koch also developed a procedure by which the specific relation between a disease agent and a disease could be shown. The steps of this experimental approach, known as Koch's postulates, are:

1. Identify the suspected disease agent in all occurrences of the disease.
2. Isolate and cultivate the agent in pure culture.
3. Cause the specific disease by inoculating the suspected agent into a healthy susceptible animal (host).
4. Recover the suspected agent from the infected host.

If an appropriate animal model is not available, Koch's postulates cannot be achieved.

T. M. Rivers modified Koch's postulates to accommodate both animal and plant viruses. The distinguishing steps in this approach are:

1. The viral agent must be found in the host's body fluids at the time of the disease or in the cells showing specific lesions.
2. The isolated viral agent must produce the specific disease in a healthy animal or plant or provide evidence of infection by showing the presence of antibodies (immunoglobulins). Antibodies are proteins that are produced by the host's immune system against the viral agent.
3. Similar material from newly infected animals and plants must in turn be capable of transmitting the disease under study to other hosts.

Early Technical Achievements

Improved techniques for the handling and study of microorganisms included the development of various preparations containing combinations of nutrients, *media,* the introduction of an effective container for such media, the *Petri dish,* and the incorporation of *agar* as a solidifying agent for media.

Chemotherapy, the use of chemicals in treatment to control microbial diseases, was introduced by Paul Ehrlich. Other advances in treatment included the discovery and isolation of the *antibiotic* penicillin by A. Fleming in 1929. Since this discovery, numerous antibiotics have been developed. (Chapter 12 presents descriptions and discussions of a wide variety of chemicals used in treatment.)

Molecular Biology, an Alliance of Specialties

Microorganisms have become an invaluable tool in unraveling complex molecular processes of life. Among the important advances were the isolation of a number of genetically different forms (mutants) of the yeast *Neurospora* (noo-RAHS-paw-rah) by G. Beadle and E. Tatum in 1941 and the discovery of the chemical nature of genetic material to be deoxyribonucleic acid (DNA) by O. Avery, C. McLeod, and M. McCarthy in 1944. With these and several other discoveries, the specialty known as *molecular biology* made its appearance.

The Growth of Organized Microbiology

Microbiology has become a significant influence in our society within a relatively short period of time. As in most other important branches of the biological sciences, microbiologists have established national and international professional organizations and developed publications for the exchange of information and to maintain ethical standards.

Some Challenges for Microbiology

Despite the years of intense study of microorganisms and their activities, numerous questions remain unanswered. Many decisions affecting the future of world populations and environments—for example, food and energy production and control of diseases and pollution—may depend on microbial activities.

V. CHAPTER SELF-TEST

Continue with this section only after you have read Chapter 1 of the text and have completed Section IV. A score of 80 percent or better is good. If your score is less than 65, reread the chapter.

Correct answers to all questions can be found at the end of this section. Write your answers in the appropriate space provided.

Matching

Select the answer from the right-hand side that corresponds to the term or phrase on the left-hand side of the question sheet. An answer may be used more than once. In some cases, more than one answer may be required.

_______ 1. An organism's resistance to disease

_______ 2. Development of new methods of disease prevention and detection

_______ 3. The study of submicroscopic obligate intracellular parasites

a. mycology

b. virology

c. microbial ecology

d. immunology

e. food microbiology

_______ 4. The study of microscopic details of microbial cells

_______ 5. Naming and classifying micro-organisms

_______ 6. Blood typing

_______ 7. The study of the principles of physics as they apply to all living matter

_______ 8. Improvements of food quality

_______ 9. Preventing heavy economic livestock losses

_______ 10. The study of hookworms and tapeworms

f. genetic engineering

g. microbial physiology

h. biophysics

i. microbial taxonomy

j. microbial morphology and ultrastructure

k. food and dairy microbiology

l. parasitology

m. veterinary microbiology

n. biotechnology

o. none of these

Matching

Select the answer from the right-hand side that corresponds to the term or phrase on the left-hand side of the question sheet. An answer may be used more than once. In some cases, more than one answer may be required.

_______ 11. Was first to describe anaerobes

_______ 12. Supported the concept of spontaneous generation

_______ 13. Showed that maggots did not arise spontaneously from decaying meat

_______ 14. Demonstrated the role of yeasts in fermentation

_______ 15. Introduced the cotton plug into bacteriology

_______ 16. Is noted for the development and use of simple microscopes

_______ 17. Showed that superheated air does not give rise to microbial growth

_______ 18. Independently observed spores

_______ 19. Was first to describe viruses

_______ 20. Was one of the first to recognize the true biological functions of microbes

_______ 21. Isolated penicillin

_______ 22. Introduced the Petri dish

_______ 23. Found agar to be an excellent solidifying ingredient for media

_______ 24. Is noted for the introduction of chemotherapy

a. A. van Leeuwenhoek

b. L. Pasteur

c. J. Lister

d. T. Schwann

e. L. Joblot

f. J. Tyndall

g. F. Redi

h. J. Needham

i. T. von Dusch

j. T. Rivers

k. A. Fleming

l. Frau Hesse

m. R. J. Petri

n. Paul Ehrlich

o. none of these

Completion Questions

Provide the correct term or phrase for the following. Spelling counts.

25. Organisms capable of growing in the absence of free oxygen are called ______________.
26. ______________ are totally dependent for their survival on the cells of higher forms of life.
27. The fusion of nuclear material from two different cells resulting in a genetically new individual is known as ______________ reproduction.
28. The majority of organisms that require free oxygen for their essential activities are known as ______________.
29. The study of yeasts belongs in the subdivision of ______________.
30. *Penicillium* would be an example of a microorganism studied in the subdivision of ______________.
31. ______________ are major sources of antibiotics.
32. One example of a protozoan disease affecting humans is ______________.
33. Viruses that invade bacteria are called ______________.
34. The first viruses to be detected were found in infected ______________.
35. The control and possible elimination of certain human genetic defects through the transfer of genetic material from one type of organism to another is one aspect of the area known as ______________.
36. In the scientific method, proving a hypothesis to be correct is the basis of a(an) ______________.
37. For an experiment to support a hypothesis effectively, it must be (a) , ______________ and it must contain adequate (b) ______________.
38. ______________ serve as important means for the transmission of knowledge among areas of the scientific community.
39. Immunization was responsible for the elimination of the viral disease ______________.
40. AIDS stands for the viral disease ______________.
41. ______________ refers to the ability to provide greater detail.
42. ______________ is called the "father" of bacteriology because of his early observations of bacteria.
43. Pasteurization was first applied to ______________.
44. The vitalist theory of fermentation considers yeasts to be the ______________ of the process.
45. Enzymes are ______________ for chemical reactions.
46. ______________ is the yeast involved in alcohol fermentation processes.
47. ______________ are disease-causing organisms.
48. ______________ are young flies.
49. ______________ are easily killed by a few minutes of exposure to boiling temperatures.
50. ______________ is a modern device used to rapidly destroy spores.
51. It incorporates ______________ to produce temperatures of 121.5° C.

52. ______________________ developed a system with which to control surgical infections.
53. The definite sequence of experimental steps by which the causal relationships between specific bacteria and disease states can be proved beyond a doubt is known as ______________________.
54. The infected individual is known as a(an) ______________________.
55. Combinations of nutrients used for microbial cultivation are called ______________________.
56. The use of chemicals for the control and treatment of microbial diseases is known as ______________.
57. With Rivers' postulates, evidence of infection must be provided in the form of either (a) ______________ or (b) ______________________.

Essay Questions

Answer the following questions on a separate sheet of paper.

58. How do viruses differ from other types of microorganisms?
59. Support the following statement: For the most part, microorganisms exhibit the characteristics common to other forms of life.
60. What are Koch's postulates? Are they applicable today?

Answers

1. d
2. d
3. b
4. j
5. i
6. d
7. h
8. e
9. m
10. l
11. b
12. h
13. g
14. b
15. i
16. a
17. d
18. f
19. o
20. b
21. k
22. m
23. l
24. n
25. anaerobes
26. viruses
27. sexual
28. aerobes
29. mycology
30. mycology
31. molds (fungi)
32. malaria
33. bacteriophages
34. tobacco plants
35. genetic engineering
36. theory

37a. pertinent
37b. controls
38. scientific publications
39. smallpox
40. acquired immune deficiency syndrome
41. resolving power
42. A. van Leeuwenhoek
43. wine
44. cause
45. specific activators or catalysts
46. *Saccharomyces cerevisiae*
47. pathogens
48. maggots
49. vegetative cells
50. the autoclave
51. steam under pressure
52. Joseph Lister
53. Koch's postulates
54. host
55. media
56. chemotherapy

57a. the specific disease in a suitable healthy animal or plant
57b. antibodies against the viral agent

58. Viruses are dependent totally on living cells for their continued survival. They are submicroscopic and do not have a true cellular appearance. In addition, viruses differ from other types of microorganisms in their patterns of development and multiplication.
59. Microbial forms of life have a definite level of organization that enables them to perform with precision the characteristic processes common to biological systems. These processes include reproduction, metabolism, growth, responding to environmental stimuli, and adjusting to environmental stresses.
60. Koch's postulates represent a definite sequence of experimental steps by which the specific relationship between a specific disease agent and a disease could be proved beyond a doubt. The concept of Koch's postulates is clearly applicable today. However, if the disease agent in question is a virus, Rivers' postulates are more appropriate.

VI. ENRICHMENT

Nobel Prizes

Louis Pasteur once said, "There are not two sciences. There is only one science and the application of science, and these two activities are linked as the fruit to the tree." No time in history has come closer to providing Pasteur's link than has the twentieth century. In a world of dramatic and rapid changes, the extraordinary growth of science and technology has the potential directly to impact and improve every aspect of life as we know it. A major contributor of these benefits has been the research scientist.

Research represents one of the most exciting and promising frontiers for all of the biological sciences. The discoveries of various scientists have generated new knowledge of natural phenomena and provided insight into problems associated with human well-being.

Perhaps the most prestigious public recognition of the contributions by scientists are the Nobel prizes. These annual awards were established by Alfred Bernhard Nobel, the inventor of dynamite, whose discovery made him an extremely wealthy industrial magnate. He died in 1896 leaving most of his accumulated fortune of $31.5 million for use in the awarding of the Nobel prizes to individuals "who, during the preceding year, shall have conferred the greatest benefit on mankind." The categories for consideration include chemistry, economics, literature, medicine or physiology, peace, and physics. A selected list of Nobel prize winners whose contributions have influenced the development of microbiology is given in Table 1-5.

Alfred Bernhard Nobel (1833-1896)

Figure 1-1

TABLE 1-5 Selected Nobel Prize Winners and Their Contributions to the Development of Microbiology

Year	*Category of Prize*	*Nobel Laureate*	*Contribution to Microbiology*
1901	Physiology or Medicine	Emil A. von Behring	Development of diphtheria antitoxin and other forms of serum therapy
1902	Physiology or Medicine	Ronald Ross	The transmission and life cycle of malaria in the human
1905	Physiology or Medicine	Robert H. Koch	Studies concerning tuberculosis and the development of Old Tuberculin, OT (the latter material was the forerunner of the preparation used today in skin testing tuberculosis)
1907	Physiology or Medicine	Charles L. A. Laveran	*Plasmodium vivax* and the role of protozoa in producing human disease
1908	Physiology or Medicine	Paul Ehrlich	The humoral theory of antibody formation and related aspects of immunity
		Elie Metchnikoff	Phagocytosis and its role in immunity
1912	Physiology or Medicine	Alexis Carrel	Studies concerned with the transplantation of blood vessels and organs
1913	Physiology or Medicine	Charles R. Richet	Anaphylaxis
1919	Physiology or Medicine	Jules Bordet	Complement fixation and immunity
1927	Physiology or Medicine	Julius Wagner-Jauregg	The use of "malaria inoculation" for the treatment of mental deterioration resulting from syphilis infection
1928	Physiology or Medicine	Charles J. Nicholle	Studies on typhus fever
1930	Physiology or Medicine	Karl Landsteiner	Discovery of human blood groups
1931	Physiology or Medicine	Otto H. Warburg	Studies concerning the nature and mode of action of respiratory (cellular) enzymes
1937	Physiology or Medicine	Albert Szent-Gyorgyi von Nagyrapolt	Studies in cellular metabolism
1939	Physiology or Medicine	Gerhard Domagk	Antibacterial effects of the drug prontosil
1945	Physiology or Medicine	Ernst B. Chain Sir Alexander Fleming Sir Howard W. Florey	The discovery and subsequent development of the antibiotic penicillin for use in the treatment of infectious diseases
1946	Chemistry	John H. Northrop Wendell M. Stanley	The preparation of enzymes and viral proteins in pure form
		James B. Sumner	Crystallization of enzymes
1946	Physiology or Medicine	Hermann J. Muller	The production of mutations by x-ray irradiation
1948	Chemistry	Arne W. Tiselius	The development of electrophoresis, and discoveries demonstrating the complex nature of serum proteins
1951	Physiology or Medicine	Max Theiler	Development of yellow fever vaccine

TABLE 1-5 Selected Nobel Prize Winners and Their Contributions to the Development of Microbiology *(continued)*

Year	*Category of Prize*	*Nobel Laureate*	*Contribution to Microbiology*
1952	Physiology or Medicine	Selman Abraham Waksman	Discovery of streptomycin
1953	Physiology or Medicine	Hans A. Krebs	Discovery of citric acid cycle
		Fritz A. Lipmann	Discovery of coenzyme A and its role in intermediary metabolism
1954	Physiology or Medicine	John F. Enders Frederick C. Robbins Thomas H. Weller	The cultivation of poliomyelitis virus in tissues other than nervous (i.e., extraneural) tissue culture
1958	Physiology or Medicine	George W. Beadle Edward L. Tatum Joshua Lederberg	Various contributions to microbial genetics
1959	Physiology or Medicine	Arthur Kornberg Severo Ochoa	Mechanisms involved in the biologic synthesis of RNA and DNA
1960	Physiology or Medicine	Sir F. Macfarlane Burnet Peter B. Medawar	Acquired immunological tolerance
1962	Physiology or Medicine	Francis H. Crick James D. Watson Maurice H. F. Wilkins	The molecular structure of DNA and its relationship to information transfer in living organisms
1965	Physiology or Medicine	Francois Jacob Jacques Monod Andre Lwoff	Regulatory processes that contribute to the genetic control of enzymes, including the "operon concept"
1966	Physiology or Medicine	Francis Peyton Rous	Regulatory processes associated with viral replication (synthesis). Discovery of tumor-causing viruses in chickens
1968	Physiology or Medicine	Robert W. Holley Har Gobind Khorana Marshall W. Nirenberg	The relationships of genetic code components in determining cellular function
1969	Physiology or Medicine	Max Delbrück Alfred D. Hershey Salvador D. Luria	Viral genetic structure and mechanisms involved in viral replication
1972	Physiology or Medicine	Albert Claude Christian de Duve George Palade	Discoveries concerning the structural and functional organization of the cell

TABLE 1-5 Selected Nobel Prize Winners and Their Contributions to the Development of Microbiology *(continued)*

Year	*Category of Prize*	*Nobel Laureate*	*Contribution to Microbiology*
1975	Physiology or Medicine	David Baltimore Renato Dulbecco Howard Temin	Discoveries concerning the interactions between tumor viruses and cellular genetic material
1976	Physiology or Medicine	Baruch C. Blumberg	Discovery of the Australia antigen and its relationship to hepatitis B virus infection
		D. Carlston Gajdusek	Demonstration of the slow-reacting viral cause of the infectious neurological disease Kuru
1977	Physiology or Medicine	Rosalyn Sussman Yallow	Development of the radioimmunoassay
1978	Chemistry	Peter Mitchell	Mechanism involved with the conversion of adenosine diphosphate (ADP) to adenosine triphosphate (ATP)
1978	Physiology or Medicine	Werner Arber Daniel Nathans Hamilton S. Smith	Discovery and application of enzymes that break large DNA molecules into smaller manageable pieces (restriction enzymes)
1980	Chemistry	Paul Berg	Pioneering experiments in genetic engineering involving the molecular biology of nucleic acids and recombinant DNA
		Walter Gilbert Frederick Sanger	Discoveries of ways to determine DNA sequences coding for proteins
1980	Physiology or Medicine	Baruj Benacerraf Jean Dausett George Snell	Various contributions uncovering the functions and activities of the genes in the histocompatibility complex
1982	Chemistry	Aaron Klug	Development of crystallographic electron microscopy and structural analysis of biologically important nucleic acid-protein complexes
1982	Physiology or Medicine	Bengt Samuelson	Biosynthesis and structures of substances active in certain allergic responses, the leukotrienes
1983	Physiology or Medicine	Barbara McClintock[a]	Discovery of mobile or transposable DNA segments
1984	Physiology or Medicine	Cesar Milstein Georges J. F. Köhler	Development of monoclonal antibody technique
		Niels K. Jerne	Theory of antibody formation
1984	Chemistry	R. Bruce Merrifield	Methodology for synthetic peptide and protein production
1987	Physiology or Medicine	Susumu Tonegawa	Genetic control mechanism in antibody formation

[a]Barbara McClintock is the seventh woman to receive a Nobel prize.

2

Survey of the Microbial World and Introduction to Classification

I. INTRODUCTION

Because of inadequate microscope procedures and methods of study, microorganisms presented difficulties to individuals attempting to classify them in the mid-eighteenth century. With the development of more sophisticated and specialized molecular and analytical techniques, basic differences and similarities of these and other forms of life could be evaluated more accurately. This chapter presents approaches to the classification schemes, computer usage in classification, and general properties of microbial groups and other forms of life.

II. PREPARATION

Knowing the following terms will be helpful to understanding the subject matter of this chapter. Refer to the glossary and the appropriate chapters of your text if you are unsure of any of them. A pronunciation guide for selected terms is provided as a learning aid.

1. algae (AL-je)
2. amino acid
3. archaeobacteria (ar-key-bak-TEE-ree-ah)
4. bacteria (bak-TEE-ree-ah)
5. *Bergey's Manual of Systematic Bacteriology*
6. carbohydrate
7. cell
8. chlamydia (klah-MID-ee-ah)
9. chloroplast
10. chromosome
11. classification
12. cyanobacterium (si-ah-no-bak-TEE-ree-um)
13. dendrogram
14. DNA homology (ho-MOL-oh-jee)
15. eubacteria (u-back-TEE-ree-ah)
16. eucaryotic (u-kar-ee-OH-tik)
17. *Escherichia* (esh-ur-EEK-ee-ah)
18. fossil
19. fungi (FUN-ji)
20. genera (JEN-er-ah)
21. genus (GEE-nus)
22. halophile (HAL-oh-file)

23. lamellae (lah-MEL-ah)
24. lysosome
25. methogen (meth-OH-gen)
26. mitochondria (mi-toe-KON-dree-ah)
27. morphology (more-FOL-oh-gee)
28. mutation (mu-TAY-shun)
29. neotype (nee-OH-type)
30. nomenclature (NO-men-klay-chur)
31. nucleoid (new-KLEE-oid)
32. nucleolus (new-KLEE-oh-lus)
33. nucleus (NEW-klee-us)
34. operational taxonomic unit
35. paleomicrobiology (pa-lee-oh-my-kro-by-OL-oh-gee)
36. phenon (FEE-non)
37. phylum (FI-lum)
38. pleomorphic (plee-oh-MOR-fik)
39. polypeptide
40. prebiotic
41. procaryotic (pro-kar-ee-OH-tik)
42. Protista (pro-TIS-tah)
43. purine (PEW-reen)
44. pyrimidine (pure-RIM-ee-deen)
45. ribonucleic acid
46. rickettsia (rik-ETT-see-ah)
47. RNA homology
48. similarity coefficient
49. species (SPEE-sheez)
50. stromatolite (STROH-mah-toe-lite)
51. taxonomy (taks-ON-oh-mee)
52. thermoacidophile
53. type stain
54. urcaryote (ur-KARE-ee-oat)
55. virus (VYE-rus)

III. PRETEST

Correct answers to all questions can be found at the end of this section. Write your responses in the appropriate space provided.

Completion Questions

Provide the correct term or phrase for the following. Spelling counts.

1. The origin of classification dates back to the ______________________.
2. ______________________ is the study of classification.
3. Reasons for the classification of organisms include (a) ________________, (b) ________________, and (c) ______________________.
4. A (a) ______________________ classification system identifies relationships between organisms on the basis of (b) ______________________.
5. (a) ______________________ classification systems are based on easily recognizable properties of (b) ______________________ organisms.
6. The system under which all forms of life have two-word names is known as the ________________.
7. A family is composed of several ______________________.

8. A(an) ______________________ represents a newly identified bacterium.
9. Possession of a structure such as a(an) ______________________ distinguishes a plant cell from an animal cell.
10. The study of the form and structure of organisms is known as ______________________.
11. A major text on bacterial classification is ______________________.
12. According to the five-kingdom approach to classification, bacteria belong to the ______________________.
13. Bacteria exhibits a(an) ______________________ form of cellular organization.
14. The presence of mitochondria, chloroplasts, and a true nucleus is a characteristic of organisms with a(an) ______________________ cellular organization.
15. Microorganisms having cell walls are (a) ______________________, (b) ______________________, and (c) ______________________.
16. Microorganisms included in the Protista are (a) ______________________ and (b) ______________________.
17. (a) ______________________ and (b) ______________________ are representative forms of fungi.
18. ______________________ do not lend themselves to the rules of classification.
19. Specific microbial groups included in the procaryotae are (a) ______________________, (b) ______________________, and (c) ______________________.
20. ______________________ are primitive procaryotes.
21. ______________________ are branched diagrams used to show relationships among clusters of bacteria.
22. One molecular approach to taxonomy involves comparisons of entire linear nucleotide arrangements and is known as (a) ______________________ or (b) ______________________.
23. Numerical taxonomy involves the use of a large number of ______________________ characteristics for their classification.
24. The recent discovery of genetically different bacteria known as ______________________ suggests that there may be a third line of evolution.
25. Comparative ribosomal RNA analyses suggest the existence of two different procaryote types: (a) ______________________ and (b) ______________________.
26. The ______________________ has been proposed as an eucaryote ancestral cell line distinct from procaryotes.

Answers

1. ancient Greeks
2. taxonomy
3a. avoiding confusion
3b. establishing criteria for identification
3c. obtaining information of how organisms evolved
4a. natural
4b. probable origins
5a. artificial
5b. known
6. binomial system of nomenclature
7. genera
8. type strain
9. cell wall
10. morphology
11. *Bergey's Manual of Systematic Bacteriology*
12. Procaryotae
13. procaryotic
14. eucaryotic
15a. algae
15b. bacteria
15c. fungi
16a. protozoa
16b. unicellular algae
17a. molds
17b. yeasts
18. viruses
19a. cyanobacteria
19b. rickettsia
19c. chlamydia
20. archaeobacteria
21. dendrograms
22a. DNA homology
22b. RNA homology
23. equally weighted
24. methanogens
25a. eubacteria
25b. archaeobacteria
26. urcaryote

IV. CONCEPTS AND TERMINOLOGY

The Importance of Biological Classification

There are several reasons for classifying organisms, namely, establishing a basis for identifying organisms, arranging similar organisms into groups, obtaining information on how different forms of life evolved, avoiding confusion, and allowing an exchange of ideas among scientists. Their systematic arrangement is called taxonomy, which involves naming and identification of organisms according to an international code of principles, rules, and recommendations.

Natural Versus Artificial Classification

Classification schemes are based on the similarities among organisms. These schemes may be *natural* (*phylogenetic*) or *artificial.* Natural systems identify relationships between organisms on the basis of their probable origin. Artificial schemes are based on easily recognizable properties of known organisms. This approach provides a practical and useful guide for unknown organism identification.

The Naming of Organisms: The Binomial System of Nomenclature

In Carolus Linnaeus's classification scheme, all organisms have two-word names, *genus* and *species* designations. Using this system, known as the *binomial system of nomenclature,* the name *Staphylococcus aureus* (staff-il-oh-KOK-kuss ORE-ee-us) first indicates the genus to which the organism belongs, then the species name. Both designations are usually in italics, and the genus always starts with a capital letter. Genus and species names are Latin terms or words from other languages to which Latin endings are added. The name of an organism frequently refers to distinctive properties, which include colonial colors, diseases it can cause, or its habitat. Or an organism may be named after its discoverer or some other individual. In bacterial taxonomy, a new identified species is designated as the type strain. A neotype is used as a replacement for a lost type strain.

Groupings in the Linnean System

In the Linnean classification system, organisms are ranked in a series of taxonomic categories that reflect their interrelationships. Beginning with the broadest one, these taxons or taxonomic ranks and their contents are indicated in Table 2-1.

TABLE 2-1 Taxonomic Ranks

Taxonomic Rank (Taxon)	*Content*
Kingdom	A group of related phyla
Phylum or Division (plants)	A group of related classes
Class	A group of related orders
Order	A group of related families
Family	A group of related genera
Genus	A group of related species
Species	Organisms of only one type

The Position of Microorganisms in the Living World

Early classifications placed most organisms into the kingdoms of Animalia and Plantae. With the discovery of microorganisms having properties of both animals and plants, it soon became apparent that additional classification schemes were necessary. The first edition of *Bergey's Manual of Systematic Bacteriology,* a major classification reference, incorporates a number of significant properties of organisms and is a useful identification tool. This publication consists of four subvolumes, each of which focuses on specific bacterial groups and which can be prepared, published, and revised independently.

The Five-Kingdom Approach

A current classification scheme proposed by R. H. Whittaker in 1969 places all cellular forms of life into the five kingdoms of Plantae, Animalia, Protista, Fungi, and Procaryotae. The organisms found in these kingdoms exhibit one of two fundamental cellular organizations, namely, the *procaryotic* (primitive) nucleus and *eucaryotic* (true) nucleus. The general features of these cellular forms are shown in Table 2-2. The general characteristics of the five kingdoms are summarized in Table 2-3.

The kingdom of Procaryotae includes a wide variety of bacteria referred to as the eubacteria as well as the cyanobacteria and other photosynthesizing organisms, agents of disease such as the chlamydia and rickettsia, and numerous beneficial and harmless forms. Research studies used to trace the evolutionary relationship between the eubacteria and the common ancestor to all existing life forms suggest the existence of another, but different, procaryote type, the more primitive *archaeobacteria.*

True Bacteria and the Archaeobacteria

Results of comparative ribosomal RNA analyses suggest that there are two different kinds of procaryotes, the true bacteria and the archaeobacteria. The archaeobacteria are believed to have descended from ancient organisms and include the methanogens, the extreme halophiles, and the thermoacidophiles. The halophiles require high salt concentrations for survival, whereas thermoacidophiles grow best in hot, acidic environments. Properties that separate the archaeobacteria from true bacteria include chemical differences in cell wall, cell membrane, ribosomes and proteins, and antibiotic sensitivities that differ from those found with other bacteria.

The discovery of archaeobacteria suggests some changes in current views concerning the evolutionary relation between procaryotes and eucaryotes. The urcaryote was proposed to represent a eucaryote ancestral cell line to explain the many biochemical and metabolic differences between procaryotes and eucaryotes.

A Consideration of Viruses

Viruses do not have the typical structures of procaryotic and eucaryotic cells. Individual virus particles possess one type of nucleic acid, DNA or RNA, are surrounded by a protein covering, require the metabolic and genetic machinery of living cells to produce more virus particles, and can undergo permanent genetic changes or mutations.

TABLE 2-2 A Comparison of Eucaryotic and Procaryotic Cells

Structure or Chemical	*Eucaryotic Cell*	*Procaryotic Cell*
Nuclear envelope	Present	Absent
Nucleolus	Present	Absent
Chromosomes	Multiple and generally linear	Single and generally circular
Mitochondria	Present	Absent
Photosynthetic system	Chlorophyll, when present, and contained in chloroplasts	May contain chlorophyll and other pigments, but not in chloroplasts
Golgi apparatus (complex), endoplasmic reticulum, lysosomes, etc.	Present	Absent
Histones	Present	Absent
Nucleosomes	Present	Absent
Ribosomes	Large (80S = 60S + 40S)[a]	Small (70S = 50S + 30S)
Membrane sterols	Present	Absent
Mitotic apparatus	Present	Absent
Flagella	Present	Present but simpler structurally
Representative microorganisms	Algae, fungi, and protozoa	All bacteria (including rickettsia, chlamydia, cyanobacteria, and the archaeobacteria)

[a]S = Svedberg units, a measure of molecular size.

TABLE 2-3 Characteristics of the Five Kingdoms

	Kingdoms				
Characteristics	*Plantae*	*Animalia*	*Protista*	*Fungi*	*Procaryotae*
Cellular organization	Eucaryotic and multicellular	Eucaryotic and multicellular	Eucaryotic, unicellular, and some colonial forms	Eucaryotic, multicellular, and unicellular	Procaryotic and unicellular
Cell wall	Present	Absent	Present with algae	Present	Present
Differentiation of tissues	Present	May be present or absent	Absent	Absent	Absent
Mode or type of nutrition	Primarily photosynthetic	Ingestive and some absorptive	Absorptive, ingestive, photosynthetic, and combinations	Absorptive	Absorptive; few are photo-synthetic
Reproduction	Generally both asexual and sexual	Generally both asexual and sexual	Asexual and sexual	Asexual and sexual with most	Asexual and rarely sexual
Motility (ability to move)	Mostly nonmotile	Motile	Both motile and nonmotile	Generally nonmotile	Both motile and nonmotile
Microbial representatives	None	None	Microscopic algae and protozoa	Molds and yeasts	Bacteria (including the rickettsia, chlamydia, and cyanobacteria)

Trends in Microbial Classification

Numerical taxonomy or Adansonian analysis takes into account numerous properties of organisms and eliminates the need to establish different values for different properties. Using appropriately programmed computers, many organisms can be compared, and similar organisms can be arranged into taxonomic groupings or *clusters*. In this approach, each organism is an *operational taxonomic unit* (OTU). Similar clusters are called *phenons*. Branched diagrams or *dendrograms* are used to show relationships between clusters.

Molecular approaches are important additions to taxonomy since they can be used to evaluate the relationship among organisms to genetic and molecular levels. Molecular approaches are listed in Table 2-4.

DNA homology values are average measurements and demonstrate similarities between closely related organisms. RNA homology values are specific for each type of RNA and are used to detect similarities between distantly related organisms.

TABLE 2-4 Molecular Approaches to Taxonomy

Approach	*Brief Description*
DNA composition	Comparisons of the overall chemical composition of DNA
DNA homology	Comparisons of the entire linear arrangement of nucleotides (nucleic acid subunits)
RNA homology	Comparisons of those portions of DNA that contain information for the formation of certain RNA types
Protein composition	Comparisons of specific proteins or their components from different organisms

V. CHAPTER SELF-TEST

Continue with this section only after you have read Chapter 2 and have completed Section IV. A score of 80 percent or better is good. If your score is less than 65 percent, reread the chapter.

Correct answers to all questions can be found at the end of this section. Write your responses in the appropriate space provided.

Completion Questions

Provide the correct term or phrase for the following. Spelling counts.

1. The study of classification is called ______________________.
2. A(an) (a) ______________________ classification system is based on easily recognizable properties of (b) ______________________.
3. Using the binomial system of nomenclature, for the microorganism *Salmonella typhi*, the first designation is its (a) ______________________ and the second one is its (b) ______________________.
4. Taxonomic orders are subdivided into (a) ______________________; classes are subdivided into (b) ______________________.
5. Fungi consist of both (a) ______________________ and (b) ______________________.
6. The study of form and structure is called ______________________.
7. According to the Whittaker scheme of classification, the members of the five biologic kingdoms are (a) ______________, (b) ______________, (c) ______________, (d) ______________, and (e) ______________.

8. A closely resembling substitute for a type strain is a(an) ____________________.

9. ____________________ have well-defined nuclei.

10. Protista exhibit a(an) (a) ____________________ type of cellular organization, whereas cyanobacteria have a(an) (b) ____________________ type.

11. The Procaryotae include the specific disease-causing bacterial groups (a) ____________________ and (b) ____________________.

12. Procaryotes are noted for the presence of ____________________ chromosomes.

13. Fungi obtain their nutrition from the environment by ____________________.

14. A permanent genetic change is a(an) ____________________.

15. With the numerical taxonomy method, properties of organisms used are given ____________________ weight.

16. A comparison of those portions of DNA that contain information for the formation of certain RNA types is ____________________.

17. (a) ____________________ values are average measurements and demonstrate (b) ____________________ between closely related organisms.

18. A newly discovered procaryotic group of organisms believed to be closely related to ancient forms is the ____________________.

19. The existence of two procaryotic groups is being proposed: (a) ____________________ and (b) ____________________.

20. Bacteria requiring high salt concentrations for their survival are ____________________.

21. According to certain research studies, the two eucaryotic structures, (a) ____________________ and (b) ____________________, may have descended from procaryotes.

22. ____________________ has been proposed as the ancestor for eucaryotic cells.

Matching

Select the answer from the right-hand side that corresponds to the term or phrase on the left-hand side of the question sheet. An answer may be used more than once. In some cases, more than one answer may be required.

Topic: Eucaryotic and Procaryotic Organization

______ 23. Mitochondria
______ 24. Nucleoli
______ 25. Lysosomes
______ 26. Chloroplasts
______ 27. Histones
______ 28. Nuclear membrane
______ 29. Nucleoid

a. associated with procaryotes
b. associated with eucaryotes
c. associated with both procaryotes and eucaryotes
d. associated with neither procaryotes nor eucaryotes

_______ 30. Mitotic spindle

_______ 31. Ribosomes

_______ 32. Flagella

_______ 33. Chromosomes

Topic: Characteristics of the Five Kingdoms

_______ 34. Differentiation of tissues

_______ 35. Procaryotic organization

_______ 36. Eucaryotic organization

_______ 37. Absorptive nutrition only

_______ 38. Photosynthetic nutrition, mainly bacteria

_______ 39. Forms of nutrition include absorption, ingestion, photosynthesis, and combinations of these

_______ 40. Cell wall present in all members

a. Animalia
b. Fungi
c. Procaryotae
d. Plantae
e. Protista
f. a, b, c *only*
g. none of the above

Essay Questions

Answer the following questions on a separate sheet of paper.

41. What is the binomial system of nomenclature?

42. How do archaeobacteria differ from the true bacteria?

Answers

1. taxonomy
2a. artificial
2b. known
3a. genus
3b. species
4a. families
4b. orders
5a. molds
5b. yeasts
6. morphology
7a. Animalia
7b. Fungi
7c. Plantae
7d. Monera
7e. Protista
8. neotype
9. eucaryotes
10a. eucaryotic
10b. procaryotic
11a. chlamydia
11b. rickettsia
12. a single
13. absorption
14. mutation
15. equal
16. RNA homology
17a. DNA homology
17b. similarities
18. archaeobacteria
19a. eubacteria
19b. archaeobacteria
20. extreme halophiles
21a. mitochondria
21b. chloroplasts
22. urcaryote
23. b
24. b
25. b
26. b
27. b
28. b
29. a
30. b
31. c
32. c
33. c
34. a,d
35. c
36. a,b,d,e
37. b,c
38. d
39. e
40. b,c,d, and certain e

41. The binomial system of nomenclature was developed by Carolus Linnaeus. According to this system, all organisms have two names, the genus and species. The genus designation always begins with a capital letter, while the species name which follows begins with a lower case letter. One example of the system would be *Yersinia* (genus) *pestis* (species).

42. The archaeobacteria are truly unusual microorganisms. They can be distinguished from true bacteria on the basis of several distinctive properties, including differences in cell wall, cell membrane, and ribosomal chemical composition and organization; and the unique proteins and antibiotic sensitivities of certain metabolic reactions differ from those found with other bacteria.

VI. ENRICHMENT

Rules of Nomenclature

The purpose of nomenclature is to provide a convenient system of communication to define an organism without having to list its characteristics. The most important level of such communication is the species. A species designation must convey the same meaning to all scientists. The rules for naming living organisms are contained in international codes of nomenclature. There are separate codes for animals, cultivated plants, noncultivated plants, bacteria, and viruses.

Bacterial species are named in accordance with the principles and rules of nomenclature as set forth in the Bacteriological Code.* The first principle is concerned with creating stability, avoiding or rejecting names that cause error or confusion, and avoiding the useless creation of names. Scientific names are usually taken from Latin or Greek and, regardless of their origin, are treated as Latin. The correct name of a species or higher taxonomic designation is determined by three criteria: valid publication, legitimacy of the name with respect to the rules of nomenclature, and priority of publication.

Until January 1, 1980, the priorities for names dated from May 1, 1753. Such priorities caused much confusion since early descriptions of organisms often were sketchy and were based on fewer and often different tests than are now used. In addition, strains representing species proposed later in the nineteenth century often were not available for testing or, when tested, did not exactly correspond to the published properties of the species.

Priorities for bacterial names now start as of January 1, 1980, on which date the "Approved Lists of Bacterial Names" were published in the *International Journal of Systematic Bacteriology* (IJSB). Names not on those lists lost all standing nomenclature status.

To be published validly, a new species proposal must contain the species name, a description of the species, and the designation of a type strain for the species, and the name must be published in the *IJSB*. The proposed name is automatically validly published only if the proposal is published in the *IJSB*. A validly published name is assumed to be correct unless and until it is officially challenged. Challenges are the responsibility of the judicial commission of the International Association of Microbiological Societies.

New species obviously will continue to be described and, of course, will be important to individuals with varying interests. The best way to keep abreast of new species and nomenclature changes is to refer to the *International Journal of Systematic Bacteriology*. Even though other professional journals contain descriptions of newly discovered bacteria, all new names must be listed in the *IJSB* to have valid published standing in nomenclature.

*See S. P. Lapage, R. H. A. Sneath, E. F. Lessel, V. B. D. Skerman, H. P. R. Seeliger, and W. A. Clark, *International Code of Nomenclature of Bacteria*, 1975 revision (Washington, DC: American Society of Microbiology, 1976).

3

Techniques Used in the Observation of Microorganisms

I. INTRODUCTION

Much of the equipment and materials used in biochemistry, physics, and other biological sciences is also utilized in microbiology. This chapter deals with the different types of microscopes and procedures used to study the organization, structure, and functions of microorganisms and other forms of life. Attention also is given to application of computer technology and special devices to provide more precise views of biological materials.

II. PREPARATION

Chapters 2, 3, portions of Chapter 10 concerned with virus particles, and portions of Chapter 7 concerned with bacterial parts should be read before continuing. The following terms are important for you to know. Refer to the glossary and the appropriate chapters if you are uncertain of any of them. A pronunciation guide for selected terms is provided as a learning aid.

1. acid-fast technique
2. amplitude
3. Angstrom (ONG-strum) unit
4. antibody (AN-tee-bod-ee)
5. antigen (AN-tee-jen)
6. bacteria (back-TEE-ree-ah)
7. bar marker
8. compound microscope
9. condenser
10. dark-field microscopy
11. differential staining
12. endospore (end-OH-spor)
13. energy dispersive x-ray spectroscopy
14. fluorescence microscopy
15. freeze-fracture
16. gram-negative
17. gram-positive
18. Gram stain
19. hanging drop technique
20. high-voltage electron microscopy
21. image enhancement
22. inverted backscatter scanning

23. iris diaphragm (EYE-ris DYE-ah-fram)
24. meter (ME-ter)
25. metric (MET-rik) system
26. micrometer (MY-kroh-me-ter)
27. micrometry
28. microscope objective
29. microscope ocular
30. mordant
31. nanometer (NAY-no-mee-ter)
32. numerical aperture
33. objective
34. ocular
35. photographic averaging
36. phase contrast microscopy
37. prion (PRE-ahn)
38. refraction
39. refractive index
40. resolving power
41. scanning electron microscopy
42. shadow casting
43. simple staining
44. SI system
45. smears
46. specimen contrast
47. staining
48. surface replicas
49. transmission electron microscopy
50. ultramicrotomy (UL-trah my-KROT-oh-mee)
51. virus (VYE-rus)

III. PRETEST

Correct answers to all questions can be found at the end of this section. Write your responses in the appropriate space provided.

Completion Questions

Provide the correct term or phrase for the following. Spelling counts.

1. Examples of light or optical microscopes are (a) ______________, (b) ______________, (c) ______________, and (d) ______________.
2. The ______________ is the more widely used international system to express measurement in scientific areas.
3. One micrometer consists of 1,000 ______________.
4. A micrometer is expressed by ______________.
5. The limit of visibility for the bright-field microscope is about ______________ μm.
6. Provide the missing factor in the following formula related to the properties of light waves. ______________

$$\text{frequency} = \frac{\text{velocity}}{(?)}$$

7. The distance between two successive peaks on a light wave is its ______________.
8. The (a) ______________ of a microscope is its ability to reveal (b) ______________ detail of a specimen.

9. The changing of direction by light waves as they pass at an oblique angle from one type of material into another having a different refractive index is called ________________ .
10. ________________ is the changing of objectives without major focusing adjustments.
11. The ________________ of a microscope controls light intensity.
12. Procedures used to study living organisms are (a) ________________ and (b) ________________ .
13. ________________ are thin, air-dried films of bacteria on glass slides.
14. Simple staining procedures utilize ________________ stain.
15. Simple stains can demonstrate (a) ________________ , (b) ________________ , (c) ________________ , and presence of (d) ________________ .
16. Examples of differential staining methods used in bacteriology are (a) ________________ and (b) ________________ .
17. Bacterial cells stained by the Gram stain procedure that are resistant to decolorization and retain the primary stains are called ________________ .
18. The decolorizer in the acid-fast procedure is ________________ .
19. All rickettsia are gram-________________ .
20. Dark-field microscopy can be used for the examnation of (a) ________________ , (b) ________________ , and (c) ________________ .
21. ________________ serve as sources of illumination for both transmission and scanning electron microscopes.
22. The lens systems of electron microscopes are ________________ rather than glass.
23. An electron microscope will function only if a ________________ is maintained within the microscope column.
24. Most biological specimens, because of their chemical composition, have ________________ ability to scatter or deflect electrons.
25. (a) ________________ and (b) ________________ are examples of methods used in transmission electron microscopy to increase specimen contrast.
26. (a) ________________ characteristically produces a three-dimensional view of specimens and a greater (b) ________________ .
27. Penetration of thick specimens as well as three-dimensional relationships may be achieved with ________ .
28. Precise locations of enzymes and/or their activities can be determined by ________________ .
29. Three methods used to improve the images obtained in electron microscopy are (a) ________________ , (b) ________________ , and (c) ________________ .
30. EDX means (a) ________________ and is used to determine the (b) ________________ of a specimen

Answers

1a. bright-field
1b. dark-field
1c. fluorescence
1d. phase-contrast
2. metric system
3. nanometers
4. μm
5. 0.2
6. wavelength
7. wavelength
8a. resolving power
8b. fine
9. refraction
10. parfocal
11. iris diaphragm
12a. hanging-drop
12b. temporary wet mount
13. smears
14. one
15a. shape
15b. morphological arrangement
15c. size
15d. spores
16a. Gram stain
16b. acid-fast stain
17. gram-positive
18. acid alcohol
19. negative
20a. unstained microbes
20b. hanging-drop preparations
20c. colloidal solutions
21. electrons
22. magnetic
23. vacuum
24. limited or poor
25a. shadow-casting
25b. negative staining
26a. scanning electron microscopes
26b. depth of focus
27. high-voltage electron microscopy
28. localization techniques
29a. photographic averaging
29b. optical diffraction
29c. computer processing
30a. energy-dispersive x-ray
30b. chemical (element) composition

IV. CONCEPTS AND TERMINOLOGY

Microscopes and Microscopy

Depending upon the magnification principle involved, microscopes are known as either light microscopes or electron microscopes. Light or optical microscopes may be *bright-field, dark-field, fluorescence,* or *phase-contrast* instruments. Electron microscopes are of two kinds—*transmission* and *scanning electron* microscopes. In the biological sciences, the major systems of measurements are the English system (pounds, feet, inches) and the metric system, which is more widely used internationally and for scientific work. The basic units of length (meter), volume (liter), mass (gram), and temperature (Celsius degree) and some commonly used units are listed in Table 3-1. The metric system also uses decimal counting and incorporates standardized prefixes to express multiples or subdivisions of basic metric units. Every unit of the metric system is some power of 10 (10, 100, 0.1, etc.) times the basic unit. Examples are shown in Table 3-2.

To understand microscopy, one must be aware of certain fundamental properties of light. Specific terms of value are:

1. *Amplitude:* maximum displacement of a light wave from its equilibrium position.
2. *Frequency:* the number of vibrations of a light wave that occur in 1 second. Frequency is inversely related to wavelength as shown in the formula:

$$\text{Frequency} = \frac{\text{velocity}}{\text{wavelength}}$$

The resolving power of a microscope—its ability to reveal fine details of a specimen—depends on the wavelength of the source of illumination. As a general rule, the shorter the wavelength, the greater the resolving power. Thus, using ultraviolet as the source of illumination produces fine details. Light moves through different media at varying rates. Denser materials exert a slowing effect on light. This difference in velocity is expressed in the form of a *refractive index* or *index of refraction:*

$$\eta \text{ (refractive index)} = \frac{\text{speed of light in a vacuum}}{\text{speed of light in the medium being tested}}$$

TABLE 3-1 SI Units and Some Common Prefixes

Quantity	*SI Base Unit*	*Symbol*
length	meter	m
mass	kilogram	kg
time	second	s
temperature	Kelvin	K
Prefix	***Means Multiply by***	***Symbol***
giga	10^9 = 1,000,000,000	G
mega	10^6 = 1,000,000	M
kilo	10^3 = 1,000	k
hecto	10^2 = 100	h
deca	10^1 = 10	da
deci	10^{-1} = 0.1	d
centi	10^{-2} = 0.01	c
milli	10^{-3} = 0.001	m
micro	10^{-6} = 0.000,001	μ
nano	10^{-9} = 0.000,000,001	n
pico	10^{-12} = 0.000,000,000,001	p

For example,
1 millimeter = 10^{-3} meter, or 0.001 meter
1 nanometer = 10^{-9} meter, or 0.000,000,001 meter
1 kilogram = 10^3 grams, or 1,000 grams

In symbols:
1 mm = 10^{-3} m, 1 nm = 10^{-9} m, 1kg = 10^3 g

TABLE 3-2 Some Metric Units and Useful Equivalents

Unit	*Symbol*	*Equivalents[a]*
meter	m	1 m = 30.37 in.
centimeter	cm	1 cm = 10^{-2} m = 0.39 in.
micrometer	μm	1 μm = 10^{-6} m = 0.39 x 10^{-4} in.
nanometer	nm	1 nm = 10^{-9} m = 0.39 x 10^{-7} in.
Angstrom	Å	1 Å = 10^{-10} m = 0.39 x 10^{-8} in.
liter	ℓ or L	1 ℓ = 1 dm^3 = 0.001 m^3 = 1.06 qt.
milliliter	ml or mL	1 ml = 0.001 ℓ = 0.001 qt.
degree Celsius	°C	1°C = 1 K = 1.8°F
		0°C = 273.15 K = 32°F

[a]English equivalents are approximate.

Light Microscopes

Table 3-3 lists common microscope components and their functions.

Procedures used to prepare microorganisms for examination include *hanging-drop* and *temporary wet-mount* techniques for living organisms, and *staining* techniques, which employ *smears* or thin films of microorganisms spread on the surface of a clean glass slide. Smears are heat-fixed by passing them through a hot flame, which kills the organisms and causes them to stick to the slide surface.

Staining procedures include *simple* and *differential* methods. Simple staining involves the application of one stain to a smear. Such techniques can be used to demonstrate the shape, arrangement, and size of microbes, to differentiate bacterial cells from nonliving structures, and to show the presence of bacterial spores.

TABLE 3-3 Microscope Components and Their Functions

Component	*Function(s)*
Ocular (eyepiece)	Magnification
Objectives	Magnification and gathers light rays from the specimen being viewed
Condenser	Concentrates light to provide adequate illumination
Iris diaphragm	Controls light intensity

Differential staining methods usually involve the application of more than one stain. These procedures not only can demonstrate the same type of microbial properties as obtained with simple staining procedures, but can also serve as a basis for group bacteria, and to distinguish between bacterial cells and certain bacterial structures. Examples of these procedures, together with the reagents used and results obtained, are summarized in Table 3-4.

TABLE 3-4 Differential Staining Procedures

Staining Procedure	*Primary Stain*	*Decolorizer*	*Counter-stain*	*Result*	*Color Seen*
Gram stain	Crystal violet[a]	Acetone-Alcohol or 95% ethanol	Safranin	Gram-positive Gram-negative	Purple Pink-red
Acid-fast or Ziehl-Neelsen stain	Carbol-fuchsin[a]	Acid alcohol	Methylene blue	Acid-fast Nonacid-fast	Red Blue
Spore stain (Schaeffer-Fulton technique)	Malachite green[b]	None	Safranin	Spores Vegetative cells	Green Red

[a]Gram's iodine (or another form of iodine) functions as a mordant in this procedure and forms an insoluble complex with crystal violet.
[b]These stains are steamed on the smear.

Other Microscopes

Table 3-5 compares other types of microscopes and summarizes their properties.

Electron Microscopes

These instruments have become extremely valuable research tools. Two general types are used: *transmission* and *scanning.* While most transmission electron microscopes are similar in basic design to the light microscope, certain differences exist. Table 3-6 summarizes both differences and similarities. Preparation techniques for specimens are different for electron microscope examination. Special methods are needed to increase specimen contrast or to make specimens stand out from their background. Examples of commonly used procedures include:

Shadow casting: the depositing of a thin coat of heavy metal at an angle on a specimen.

Surface replicas: metal impression of a specimen's surface.

Electron staining: applications of solutions containing heavy metals.

Thin sectioning (ultramicrotomy): thin slicing or sectioning of specimens.

TABLE 3-5 Microscope Comparisons

Microscope[a]	*Source of Illumination*	*General Magnification Range*	*Representative Specimen Preparation Techniques*	*Selected Applications*
Bright light	Visible light	1,000-2,000	Living, dead, stained, and unstained	Identification of cell types, cell structure, and function study. Diagnosis of disease.
Dark field	Visible light	1,000-2,000	Living and generally unstained	Identification of cell types, study of certain cellular activities. Diagnosis of certain diseases.
Fluorescence	Ultraviolet	1,000-2,000	Living and generally unstained	Identification of cellular components, study of cellular activities. Diagnosis of certain disease states.
Phase-contrast	Visible light	1,000-2,000	Living, unstained	Identification of cellular components. Estimations of cellular materials concentration. Study of cellular structures and functions.
Transmission electron microscope (TEM)	Electrons	100,000-500,000+	Dead, stained, coated with thin layers of heavy metals, and ultrathin sectioning	Study of cellular ultrastructure and function. Identification of cell types, viruses, and other microbes. Diagnosis of diseases and disease processes.
Scanning electron microscope (SEM)	Electrons	100,000-500,000+	Dead, coated with thin layers of heavy metals	Study of the structures, arrangements, and development of tissues, organs, etc. Correlation of results with those of other techniques in building conceptual insights into structures.
Scanning transmission electron microscope (STEM)	Electrons	100,000-500,000+	Similar to those for TEM and SEM	Similar to those of the TEM and SEM plus chemical microanalyses (energy dispersive x-ray) and computerized measurements of structures.

[a]As technology develops further, the magnification ranges and other properties of these instruments will no doubt be improved.

TABLE 3-6 Differences Between the Ordinary Light Microscope and a Typical Transmission Electron Microscope

Property or Procedure	*Light Microscope*	*Transmission Electron Microscope*[a]
Source of radiation for image formation	Visible light	Electrons
Medium through which radiation travels	Air	Vacuum (approx. 10^{-4} mm mercury)
Specimen mounting	Glass slides	Thin films of collodion or other supporting material on metal grids
Nature of lenses	Glass	Magnetic fields or electrostatic lenses
Focusing	Mechanical (i.e., raising or lowering objectives)	Electrical (i.e., the current of the objective lens coil is changed)
Magnification adjustments	Changing objectives	Electrical (i.e., changing current of the projector lens coil)
Major means of providing specimen contrast	Light absorption	Electron scattering

[a]Most of these properties are also true for scanning instruments. Specimens, however, are supported on round metal stubs.

Freeze-fracture: quick freezing of specimens, which are physically split or cleared. Exposed surfaces are then covered with a thin layer of metal for replica formation. Techniques for scanning electron microscopy specimens largely involve the depositing of heavy metals onto specimen surfaces.

Localization techniques: procedures used to find the cellular sites of enzymes and their activities or to follow the formation of specific cellular structures. Techniques include combining antibody molecules with an electron-opaque, iron-containing protein. Such preparations are then applied to antigen-containing specimens, which are detected by a specific antibody-antigen reaction.

Inverted backscatter scanning: a technique used in scanning electron microscopy (SEM) which involves exposing a specimen section (slice) to one of several heavy-metal electron stains followed by examining the preparation with a backscatter detector. The detector can distinguish between the atoms of heavy-metal stains which penetrate into deeper portions of the specimen. By reversing the polarity of the SEM, the image of the specimen is reversed, thus giving a view of the specimen comparable to that seen under a light microscope.

Electron Microscopic Image Enhancement

Electron microscopic image enhancement involves techniques to remove features of a specimen that do not contribute to its specific organization. Three methods used to improve the images of viruses and structures of both procaryotic and eucaryotic cells are briefly described in Table 3-7.

TABLE 3-7 Methods for Image Enhancement

Method	*Brief Description*	*Features and/or Advantages*
Photographic averaging	The production of a composite (average) image on a single photographic plate by accurately superimposing the images of a number of similar images.	Improves the general features of specimens and eliminates nonspecific detail.
Optical diffraction	An electron micrograph is used as a diffraction device to change the path of light waves passing through it and to pass the light through the specimen image and to focus it on a screen to produce a diffraction pattern.	Pattern produced is directly related to the structural detail of the image. Unnecessary and interfering details are eliminated.
Computer processing	A computer changes the visual image into an electrical image to perform a mathematical analysis and to produce an average image. The computer detects all parts having similar parts and orientation.	All nonessential information is removed by a computer programmed to do so.

V. CHAPTER SELF-TEST

Continue with this section only after you have read Chapters 2, 3, portions of Chapter 7 concerned with bacterial parts, and portions of Chapter 10 concerning virus particles. A score of 80 percent or better is good. If your score is less than 65 percent, reread the chapter.

Correct answers to all questions can be found at the end of this section. Write your responses in the appropriate space provided.

Completion Questions

Provide the correct term or phrase for the following. Spelling counts.

1. The basic unit for length in the metric system is the ______________________.
2. The designation for a micrometer is (a) ______________ and for a nanometer is (b) ______________ .
3. The maximum displacement of a light wave from its equilibrium position is called its ______________ .
4. The ability of a microscope to reveal fine details is called ______________________.
5. A bright-field microscope with one lens is called a(an) ______________________ microscope.
6. The application of one dye solution to a smear is representative of a(an) ______________________ staining procedure.
7. Gram-positive organisms stain ______________________.
8. Acid-fast organisms stain ______________________.

Matching

Select the answer from the right-hand side that corresponds to the term or phrase on the left-hand side of the question sheet. An answer may be used more than once. In some cases, more than one answer may be required.

Topic: Compound Microscope Parts and Functions

______ 9. Coarse adjustment knob	a.	magnification
______ 10. Iris diaphragm lever	b.	concentration of illumination
______ 11. Ocular	c.	controls brightness
______ 12. Objective (oil immersion)	d.	fine focusing adjustment
______ 13. Nosepiece	e.	none of these

Topic: Microscopy and Procedure Applications

______ 14. Spore staining is applicable	a.	bright-field microscopy
______ 15. Ultraviolet light is used for illumination	b.	fluorescent microscopy
______ 16. Demonstration of viruses	c.	scanning electron microscopy
______ 17. Specimens are generally dead	d.	dark-field microscopy
______ 18. Shadow-casting	e.	transmission electron microscopy
______ 19. Inverted backscatter scanning		

Topic: Metric System

______ 20. 19 m	a.	2 mm
______ 21. 1,900 μm	b.	100 μm
______ 22. 2000 Å	c.	190 pm
______ 23. 1,000,000 nm	d.	10 nm
______ 24. 20,000 mm	e.	none of these
______ 25. 10,000 pm		

Topic: Differential Staining Procedures

______ 26. Carbolfuchsin

______ 27. Malachite green

______ 28. Methylene blue

______ 29. Acid alcohol

______ 30. Crystal violet

______ 31. Safranin

a. counterstain in the gram reaction

b. decolorizing agent in the acid-fast procedure

c. primary stain in the acid-fast procedure

d. counterstain in the acid-fast procedure

e. none of these

True or False

Indicate all true statements with a "T" and all false statements with an "F".

______ 32. Energy-dispersive x-ray techniques employ antibody-tagged preparations to determine the chemical composition of a specimen.

______ 33. Ultrathin sectioning is used to study thick specimens.

______ 34. Image enhancement makes it possible to penetrate thicker specimens.

______ 35. Specimens for electron microscopy study are generally dead.

______ 36. Electron scattering is a major means of providing specimen contrast with the light microscope.

Identification Questions

37. What type of cellular organization is shown in Figure 3-1? ______________________

38. Give the technique used to obtain the view of the specimen shown in Figure 3-1. ______________

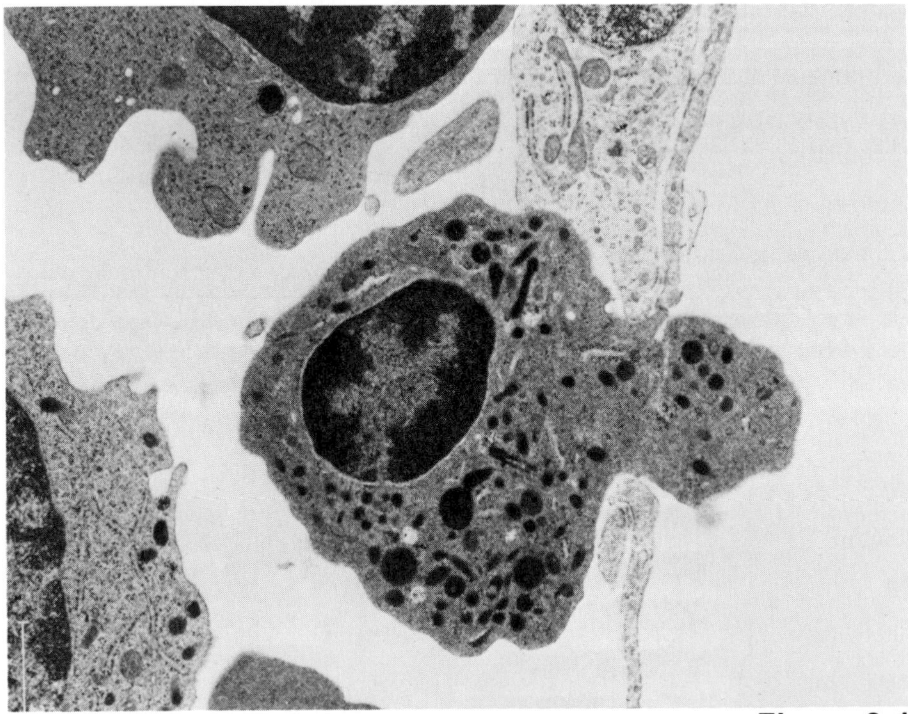

Figure 3-1

Essay Questions

Answer the following questions on a separate sheet of paper.

39. What is the purpose of the bar marker usually found near the bottom of photomicrographs?

40. What distinguishes a bacterial differential staining procedure from a simple staining procedure?

Answers

1. meter
2a. μm
2b. nm
3. amplitude
4. resolving power
5. simple
6. simple
7. purple
8. red
9. e
10. c
11. a
12. a
13. e
14. a
15. b
16. c and e
17. c and e
18. e
19. c
20. e
21. e
22. e
23. e
24. e
25. d
26. c
27. e
28. d
29. b
30. e
31. a
32. F
33. T
34. F
35. T
36. F
37. eucaryotic
38. thin or ultrathin sectioning

39. The bar marker usually found on photomicrographs functions as a size reference. The length of the marker represents the length of a metric unit at the particular magnification of the photomicrograph.

40. Bacterial differential staining procedures enable the investigator to categorize or to divide nearly all bacteria into major groups on the basis of differences in color (tinctorial) reactions. In addition, these procedures also can be used to distinguish between bacterial cells and certain of their structures. With simple staining procedures, all organisms stain the same color and thereby cannot be distinguished from one another on the basis of color.

VI. ENRICHMENT

Atomic Peeling of Cells

Analytical Ion Microscopy (AIM) is a highly sensitive technique used to determine the chemical content of cells. Even trace amounts can be detected. In this method, the surface of a 1-μm thick biological section (slice) is bombarded over a relatively large area of the specimen with an ion beam. As a result of this bombardment, the outermost parts of the specimen are progressively removed at a rate of about 1 to 10 atom layers per second, resulting in the specimen's slow erosion. Among the atoms that are released in this process, some are emitted as electrically charged particles and are chemically characteristic of a particular surface area of the specimen. The charged particles (ions) are collected, focused, and electronically analyzed to identify the chemical elements present.

PART I
MICROBIOLOGY TRIVIA PURSUIT
(Chapters 1 through 3)

This section examines your attention to detail, and your fact-gathering ability. While there are *four* challenging Parts to this section, a certain number of questions must be answered in each Part before you proceed up the MICROBIOLOGY TRIVIA PURSUIT trail. The answers and directions for continuing are given at the end of each Part. Try Part 1.

PART 1

Completion Questions

Provide the correct term or phrase for the following. Spelling counts.

1. What did Anton van Leeuwenhoek call the tiny creatures he saw in his microscopes? ________________
2. What is the formal or currently accepted name of the AIDS virus? ________________
3. What type of microscope is shown in Figure I-1? ________________

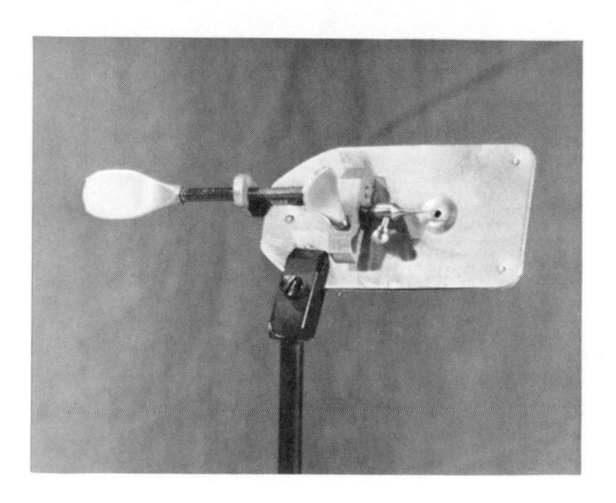

Figure I-1

4. (a) ________________ and (b) ________________ introduced the use of cotton plugs for bacteriological culture flasks and tubes.
5. The ________________ view of the basis of fermentation holds that yeasts were by-products of fermentation.

Directions

Check your responses with the *Answers* section, and add up your score. Enter the number of correct answers in the space provided.

Total Correct Answers for Part 1: ____________.

If your score was 4 or higher, proceed to Part 2.

> ***Answers***
>
> 1. Little beasties; 2. human immunodeficiency virus (HIV); 3. a simple microscope; 4a. F. Schröder; 4b. T. von Dusch; 5. nonvital.

PART 2

Completion Questions

Provide the correct term or phrase for the following. Spelling counts.

6. The German physiologist ______________________ clearly demonstrated the role of yeasts in alcoholic fermentation.
7. Intermittent sterilization is also known as ______________________.
8. ______________________ employs the use of chemicals for the treatment of diseases without damaging the infected individual.
9. The procedure to show the causal role of viruses in diseases was established by ______________________.
10. ______________________ of a light wave refers to the number of vibrations that occur in 1 second.

Directions

Check your responses with the *Answers* section, and add up your score. Enter the number of correct answers in the space provided.

Total Correct Answers for Part 2: ____________.

If your score was 4 or higher, proceed to Part 3.

> ***Answers***
>
> 6. T. Schwann; 7. tyndallization; 8. chemotherapy; 9. T. M. Rivers; 10. Frequency.

PART 3

Excellent! You are doing very well. Here is a challenging Part 3.

Completion Questions

Provide the correct term or phrase for the following. Spelling counts.

11. 1,000,000 nm = ______________________ μm.
12. Identify the electron microscopy specimen technique with the bacterial cells shown in Figure I-2.

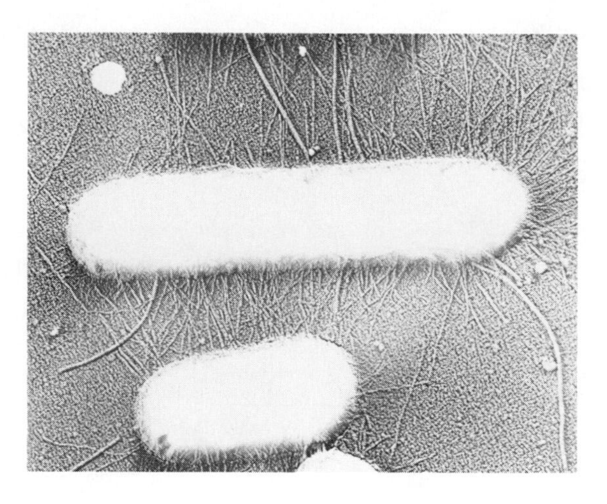

Figure I-2

13. A close internal relationship between a cell and internally contained second cell is known as __________ .
14. The ______________________________ was used for vaccination against smallpox.
15. ______________________________ is another term for worms.

Directions

Check your responses with the *Answers* section, and add up your score. Enter the number of correct answers in the space provided.

Total Correct Answers for Part 3: ______________ .

If your score was 5, proceed to Part 4.

Answers

11. 1,000; 12. shadow-casting; 13. endosymbiosis; 14. vaccinia; 15. Helminths.

PART 4

You have come a long way. Now for the last and final challenge.

Completion Questions

Provide the correct term or phrase for the following. Spelling counts.

16. (a) ______________________________ and (b) ______________________________ are noted for the invention of the compound microscope.
17. The particular systems designed by L. Pasteur to disprove spontaneous generation were called ________ .
18. The ______________________________ is the sterilizing apparatus that incorporates steam under pressure and can reach temperatures of 121.5° C and higher.

19. The bacterium used by Robert Koch to show the role of bacteria as agents of disease was (a) ________ , the cause of (b) ____________________ .

20. ____________________ are the sources of the agar used as the solidifying agent in culture media.

Directions

Check your responses with the *Answers* section, and add up your score. Enter the number of correct answers in the space provided. Then add all your scores for Parts 1 through 4 together and find your score on the *Performance Score Scale.*

Total Correct Answers for Part 4: __________ .

Total Correct Answers for Parts 1 through 4: __________ .

16a. Zacharias Janssen; 16b. Hans Janssen; 17. swan-necked or S-shaped flasks; 18. autoclave; 19a. *Bacillus anthracis*; 19b. anthrax; 20. Algae.

PERFORMANCE SCORE SCALE

Number Correct	*Ranking*
12	You should have done better
13	Better
14	Good
16	Excellent
18	Outstanding! Keep it up

Note that each major section of this Study Guide has a *MICROBIOLOGY TRIVIA PURSUIT* section. If you did not do as well as you expected, there are many more opportunities ahead.

4

Bacterial Growth and Cultivation Techniques

I. INTRODUCTION

Bacterial growth is the orderly increase in cell size leading to cell division. Growth requires processing of compounds from the environment for two purposes: to obtain energy and to form new cellular material. Different types of bacteria have developed their own individual methods for meeting these requirements. Bacterial growth is best understood in laboratory culture where conditions can be controlled by the investigator. This chapter examines the nature and importance of these nutrients. Since each bacterial type has specific nutritional requirements, it is essential that the proper nutrients be provided when bacteria are cultured.

The study of bacterial growth, even under the controlled conditions of the laboratory, requires special procedures designed to allow accurate examination of these very small creatures. This chapter discusses some of these methods and shows why they are especially suited for the study of bacterial growth. Every method has limitations to its use and appropriate precautions are considered here so that the results obtained can be properly interpreted.

II. PREPARATION

Read Chapter 4 before continuing. The following terms are important for understanding this material. Refer to the glossary and the appropriate chapters if you are uncertain about the meaning of any of them. A pronunciation guide for selected terms is provided as a learning aid.

1. aerobic
2. aerotolerant anaerobes
3. agar
4. aseptic (a-SEP-tik) technique
5. autotrophy
6. capneic (cap-NAY-ik)
7. cardinal temperatures
8. CFU
9. complex medium
10. continuous culture
11. death phase
12. differential media
13. direct count
14. enrichment
15. facultative (fak-uhl-TAY-tiv) anaerobes
16. generation time
17. growth medium
18. heterotrophy (HET-er-oh-tro-fee)

19. hypertonic
20. hypotonic
21. *in vitro*
22. *in vivo*
23. isotonic
24. lag phase
25. log phase
26. mesophile (ME-zoh-fil)
27. microaerophilic
28. PFU
29. plasmolysis (plaz-MOL-ee-sis)
30. plasmoptysis (plaz-MOP-toh-sis)
31. pour plates
32. psychroduric
33. psychrophiles (SI-kro-fil)
34. pure culture
35. selective media
36. stationary phase
37. streak plate
38. strict anaerobes
39. synthetic medium
40. thermoduric
41. thermophile (thur-MOE-fil)
42. turbidity (tur-BID-e-tee)
43. viable count

III. PRETEST

Correct answers to all questions can be found at the end of this section. Write your responses in the appropriate space provided.

Completion Questions

Provide the correct term or phrase for the following. Spelling counts.

1. The growth of bacteria in culture is called ______________________.
2. The growth of pathogenic bacteria in a diseased person is called ______________________.
3. The method for growing bacteria by constantly adding fresh medium is called ________________.
4. This growth method keeps all cells in the ______________________ phase.
5. The three basic nutritional classifications of bacteria are (a) ______________, (b) ______________, and (c) ______________________.
6. Different microorganisms require varying amounts of oxygen. (a) ______________________ are killed by oxygen; (b) ______________________ require an atmosphere of 20 percent oxygen; (c) ______________________ require low levels (about 2 percent) of oxygen; (d) ______________________ grow equally well in the presence or absence of oxygen; and (e) ______________________ grow better in the presence of oxygen, but will grow in the absence of oxygen.
7. In addition to requiring low levels of oxygen, many bacteria must have higher levels of (a) __________. These bacteria are called (b) ______________________ and many of them are important as causative agents of (c) ______________________.

8. Microorganisms which survive high temperatures, even though they may not be growing, are called (a) ____________________ . Members of the genus *Bacillus* which form (b) ____________________ are good examples.
9. An environment with a higher concentration of H^+ ions than water is called (a) ____________________ . One with a lower concentration of H^+ ions than water is called (b) ____________________ .
10. Organisms that grow best below 20° C are called (a) ____________________ . Those that grow from 20° C to 45° C are (b) ____________________ . Those that grow above 55° C are (c) ____________________ .
11. A medium or solution having an osmotic strength lower than that of the cells suspended within it is called (a) ____________________ . If the osmotic difference is too great, (b) ____________________ can occur. Conversely, media with high osmotic strengths are (c) ____________________ and can cause (d) ____________________ .
12. When a growth medium is being designed, the two major elements which are considered are (a) ____________________ and (b) ____________________ .
13. Media in which all components are precisely known are called (a) ____________________ , or (b) ____________________ media.
14. There are three basic procedures for growing and separating bacteria on solidified agar media: (a) ____________________ , (b) ____________________ , and (c) ____________________ .
15. These procedures (question 14) are used for the preparation and maintenance of ____________________ .
16. ____________________ media allow the growth of only certain types of bacteria.
17. ____________________ media allow many types of bacteria to grow, but with distinct appearances.
18. The growth of bacteria in the absence of oxygen requires special methods to create (a) ____________________ . Three devices used to obtain this condition are (b) ____________________ , (c) ____________________ , and (d) ____________________ .
19. The four phases of a general bacterial growth curve are called (a) ____________________ , (b) ____________________ , (c) ____________________ , and (d) ____________________ .
20. The ____________________ procedure is used to determine the number of living cells present at any time.
21. This determination uses (a) ____________________ plates or (b) ____________________ plates.
22. Bacterial growth can be measured by the ____________________ of the medium resulting from cell growth.
23. One can use the microscope for a (a) ____________________ or use the (b) ____________________ for an electronic determination of cell density.

Answers

1. *in vitro*
2. *in vivo*
3. continuous culture
4. log
5a. heterotrophy
5b. autotrophy
5c. hypotrophy
6a. strict anaerobes
6b. aerobes
6c. microaerophiles
6d. aerotolerant anaerobes
6e. facultative anaerobes
7a. carbon dioxide
7b. capneic
7c. disease
8a. thermoduric
8b. endospores
9a. acidic
9b. basic
10a. psychrophiles
10b. mesophiles
10c. thermophiles
11a. hypotonic
11b. plasmoptysis
11c. hypertonic
11d. plasmolysis
12a. carbon
12b. nitrogen
13a. synthetic
13b. chemically defined
14a. streak plate
14b. pour plate
14c. spread plate
15. pure cultures
16. selective
17. differential
18a. anaerobiosis
18b. candle jar
18c. GasPak
18d. Brewer jar
19a. lag
19b. log
19c. stationary
19d. death
20. viable count
21a. spread
21b. pour
22. turbidity
23a. direct count
23b. Coulter counter

IV. CONCEPTS AND TERMINOLOGY

Nutritional Categories

Complex subjects are often easier to understand if they are divided into smaller units. The division for bacteria (and most microorganisms) is made on the basis of the nature and source of carbon molecules they use to synthesize cell material. Let us consider the nature of the carbon supply first. Those organisms that use organic molecules are called *heterotrophs.* Those that use CO_2 (or HCO_3^-) are called *autotrophs.* The third and final group contains organisms called *hypotrophs,* which are heterotrophs that live inside other living organisms and parasitize them.

Gas Relations

The two most important gases to consider are oxygen (O_2) and carbon dioxide (CO_2). Microorganisms span the range from requiring O_2 as 20 percent of the atmosphere to being killed by even traces of O_2. *Obligate aerobes* require O_2 in high concentrations, usually near the 20 percent level present in the atmosphere. *Microaerophiles* also must have O_2 for growth, but in low levels (about 2 percent). *Facultative anaerobes* will grow either in the presence or absence of O_2, but they always grow better with O_2. When they grow anaerobically, they change their metabolism extensively. *Aerotolerant anaerobes* also grow either in the presence or absence of O_2, but they grow at the same rate in either instance. *Strict anaerobes* grow only in the complete absence of O_2 and are killed even by traces of O_2.

The relation of organisms to CO_2 is less clear. Autotrophs obviously require CO_2. Heterotrophs also require some CO_2 for growth. Organisms that grow better when the CO_2 level is increased are called *capneic.* It is usually difficult to accurately determine if an organisms is capneic. These organisms often occur below the surface in liquid medium where CO_2 is more concentrated and O_2 is less concentrated. This growth pattern can easily be misinterpreted as indicating that an organism is microaerophilic. Many clinical incubations are now done in incubators with a CO_2-enriched atmosphere to ensure that CO_2 limitation is not a problem in growth. This raises the question of the danger of CO_2 levels which are too high, but it has been found that very few organisms are harmed by high CO_2 concentrations. Finally, it should be noted that heterotrophic organisms produce CO_2 as a by-product of their catabolism.

Temperature Relations

A given organism has a fairly constant relation to temperature. This relation is most simply seen by examining the *cardinal temperatures* of an organism. There are three cardinal temperatures: minimum (the lowest temperature at which growth will occur), optimum (the temperature at which growth occurs fastest), and maximum (the highest temperature at which growth will occur).

It is important to distinguish between the temperature requirements and the temperature tolerance of an organism. Organisms that grow best at cold temperatures (less than 20° C) are called *psychrophilic,* whereas those that survive these temperatures without growing or dying are called *psychroduric.* Organisms that grow best at high temperatures (greater than 50° C) are called *thermophilic,* whereas those that survive these temperatures without growing or dying are called *thermoduric.*

Osmotic Relations

All microorganisms are affected by the osmotic strength of their environments. The osmotic strength is determined by all solutes present. Environments that have osmotic pressures lower than the cell are called *hypotonic;* those with pressures equal to the cell are *isotonic;* and those with pressures greater than the cell are called *hypertonic.* These relations are important because they may cause water to flow into or out of a cell. To function properly, a cell must contain about 80 percent water. If the cell is in a hypotonic environment, water will enter the cell as pressures seek to equalize. If the initial difference is very large, then too much water will enter the cell. The additional water creates a hydrostatic pressure inside the cell. The cell has limited strength and may suffer *plasmoptysis* or lysis if the pressure is too great.

Many cells contain cell walls, which are rigid layers external to the cell membrane. This layer may resist the osmotic pressure created from the hypotonic conditions and prevent the occurrence of plasmoptysis. Conversely, if the cell is in a hypertonic environment, then water will flow out of the cell to equalize pressures. Again, the amount of water flow is proportional to the initial pressure difference between the inside and outside of the cell. If the flow is too extreme, the cell undergoes *plasmolysis* and collapses. Therefore, it is important that a cell be isotonic with its environment (or nearly so) in order to minimize water flow into or out of the cell.

Culture Media

The laboratory growth of microbial cultures depends on their being supplied with the necessary nutrients. It is usually not desirable to simply supply an organism with all conceivable nutrients since some compounds which are nutrients for one organism may be poisons for another. Therefore, it is most often desirable—for example, the isolation of an organism from a natural sample—to create a medium that favors the growth of a specific organism or group of organisms at the expense of other organisms in the sample. Media can be made selective by a number of means—by adding specific inhibitors or by adjusting the pH or osmotic conditions to restrictive levels, or by using unusual carbon sources, for example.

Sometimes, being able to distinguish two different organisms on a culture medium is desirable. Media designed for this purpose are called *differential* media. The most common type of differential medium contains a known sugar (mannitol, for example) and a pH indicator that will change colors if acid is produced from the sugar. The organism that ferments the sugar can then easily be distinguished from the organism which does not. Note that there are other ways in which a medium can be made differential.

The main difference between differential and selective media is that differential media are designed to produce different growth responses but not necessarily limit growth, whereas selective media are designed to limit the types of organisms that can grow, without necessarily showing a differential effect. In a strict sense, all media (and growth conditions) are selective, since it is not possible to devise a culture in which all organisms can grow. For example, aerobes and anaerobes generally will not grow together.

To be useful in the laboratory, culture media must not only be designed to contain the correct components, but they must also be assembled properly and prepared in such a way that they can be safely stored without damage. There are several companies which specialize in the manufacture of prepackaged dehydrated growth media which can be made ready for use by the addition of water followed by a sterilization procedure.

A final important concept in the formulation of microbiological growth media is the distinction between chemically defined media (also called synthetic) and complex media. The most important difference is that in chemically defined media, the exact amount of each component is known. In contrast, complex media usually contain natural extracts in which the general composition (protein, carbohydrate, lipid, etc.) may be known, but the exact chemical concentrations are variable from batch to batch. As may be expected, bacterial growth and metabolic characteristics are more reproducible when chemically defined media are used, but it is often much easier to obtain rapid growth with complex media. Therefore, the choice at any given time depends on the precise objectives of the experiment.

Plating Methods

Whatever type of medium is used, the most useful culturing procedures for isolating and purifying bacteria involve solidifying the medium with agar in a Petri dish or plate. The preparation of pure culture is essential so that the experimenter can be certain that the observed growth effects are the result of one specific type of organism. There are three methods for using the solidified medium for bacterial isolation: streak plate, pour plate, and spread plate. The three methods differ as to particulars, but all work on the same principle—physical separation on a solid surface, leading to the formation of individual colonies. It is usually necessary to repeat the physical separation several times to be sure that each colony arose from only one cell and therefore represents a pure culture.

Anaerobic Methods

The principles for growing anaerobes are the same as those for growing aerobes except that O_2 must be excluded. This exclusion is not always easy since the atmosphere is 20 percent O_2. Elaborate chambers, closets, and hoods are used to allow culture inoculations and transfers in the absence of O_2. But the inoculated media must be incubated anaerobically as well, so several different devices have been invented, all of which require placing the inoculated plates in a large jar and removing the O_2 from the atmosphere of the jar. This O_2 removal is done either biologically, physically, or chemically. The most successful method currently available is the GasPak, which removes O_2 chemically while also supplying CO_2 to aid the growth of any capneic organisms.

Phases of Culture Growth

When a flask or other closed vessel containing liquid medium is inoculated with a culture, cells begin to grow. This growth occurs in four stages: *lag, log, stationary,* and *death.* It should be emphasized that the observed growth phases are the result of the *population* or organisms in the culture and do not necessarily represent any individual cell in the culture. Let us consider the four phases in more detail. During the *lag* stage, the newly inoculated cells are adjusting to the conditions in the culture. If there are different nutrients in the culture than were found in their previous environment, then the cells may spend time and energy synthesizing new enzymes necessary for growth in this particular medium. The activity of the cells in lag phase can be summarized as preparing for rapid growth. During the *log* phase of the growth response, cells are growing and dividing at their maximum rate. The growth pattern is exponential (or logarithmic) because bacterial cells divide by binary fission; that is, one cell divides to become two, each of which can then divide again. The time needed for the number of cells to double is called the *generation time* or doubling time of the population. After an extensive period of rapid, exponential growth, the cells pass into the *stationary* phase. Because they have consumed much of the food and excreted several toxic waste products, their growth rate slows down and some cells begin to die. If the cells remain under these conditions long enough, many more die—the *death* phase.

Measurement of Growth

Accurate determinations of the number of bacteria present requires special techniques suited to the small size of these organisms. We have already discussed the spread plate and the pour plate methods. In addition, there are four methods for measuring bacterial growth:

1. *Mass:* portions (aliquots) can be removed from a culture, centrifuged to separate cells from medium components, dried, and weighed on a sensitive balance.
2. *Optical density:* as the number of cells in a bacterial culture increases, the culture becomes cloudy or turbid. This turbidity can be determined quantitatively with colorimeters, spectrophotometers, or nephelometers.

3. *Direct count:* samples can be removed from the culture and examined in the microscope. The number of cells in a given area or volume on a slide can be determined and multiplied by a factor to relate the number to the whole culture.

4. *Coulter counter:* if the culture has no suspended particles other than bacterial cells, samples can be passed through this sensitive electronic device, which can determine the exact number of particles (cells) quickly and accurately.

All four of these procedures can be used in addition to the plating methods to create a growth curve for a bacterial culture. It is simply necessary to remove samples from the culture several times at appropriate intervals and analyze them with one or more of the methods described. Note that the optical density and Coulter counter methods provide answers in a few minutes; mass and direct count, somewhat more slowly (one to a few hours); and the plating methods take at least two days.

V. CHAPTER SELF-TEST

Continue with this section only after you have read Chapter 4. A score of 80 percent is good. If your score is less than 65 percent, reread the chapter.

Correct answers to all questions can be found at the end of this section. Write your responses in the appropriate space provided.

Matching

Select the answer from the right-hand side that corresponds to the term or phrase on the left-hand side of the question sheet. An answer may be used more than once. In some cases, more than one answer may be required.

______ 1. Requires small amounts of O_2

______ 2. Organism with a low optimum temperature

______ 3. May cause plasmoptysis

______ 4. Used to favor one type of organism

______ 5. Removal of O_2 from culture or plates

______ 6. Grows equally well with or without O_2

______ 7. Period of rapid cell growth

______ 8. Distilled water

______ 9. Time of cell activity with no cell division

______ 10. A medium with high osmotic pressure

______ 11. Culturing method

______ 12. Grows better with O_2

______ 13. Used for estimating density of bacterial population in liquid medium

a. optical density
b. hypotonic
c. thermophile
d. pour plate
e. hypertonic
f. log phase
g. lag phase
h. aerobe
i. microaerophile
j. aerotolerant anaerobe
k. facultative anaerobe
l. psychrophile
m. anaerobiosis
n. selective medium
o. viable count
p. enrichment

______ 14. Time for measurement of generation time

______ 15. The response when an organism is placed in new culture medium

______ 16. Culture procedure designed to increase one bacterial type relative to others

______ 17. Measurement of the number of living cells

Completion Questions

Provide the correct term or phrase for the following. Spelling counts.

18. (a) ______________ organisms require increased CO_2 levels. These organisms often grow where (b) ______________ grow, causing confusion.

19. Some bacteria are harmed by gaseous (a) ______________ which is required by most other organisms for growth. These bacteria often convert the gaseous compound to an even more dangerous poison, (b) ______________.

20. The use of CO_2 for cellular carbon is called ______________.

21. A Brewer plate is used to create ______________.

22. The cardinal temperatures are the (a) ______________, (b) ______________, and (c) ______________ temperatures for growth.

23. A(An) (a) ______________ has an optimum temperature lower than that of a(an) (b) ______________.

24. Some microorganisms survive freezing conditions but are still able to grow when the environment is warmed. These organisms are called ______________.

25. A(An) ______________ medium is designed to show different responses of different organisms.

26. Many organisms are found in environments with pH values below 7 or somewhat above 7. Careful regulation of the pH in a growth medium is one way to create a (a) ______________ which favors the growth of these organisms—for example, (b) ______________ ______________, the causative agent of cholera.

27. Water flows into cells which are in a (a) ______________ environment and out of cells which are in a (b) ______________ environment. If the external ionic concentration is too low, compared to the internal, the cells may incorporate so much water that they burst. The (c) ______________ ______________ is a structure which protects against this kind of danger.

28. In addition to providing the necessary carbon, nitrogen, and energy sources, properly constructed growth media include required ______________ or their precursors.

29. Growth of microorganisms is more reproducible when (a) ______________ media are used, but growth is often faster on (b) ______________ media.
30. All manipulations with bacterial cultures are done with ______________ to avoid contamination.
31. The use of ______________ ______________ is necessary to allow certain determination that a given experimental result was caused by one specific type of organism.
32. The ______________ phase of a bacterial culture occurs as nutrients are depleted.
33. The number of bacteria in a culture is changing during both the (a) ______________ and the (b) ______________ phases.
34. A culture method in which medium is added and removed at a constant rate is ______________ .
35. Spectrophotometers are used to measure the ______________ of a culture.
36. The number of bacteria in a culture can also be determined electronically with a(an) ______________ .

Essay Questions

Answer the following questions on a separate sheet of paper.

37. What is the difference between differential media and selective media?
38. Growth curve experiments that use the direct count method often give different results from those using the spread plate method. Why?

Answers

1. l
2. l
3. e
4. n
5. m
6. j
7. f
8. b
9. g
10. e
11. d
12. k
13. a
14. f
15. g
16. n
17. o

18a. capneic
18b. microaerophiles
19a. oxygen
19b. superoxide
20. autotrophy
21. anaerobiosis
22a. minimum
22b. optimum
22c. maximum
23a. psychrophile (or mesophile)
23b. mesophile (or thermophile)
24. psychroduric
25. differential
26a. selective medium
26b. *Vibrio cholerae*
27a. hypotonic
27b. hypertonic
27c. cell wall
28. vitamins
29a. synthetic (or chemically defined)
29b. complex
30. aseptic technique
31. pure cultures
32. stationary
33a. log
33b. death
34. continuous culture
35. turbidity
36. Coulter counter

37. Selective media either inhibit or limit the growth of certain organisms without affecting others. Differential media distinguish between organisms growing on the same media, usually on the bases of a growth reaction or a fermentation reaction.
38. Direct count methods are used to detect all cells present, both alive and dead. The spread plate procedure counts only living cells, thereby usually giving a slightly smaller number.

VI. ENRICHMENT

Microbiological Assay

One example of the wide impact of bacterial growth studies is the use of bacterial cultures in a procedure called *microbiological assay*. This valuable procedure is used to determine the exact amount of a growth factor such as a vitamin in solution or sample of food. These components are usually present in levels so low that chemical assay procedures are not sensitive enough to detail them. The microbiological procedure is based on the simple principle that the bacteria being used have an absolute requirement for the compound being assayed; that is, without the compound, the bacteria will not grow. A carefully measured amount of the solution containing the growth factor is added to the bacterial culture. The bacterial culture is allowed to grow until it reaches stationary phase. The number of bacterial cells is determined. Since the number of cells formed is limited by the amount of growth factor added, the amount of the growth factor in the sample can be calculated.

5

Microbial Metabolism and Cellular Regulation

I. INTRODUCTION

Chapter 4 described the basic methods used for studying the growth of microorganisms with an emphasis on the wide variety of nutritional characteristics displayed by different organisms. Chapter 5 focuses on some of the details of cell metabolism which underlie the nutrition and growth of microorganisms. It is now clearly understood that basic life processes reflect chemical activities within the cells. Since the most important molecules are organic (that is, contain carbon), an understanding of basic organic chemistry is essential. The study of biologically important organic chemical reactions is called **biochemistry.** Cells also have the ability to regulate their activities precisely; that is, under a certain set of conditions, only certain abilities are expressed. When environmental conditions change, the cell changes its activities. This self-regulation prevents the cells from wasting valuable energy-synthesizing components that may not be needed.

The mechanisms by which the cells synthesize necessary components and perform their regulation are now understood quite well at the molecular level. In fact, the study of regulation mechanisms has been of major value in understanding the fundamental cell processes.

II. PREPARATION

Read Chapter 5 before continuing. The following terms are important for understanding this material. Refer to the glossary and the appropriate chapters if you are uncertain about the meaning of any of them. A pronunciation guide for selected terms is provided as a learning aid.

1. activation energy
2. active site
3. allosteric (al-oh-STEER-ik)
4. anabolism
5. anaerobic respiration
6. apoenzyme
7. ATP
8. carbon dioxide fixation
9. catabolism
10. catalysts
11. chemiosmosis
12. chromotography (kroh-mah-TOG-raff-ee)
13. citric acid cycle (Krebs cycle)
14. coenzyme A
15. cofactor
16. competitive

17. cytochromes
18. denatured
19. electron transport
20. FAD
21. feedback inhibition
22. fermentation (fer-men-TAY-shun)
23. fermentation tests
24. glycerol
25. glycolysis (gli-KOL-ee-sis)
26. heterofermentative
27. homofermentative
28. manometry (mah-NOM-ee-tree)
29. NAD
30. oxidative phosphorylation
31. photosynthetic phosphorylation
32. proton
33. pyruvate (PIE-roo-vait)
34. quinones
35. radioisotopes
36. respiration
37. respiratory chain
38. substrate
39. substrate phosphorylation

III. PRETEST

Correct answers to all questions can be found at the end of this section. Write your responses in the appropriate space provided.

Completion Questions

Provide the correct term or phrase for the following. Spelling counts.

1. Energy released during catabolism is trapped by the cell by the formation of ______________ .
2. Enzymes function as (a) ______________ . They lower the (b) ______________ ______________ of their specific reaction.
3. Conjugated enzymes are composed of two parts, the (a) ______________ and the (b) ______________ .
4. The substrate of a reaction combines with the ______________ ______________ of the enzyme.
5. When an enzyme loses its precise shape, it is said to be ______________ .
6. Oxidations and reductions always occur together. An atom is oxidized if it (a) ______________ one or more electrons and is reduced if it (b) ______________ one of more electrons.
7. A cell's metabolism can be divided into two parts, (a) ______________ and (b) ______________ .
8. The sequence of biochemical reactions that converts glucose to pyruvate is called ______________ .
9. To obtain the maximum energy, glycolysis is usually followed by the (a) ______________ and (b) ______________ , oxidizing the original glucose to CO_2.
10. Before pyruvate enters the Krebs cycle, it is converted to acetyl- ______________ .
11. A catabolic process in which the electrons from the starting organic compound are ultimately accepted by another organic compound is called ______________ .
12. Bacteria that catabolize glucose to lactic acid are called ______________ .

13. Bacteria that catabolize glucose to a mixture of lactic and other acids are called ________ .
14. ATP is formed from these processes via ________ .
15. Autotrophs obtain carbon for cell material from the process of ________ .
16. This process requires (a) ________ and (b) ________ .
17. Respiration is normally thought of as involving O_2, but some bacteria use inorganic electron acceptors other than O_2 in a process called ________ .
18. The formation of ATP via the electron transport chain involves the formation of a(an) (a) ________ gradient which is used to make ATP through the process of (b) ________ .
19. (a) ________ are heme-containing proteins involved in electron transport. (b) ________ are non-protein electron carriers in this chain.
20. A(An) (a) ________ inhibitor interferes with the activity of an enzyme because it is similar in shape to the enzyme's (b) ________ .
21. Many enzymes can have their activity altered by combining with a specific ________ effector.
22. The gas exchange activity of a culture can be used to investigate its activity by the technique of ________ .
23. More detailed physiological information can be obtained by using (a) ________ , (b) ________ , or (c) ________ .

Answers

1. ATP
2a. catalysts
2b. activation energy
3a. apoenzyme
3b. cofactor
4. active site
5. denatured
6a. loses
6b. gains
7a. catabolism
7b. anabolism
8. glycolysis
9a. Krebs cycle
9b. electron transport chain
10. CoA
11. fermentation
12. homofermentative
13. heterofermentative
14. substrate phosphorylation
15. CO_2 fixation
16a. NADPH
16b. ATP
17. anaerobic respiration
18a. proton
18b. chemiosmosis
19a. Cytochromes
19b. quinones
20a. competitive
20b. substrate
21. allosteric
22. manometry
23a. fermentation tests
23b. redox activity
23c. radioisotopes

IV. CONCEPTS AND TERMINOLOGY

Unity of Biochemistry

By the end of the nineteenth century, the basic principles of chemical reactions and atomic construction were fairly well understood. Since that time, an enormous amount of work has been done demonstrating that basic life processes in all organisms can be understood in chemical terms. The study of chemical reactions in organisms is called *biochemistry*. Because of the relative simplicity of microorganisms and the ease of growing them, they have been used extensively in discovering biochemical secrets. Recent technological advances have made it possible to examine eucaryotic systems in more detail than before. Surprising differences have been found in the genetic organization and expression of eucaryotes, although fundamental biochemical activities are quite similar to those seen in bacteria.

Enzymes

Enzymes are proteins that act as catalysts in specific cellular processes. *Catalysts* are compounds that speed up a reaction without being created or destroyed in that reaction. Most biochemical reactions must be catalyzed because they would occur much too slowly otherwise. Enzymes are specific, that is, a given enzyme will catalyze only one reaction. Since the cell requires many reactions, many different enzymes must be present.

One way in which enzymes may speed up reactions is by bringing two reactant molecules together so that chemical bonds can be formed between them. This association is often termed "lock-and-key" since the reactant molecules and enzyme molecule must perfectly match each other in three-dimensional shape. The specific shape of an enzyme is analyzed in terms of its secondary, tertiary, and quaternary structures, which are determined by its primary structure. If these specific three-dimensional features are changed, the enzyme is said to be *denatured* and no longer functions catalytically. Denaturation can be caused by excess heat or an extreme pH.

All reactions have an activation energy requirement which must be met before the reaction can occur. This requirement is often described as a barrier to the reaction.

Catalysts function by lowering the amount of activation energy needed. It is important to understand that catalysts do *not* add energy to a reaction; they only function by lowering the amount of activation energy needed.

Subdivision of Metabolism

To simplify analysis of bacterial activities, the biochemical processes are usually divided into two reactions: *catabolism* and *anabolism.* Catabolic reactions degrade substrate molecules, releasing energy which is trapped in the form of ATP within the cell. The key steps that actually release energy and lead to ATP formation are oxidation-reduction reactions (often referred to more simply as redox reactions). Anabolic reactions are the steps a cell follows to synthesize cellular components necessary for growth (proteins, nucleic acids, lipids, carbohydrates). This synthetic process consumes most of the energy released during the catabolic processes by using ATP. In simple terms, catabolism is the creation of useful energy (ATP) by the cell and anabolism is the expenditure of ATP to build cell material. As convenient and useful as this distinction is, one must remember that the separation of catabolism and anabolism does not occur so neatly in a living cell. As we shall see, all cellular processes are interconnected, a fact that must be appreciated for a full understanding of cellular biochemistry.

Glucose Metabolism

Glycolysis

Glucose is a key substrate for many types of bacteria. The sequence of reactions that causes its initial degradation in most organisms is called glycolysis. Glycolysis can be briefly summarized as the conversion of glucose to two molecules of glyceraldehyde-phosphate, which are in turn oxidized to pyruvate. This pathway creates two molecules of ATP and reduces 2 NAD to NADH. The catabolic and anabolic aspects will now be considered separately.

Catabolic

There are two possibilities again: fermentation and respiration.

1. *Fermentation.* The simplest process is the reduction of the two pyruvate molecules to lactic acid by the NADH. This is a balanced reaction in that the NADH coenzyme molecules are reoxidized and available to accept electrons for more glucose molecules. The net energy yield from this fermentation is 2 ATP. These ATP molecules are created during substrate reactions and the process is therefore called *substrate phosphorylation.* There are other possible fermentations in which the pyruvate is converted to products other than lactic acid, such as ethanol, butanol, or acetone. But all these fermentations yield the same ATP and all perform the essential reoxidation of NADH by reducing pyruvate or a derivative of it.

2. *Respiration.* In this process, the pyruvate molecules created during glycolysis are not reduced but rather are further oxidized. The first step is the oxidative removal of CO_2 coupled to the reduction of NAD to NADH and the incorporation of a molecule of coenzyme A. The organic product of this reaction is acetyl-CoA. This compound is then ready to enter the Krebs cycle.

Krebs Cycle

The acetyl-CoA combines with oxaloacetate to create citric acid. The six-carbon citric acid undergoes a series of oxidations and CO_2 removals to regenerate the oxaloacetate that was present at the start of the cycle. This cyclic series of reactions accomplishes the complete oxidation of the acetyl-CoA to CO_2 while reducing 3 NAD to 3 NADH and 1 FAD to FADH per acetyl-CoA that entered the cycle. (FAD is an electron-carrying coenzyme similar to but different from NAD.) Two molecules of GTP are also created in the Krebs cycle. These GTP molecules are energetically equivalent to ATP and are easily converted to ATP. At the end of the cycle, there have been created 10 NADH, 2 FADH, and 4 ATP from the combined action of glycolysis and the Krebs cycle.

Electron Transport

The goal of catabolism is the creation of ATP. It is also essential that NADH and FADH molecules be reoxidized so they will be available to oxidize more substrate. The summary of reactions through the glycolysis and the Krebs cycle shows the creation of only 4 ATP and 10 NADH and FADH. The electron transport chain is used to solve both problems at the same time: the creation of more ATP and the reoxidation of reduced coenzymes. Electron transport is a series of redox reactions starting with reduced coenzyme and ending with an inorganic compound such as O_2. The key intermediates in the series are proteins called cytochromes, which contain iron atoms that actually undergo the redox changes. As electrons are passed along the electron transport chain, ATP is made from ADP and phosphate. This ATP synthesis is called *oxidative phosphorylation.* The total effect of electron transport, therefore, is to reoxidize the reduced coenzymes and create ATP.

The oxidation of NADH allows the synthesis of 3 ATP while the oxidation of FADH allows only 2 ATP to be made. Therefore, the total result of oxidizing one glucose to 6 CO_2 through glycolysis, the Krebs cycle, and electron transport is the creation of 38 ATP as follows:

glycolysis:	2 ATP, 2 NADH
Krebs cycle:	2 ATP, 8 NADH, 2 FADH
Total:	4 ATP, 10 NADH, 2 FADH

4 ATP + 30 ATP (from 10 NADH) + 4 ATP (from 2 FADH) = 38 ATP

These calculations make it clear that respiration allows much more ATP synthesis than does fermentation. It is important to understand this difference in chemical terms. All energy-yielding reactions in biochemistry are redox reactions. The amount of energy released for possible ATP synthesis is proportional to the amount of oxidation that occurs. Fermentation is only a partial oxidation of glucose and therefore only a small amount of ATP is made. Respiration is the complete oxidation of glucose and therefore much more ATP is made.

Anabolism

The biosynthetic consequences of glucose metabolism will now be considered. Remember that this distinction between catabolism and anabolism is only done to simplify the analysis. Both features of metabolism occur simultaneously.

1. *Glycolysis.* Compounds are withdrawn from two places in this pathway for use in biosynthesis. The first of these compounds is glyceraldehyde-phosphate, which can be converted to glycerol, the backbone of phospholipid molecules in membranes. The second important intermediate is pyruvate, which is used as the starting material in the synthesis of several amino acids, including alanine and aspartate.

2. *Krebs Cycle.* Acetyl-CoA, the compound that initiates the cycle, is the fundamental unit in fatty acid synthesis. Fatty acids are joined to glycerol to make phospholipids. One of the intermediate Krebs cycle compounds, α-keto-glutarate, is the starting compound in the synthesis of glutamate and glutamine.

There are other anabolic offshoots from glycolysis and the Krebs cycle, but the examples given here should demonstrate that the catabolic and anabolic functions of a cell are closely related.

Nonglucose Fermentation

Glucose was used in the preceding example because it is a central compound in the metabolism of most organisms and its biochemistry is well understood. However, it is important to realize that many compounds other than glucose can be fermented and respired. For example, the Strickland reaction is a fermentation of two amino acids performed by anaerobic bacteria of the genus *Clostridium.*

Anaerobic Respiration

The foregoing example considered O_2 respiration because this is the most common and best understood type. However, several bacteria can use inorganic compounds other than O_2 as the terminal electron acceptor. The compounds used are nitrate (NO_3^-), sulfate (SO_4^{2-}), and carbon dioxide (CO_2). These respirations only occur in the absence of O_2. Of course, the details of the reactions differ from those of O_2 respiration. But the general principle of coenzyme oxidation and ATP synthesis through oxidative phosphorylation are the same in all types of respiration.

Enzyme Regulation

The environment of a bacterial cell in nature is quite variable. This variation occurs in factors such as temperature, pH, and chemical components. It therefore makes sense that a bacterium would be able to change its metabolic activities in response to environmental variations. Most of these changes will be reflected in the cell's enzyme activities, since the specifically catalyzed reactions are the main functions of a cell. One very important example of this regulation is called feedback inhibition.

Feedback Inhibition

In order to grow, a cell needs a certain amount of specific components (amino acids, nucleotides, and so forth). There are only two ways in which these compounds can be obtained: they are synthesized by the cell from other compounds, or they can be incorporated from the environment by the cell. The synthesis of these compounds requires energy. Therefore, if a cell is suddenly supplied with one of these compounds from outside, energy would be saved if the cell used the externally supplied material and stopped making the compound itself. It is now clear that in many cases cells are capable of exactly this kind of control. Consider the example of the amino acid histidine. The pathway for making histidine involves ten enzymatic steps. The mechanism of control involves the combination of the product of the pathway (histidine) with the first of the ten enzymes. This combination, called *allosteric,* causes the enzyme to become inactive or inhibited. Since the end product reacts with the first enzyme, this process is called *feedback inhibition.* This mechanism allows very quick and effective control. When there is much histidine supplied from the outside, the enzyme is inhibited and histidine synthesis stops. If the histidine supply is removed, histidine synthesis resumes.

Metabolic Methods

Since microorganisms are so small, detailed studies are quite difficult. Usually an investigator wishes to examine a physiological process of a bacterial type in culture. For these single-celled organisms, physiology is almost the same as biochemistry. There are four major types of methods used in these investigations.

Manometry

This procedure allows sensitive determination of the gas exchanges brought about by a culture. The rate of respiration can be measured, for example, by examining the rate of O_2 consumption or CO_2 evolution.

Fermentation Tests

Different types of bacteria have different enzymatic abilities. Therefore, bacteria can sometimes be distinguished by the identity of the sugars they can ferment. Fermentation usually involves acid or gas production. Simple pH indicators and inverted tubes in liquid culture are used for determining the fermentation of a compound.

Redox Methods

As has been noted, energy-yielding reactions are redox reactions. It is possible to use redox dyes in a bacterial culture that will change color if they are reduced. The reduction of these dyes in the presence of a specific compound, for example, succinate, shows whether or not the bacteria can catabolize that particular compound.

Radioisotopes

The recent availability of radioactive compounds has greatly increased knowledge of microbial physiology. These compounds allow examination of specific biochemical reactions in great detail. The use of radioisotopes can often be coupled with other techniques such as manometry in the same experiment to enhance the analysis.

V. CHAPTER SELF-TEST

Continue with this section only after you have read Chapter 5. A score of 80 percent is good. If your score is less than 65 percent, reread the chapter.

Correct answers to all questions can be found at the end of this section. Write your responses in the appropriate space provided.

Matching

Select the answer from the right-hand side that corresponds to the term or phrase on the left-hand side of the question sheet. An answer may be used more than once. In some cases, more than one answer may be required.

_____ 1. NO_3^-
_____ 2. Noncompetitive inhibition
_____ 3. Involves cytochromes
_____ 4. Occurs only in fermentations
_____ 5. Allows rapid control of enzyme activity
_____ 6. Oxidative phosphorylation
_____ 7. SO_4^{2-}
_____ 8. Reoxidizes coenzymes with the reduction of an organic compound
_____ 9. Reoxidizes coenzymes with the reduction of an inorganic compound
_____ 10. Important in both catabolic and anabolic processes
_____ 11. Regulatory site of an enzyme
_____ 12. Coenzyme

a. anaerobic
b. radioisotopes
c. regulatory protein
d. fermentation
e. substrate phosphorylation
f. Krebs cycle
g. electron transport
h. feedback inhibition
i. NAD
j. allosteric
k. respiration
l. oxidation
m. anabolism

Completion Questions

Provide the correct term or phrase for the following. Spelling counts.

13. (a) ____________ is the production of ATP, whereas (b) ____________ consumes ATP.

14. The usual way in which energy is transferred from ATP is by the addition of a phosphate group to an organic molecule in a process called ____________.

15. Most reactions require the lowering of their (a) ____________ ____________ before they can proceed. In biological systems, this is brought about by the action of (b) ____________.

16. Each enzyme is ____________, that is, it will catalyze only one reaction.

17. An apoenzyme is one which is missing its ____________.

18. The portion of the enzyme which combines with the (a) ____________ is called the (b) ____________ ____________.

19. If an enzyme encounters temperatures which are too high or pH levels which are too far from its optimum, it may become ____________.

20. A molecule that accepts electrons is called the (a) ____________; a molecule that loses electrons is called the (b) ____________ agent.

21. (a) ____________ is a more complete oxidation of the initial substrate than is (b) ____________.

22. ____________ is the complete oxidation of substrate in the absence of oxygen.

23. (a) ____________, (b) ____________, or (c) ____________ can serve as electron acceptors in this process.

24. Electron transport serves the dual purpose of synthesizing (a) ____________ and reoxidizing (b) ____________.

25. During oxidative phosphorylation, cells create a(an) (a) ____________ gradient which, in turn, performs chemiosmosis, leading to the synthesis of (b) ____________.

26. Three types of electron carriers involved in electron transport are (a) ____________, (b) ____________, and (c) ____________.

27. An inhibitor which binds at an enzyme's active site is called (a) ____________, while a(an) (b) ____________ inhibitor binds to the enzyme at a location other than the active site.

28. When a(an) (a) ____________ effector combines with an enzyme, it changes the enzyme's activity. This reaction is an example of (b) ____________ ____________, commonly used by cells to control their reactions.

29. Four procedures used in physiological investigations of bacteria are (a) ____________, (b) ____________, (c) ____________, and (d) ____________.

Essay Questions

Answer the following questions on a separate sheet of paper.

30. Give one example in which anabolism and catabolism may be closely related.
31. Explain how feedback inhibition is important in allowing a cell to utilize energy efficiently.

Answers

1. a, k
2. h, j
3. g, k
4. e
5. h, j
6. g, k
7. a, k
8. d
9. k
10. f
11. j
12. i

13a. catabolism
13b. anabolism
14. phosphorylation
15a. activation energy
15b. enzymes
16. specific
17. cofactor
18a. substrate
18b. active site
19. denatured
20a. oxidizing agent
20b. reducing agent

21a. respiration
21b. fermentation
22. anaerobic respiration
23a. nitrate
23b. sulfate
23c. carbon dioxide
24a. ATP
24b. NADH
25a. proton
25b. ATP
26a. cytochromes
26b. quinones

26c. iron-sulfur proteins
27a. competitive
27b. noncompetitive
28a. allosteric
28b. feedback inhibition
29a. manometry
29b. fermentation tests
29c. redox methods
29d. radioisotopes

30. Glycolysis serves both catabolic and anabolic purposes. This pathway is used in both the fermentation and respiration of glucose. Intermediates of glycolysis are also used in starting compounds for several biosynthetic processes, for example, lipid synthesis.
31. Feedback inhibition allows a cell to stop performing reactions which consume energy, but are not needed at that moment. Since this control process is very rapid, the cell does not waste energy.

VI. ENRICHMENT

Cellular Processes

This chapter has presented some of the large number of different reactions a cell can perform. Our understanding of these biochemical and molecular activities has increased enormously during the past 20 years. In fact, the detailed knowledge we have of specific processes sometimes makes it difficult to remember that all these reactions occur in the same cell, often simultaneously. The cellular processes described in this chapter—energy generation and regulation—each requires detailed study to be understood. But now that you have a basic knowledge of the mechanisms, it would be very valuable for you to examine these subjects again, this time with an eye toward seeing all the ways in which they interact. Remember, the cell is the basic unit of life, and the most meaningful understanding will result when cellular metabolism as a whole is considered.

6

Microbial Genetics and Its Applications

I. INTRODUCTION

As described in Chapter 5, a cell's enzymes control what reactions it can carry out. Chapter 6 considers how enzymes are made and, most importantly, how the information for this synthesis is transmitted from one cell to its offspring. It is now clear that microorganisms can exchange genetic information with each other. The mechanisms of genetic exchange and recombination have been of interest since Gregor Mendel's pioneering experiments in the nineteenth century. Since World War II, much work has been done with bacteria, revealing most of the currently known details of molecular biology. Eucaryotes, being more complicated, have begun receiving detailed molecular attention only recently. Although the fundamental principle of the genetic code has been found to be the same between procaryotes and eucaryotes, the methods of organization and expression are much more complicated in eucaryotes.

A recent and controversial application of bacterial genetic studies is the area of genetic engineering. This field is concerned with creating combinations of genes that have never existed before in nature. These procedures have been used industrially to produce quantities of human proteins with bacteria. The economic and medical implications are great, but many fear that accidental or deliberate misuse of this technology could cause great harm. Scientists performing these experiments have taken the lead in considering safety, and no damage, accidental or deliberate, has occurred.

II. PREPARATION

Chapter 6 should be read before continuing. The following terms are important for you to know. Refer to the glossary and the appropriate chapters if you are uncertain of any of them. A pronunciation guide for selected terms is provided as a learning aid.

1. alkylating agents
2. anticodon
3. auxotroph (AWK-soh-troff)
4. bacteriocin
5. competent
6. conjugation (kon-ju-GAY-shun)
7. constitutive
8. cytoplasmic inheritance
9. DNA hybridization (hi-brid-ee-ZAY-shun)
10. evolution
11. exons
12. F^+
13. F^-
14. F'

15. gene amplification
16. gene manipulation
17. gene pool
18. genetic pool
19. genetic engineering
20. genetic mapping
21. Hfr
22. inducible
23. introns
24. kappa particles
25. lysogenic (lye-so-GEN-ik) conversion
26. mRNA
27. mutagenic agent
28. mutation
29. nonsense triplets
30. one gene, one enzyme
31. operator
32. operon
33. percent G + C composition
34. plasmid (PLAZ-mid)
35. prototroph
36. recombination
37. regulator
38. replica plating
39. repressible
40. restriction endonuclease
41. ribosome
42. RTF
43. selection
44. semiconservative replication
45. transcription
46. transduction
47. transfection
48. transformation
49. translocation
50. transposon (tranz-POH-zon)
51. triplet code
52. tRNA
53. U-tube
54. wild type

III. PRETEST

Correct answers to all questions can be found at the end of this section. Write your responses in the appropriate space provided.

Completion Questions

Provide the correct term or phrase for the following. Spelling counts.

1. The Hershey-Chase experiment showed that genetic information was exclusively in (a) ______________ , and not in (b) ______________ .
2. The DNA of eucaryotes is covered with proteins called ______________ .
3. The DNA of an organism comprises its (a) ______________ , while the list of all genetic features of an organism is its (b) ______________ .
4. RNA differs from DNA by having only one (a) ______________ , the sugar (b) ______________ instead of deoxyribose, and the base (c) ______________ instead of thymine.
5. The type of RNA which contains information for making proteins is ______________ .

6. New DNA molecules contain one half old material and one half new. Their synthesis is therefore called ______________________ replication.
7. The genetic information of a cell is contained in its DNA. The first step in using this information is the creation of (a) ______________________, a process called (b) ______________________.
8. The last step is the synthesis of proteins; this is called (a) ______________________ and occurs with the help of (b) ______________________.
9. Anticodons are found in ______________________ molecules.
10. Several genes can be organized in a single regulation group called a(an) ______________________.
11. Genes in these groups are either (a) ______________________ or (b) ______________________ as needed by the cell.
12. Enzymes that are always produced by the cell under all conditions are called ______________________.
13. Enzymes that are produced only when certain conditions exist are called ______________________.
14. The genes of eucaryotes contain coding regions called (a) ______________________ and noncoding regions called (b) ______________________.
15. The changing abilities of a species of organism is called ______________________.
16. The accepted explanation today is that these changes occur by a process known as (a) ______________________, as suggested by (b) ______________________.
17. A change in a gene is called a(an) ______________________.
18. If changed genes from different organisms associate, it is called (a) ______________________. One way to induce these changes is through the use of mustard gas, which is a(an) (b) ______________________.
19. Wild-type strains often have the ability to grow on a minimal medium and are called (a) ______________________. If this ability is lost, the organism becomes a(an) (b) ______________________.
20. The ______________________ theory was proposed to explain the nutritional defects in laboratory cultures.
21. The ______________________ method is an indirect selection procedure used to isolate bacterial colonies with specific nutritional deficiencies or virus sensitivity.
22. The first known method of genetic transfer between bacteria was (a) ______________________. This process requires the recipient cell to become (b) ______________________ through the action of a protein called (c) ______________________.
23. The other two methods of genetic transfer between bacteria are (a) ______________________, which requires cell-to-cell contact, and (b) ______________________, which involves viruses.
24. Small pieces of DNA other than the chromosomes in a cell are called ______________________.
25. ______________________ is the phenomenon in which a bacterium synthesizes a toxin only when it has been infected by a virus.
26. The determination of genes on a chromosome is called ______________________.

27. Bacterial taxonomy is now helped by two molecular techniques called (a) ______________________________ and (b) ______________________________.

28. A newly discovered enzyme important in genetic engineering is ______________________________.

29. The addition of several copies of extra DNA into an organism causes the rapid production of the gene product. This process is called ______________________________.

Answers

1a. DNA
1b. protein
2. histones
3a. genome
3b. genotype
4a. strand
4b. ribose
4c. uracil
5. mRNA
6. semiconservative
7a. mRNA
7b. transcription
8a. translation
8b. ribosomes
9. tRNA
10. operon
11a. induced
11b. repressed
12. constitutive
13. inducible
14a. exons
14b. introns
15. evolution
16a. natural selection
16b. Charles Darwin
17. mutation
18a. recombination
18b. alkylating agent
19a. prototrophs
19b. auxotroph
20. one gene, one enzyme
21. replica plate
22a. transformation
22b. competent
22c. competence factor
23a. conjugation
23b. transduction
24. plasmids
25. Lysogenic conversion
26. genetic mapping
27a. percent G + C composition
27b. DNA hybridization
28. restriction endonuclease
29. gene amplification

IV. CONCEPTS AND TERMINOLOGY

Genetic Information

Molecular biology is the study of information transfer. The ultimate store of information in a cell is in the sequence of bases in its DNA. This information has several uses. We will consider the transfer of information from DNA to the structure of proteins. The synthesis of proteins is of special importance, as enzymes are proteins and actually carry out most of the specific reactions of a cell.

The first step is the creation of a strand of messenger RNA (mRNA) from a portion of the DNA. This process is called *transcription.* The sequence of bases in the mRNA is a direct reflection of the base sequence of the DNA.

As was discussed in Chapter 5, the three-dimensional shape of an enzyme and, therefore, its catalytic specificity, are determined by the sequence of amino acids in the protein (the primary structure of the protein). The next step in protein synthesis is called *translation,* which is the transfer of information from the base sequence in mRNA to the amino acid sequence in the protein. This translation occurs on ribosomes and involves another type of RNA called transfer RNA (tRNA). Each tRNA combines with one specific amino acid and has a specific anticodon, a group of three bases on the tRNA molecule. The amino acid attaches to the tRNA at a site other than the anticodon. The mRNA can be logically divided into groups of three bases called *codons.* The function of the ribosome is to place the mRNA and the tRNA molecules in the right position so the anticodon and the codon can align properly. A given anticodon and codon will only pair with each other if the sequences are complementary. If a complementary pair is formed, then the amino acid is released from the tRNA and attaches to the growing protein. This process is repeated until all the codons on the mRNA have been paired with their anticodons. For the synthesis of a protein with 100 amino acids, this process must be repeated 100 times.

Let us review this process in terms of information transfer. The information contained in the piece of DNA in our example is the structure of a specific enzyme. An enzyme's specificity is in its three-dimensional shape, which is determined by its primary structure (see Chapter 5). The information in the DNA is in the form of its base sequence. This information is transcribed to the base sequence in the mRNA made from this DNA. The information is then translated to the amino acid sequence in the protein. It is important to understand the mechanism of this process, but it is also important to keep in mind its purpose—information transfer.

Eucaryotes follow the same basic steps of transcription as do procaryotes, but there are some very significant organizational differences in transcription. Most notable is that almost all genes in eucaryotes contain sequences which are not used for the synthesis of proteins. These noncoding regions, called *introns,* are included in the transcription of a molecule of precursor mRNA. The introns are removed from the precursor, resulting in the formation of a mature mRNA consisting of the coding regions, called *exons.*

Operons

Feedback inhibition is sometimes not enough, even though it is quick and effective. If the cell continues to make the enzymes, it is expending a great deal of energy. This energy is saved by controlling enzyme synthesis through *operons,* clusters of genes containing a region of DNA called an operator. In the histidine example, there is also a regulatory protein made by another gene. This regulatory protein is normally inactive. However, in the presence of excess histidine, it becomes activated and binds to the operator. When this occurs, the genes in the operon cannot be transcribed into mRNA. In the absence of the specific mRNA, the enzymes for the histidine pathway cannot be made. The system is said to be *repressible.* The combination of feedback inhibition and the operon allows complete control over the synthesis of histidine.

The operon concept also works in the reverse manner. That is, an operon that is normally not transcribed is transcribed in the presence of a specific compound. This system is said to be *inducible.* The inducible system usually functions with operons related to catabolism (e.g., the synthesis of enzymes to degrade a rarely encountered energy source). The repressible system relates to anabolic processes such as histidine synthesis.

Genetic Variation

The nature of the changes that occur in populations has been questioned for hundreds of years. This questioning led to the currently accepted theory of evolution, which explains small population changes as well as the development of different species. It is noteworthy that this evolutionary theory arose and indeed became quite well accepted without direct experimental evidence, which is difficult to obtain. Much support for the theory has come from a series of observations of microorganisms that are consistent with the theory. These observations serve as circumstantial evidence to show that the evolutionary theory is correct. The theory of evolution may be summarized by saying that mutations occur randomly and that natural selection eliminates organisms that carry detrimental mutations.

Genetic Recombination

Once mutations are created and incorporated in a population, further genetic variation occurs through the process of *recombination.* Recombination is the transfer of genetic material (DNA) from a donor organism to a recipient, followed by the chemical rearranging of donor and recipient DNA into a new molecule (or sets of molecules in higher organisms). One advantage of this process is that different mutations that occur in different organisms can be brought together. The result is that a larger number of combinations is possible than if the DNA were retained by each cell. This larger number of combinations increases the possibility of evolutionary change.

As with most cellular processes, the transfer of genetic material is most easily studied in microorganisms. One of the earliest studies by Beadle and Tatum with the fungus *Neurospora* showed that wild-type (natural) strains could form all of their necessary biosynthetic components from a simple defined medium. These organisms are called *prototrophs.* Mutants were created that were found to have specific nutritional defects. These organisms, called *auxotrophs,* were able to grow if their specific deficiency were overcome by adding the required nutrient—amino acids, purines, pyrimidines, or vitamins—to the medium. Further detailed study showed that each auxotroph had a deficiency in just one biochemical step. This result led to the one gene, one enzyme theory, which simply says that each unit of genetic information is responsible for the synthesis of one enzyme.

Mechanisms of Genetic Transfer in Bacteria

Even though this important result was obtained with a fungus, almost all detailed advances since then have been made with bacteria. One of the major reasons is that bacteria are *haploid,* meaning that they have only one type of DNA molecule per cell. In a haploid organism, a recombinant DNA would be expressed immediately. In fungi, which are diploid, a defect would not be seen until the gene became incorporated in both DNA molecules.

The three mechanisms of DNA transfer in bacteria will be presented in the order of their discovery.

Transformation

This process is the incorporation of a small piece of DNA by a recipient cell. Even small pieces of DNA are fairly large, so the recipient cell must be prepared to accept the DNA. This state of preparedness is called *competence.* Cells are competent only for short periods of time during their growth, although it is now possible to increase the competence of a bacterial population by appropriate chemical treatment. Transformation experiments with pneumonia-causing bacteria and mice were essential in demonstrating that DNA is the genetic information material.

Conjugation

This process is the most complicated and probably the most common in nature. The key feature is that contact between donor and recipient cells is required. The donor cells contain small pieces of DNA other than the cell chromosome called *F factors.* These F factors are a type of plasmid. The F factor causes the transfer of DNA to the recipient cell. If the F factor is integrated into the chromosome, chromosomal DNA is also transferred to the recipient. Experiments with conjugation have led to the formation of genetic maps of bacterial chromosomes, showing the physical relation of genes along the circular chromosome.

An additional result from conjugation studies is the discovery of plasmids. A *plasmid* is a small piece of DNA. Two types of natural plasmids that have been studied in detail are the resistance transfer factors (RTF) and those that produce bacteriocins.

RTF plasmids contain the genes for resistance to several (usually two to five) antibiotics. When one of these plasmids is transferred to a cell, that recipient immediately becomes resistant to two to five antibiotics at once. The medical consequences of this multiple resistance are quite significant.

Bacteriocins are compounds made by bacteria that are lethal only to closely related bacteria. These compounds vary widely in their chemical nature and mode of action, but all are made from plasmids. Transfer of these plasmids from one bacterium to another is common.

Another use of plasmids, to be discussed shortly, is their involvement in genetic engineering.

Transduction

This process of DNA transfer involves the action of bacterial viruses called *bacteriophages* or *phages.* When a bacterium is lysogenized by a temperate phage (see Chapter 10), the DNA of the phage is incorporated into the cell's chromosome. This process is somewhat similar to that of F particles; some scientists even call the lysogenic phage an episome. When the phage emerges from its lysogenic state and begins to produce new phage, some of the chromosomal DNA may be incorporated into the new phage particles. These phages have usually lost some of the phage DNA and are therefore referred to as defective. When one of these defective phages infects a new bacterium, the bacterial DNA from the first bacterium enters the recipient cell where it is available for recombination and expression.

There are two types of transduction: specialized and generalized. *Specialized transduction* is the transfer of the same small piece of DNA from the donor every time transduction occurs. *Generalized transduction* occurs when all portions of the bacterial chromosome have an equal chance of being transferred in each transduction.

Genetic Engineering

This exciting new area of bacterial genetics involves the creation of specific combinations of genes. That is, the investigator determines which combinations will occur. One type of new combination being made is the merger of the DNA from higher organisms with bacterial DNA—for example, the insertion of the gene to make the animal protein insulin into a plasmid in *E. coli.* The goal is to allow the simple bacterial system to make huge amounts of insulin relatively cheaply and quickly, with obvious medical benefit.

This new technology depends on specific enzymes called *restriction endonucleases,* which break the DNA in specific, desired places. The new DNA similarly treated can then be inserted with other enzymes. This new combination is usually done on a bacterial plasmid. Plasmids have several advantages for this process: they are stable, small, and can be introduced into bacteria more easily than other forms of DNA.

It is necessary, however, to be extremely cautious when making these recombinants. Even with such precautions, it is always possible that accidents can occur and a *chimera,* or monster, might be created. Some chimeras could cause serious epidemics if they got out of control. The United States National Institute of Health has formulated careful guidelines concerning the conduct of genetic engineering research.

V. CHAPTER SELF-TEST

Continue with this section only after you have read Chapters 5 and 6. A score of 80 percent or better is good. If your score is less than 65 percent, reread the chapters.

Correct answers to all questions can be found at the end of this section. Write your responses in the appropriate space provided.

Matching

Select the answer from the right-hand side that corresponds to the term or phrase on the left-hand side of the question sheet. An answer may be used more than once. In some cases, more than one answer may be required.

______	1. Transformation	a. competence
______	2. Toxin production	b. Darwin
______	3. Breaks DNA in specific sites	c. Tatum and Beadle
______	4. Replica plating	d. the Lederbergs
______	5. Used mice and pneumonia bacteria	e. Davis
______	6. F factor	f. Griffith
______	7. Restriction endonuclease	g. RTF
______	8. U-tube	h. conjugation
______	9. Made from plasmids	i. bacteriocin
______	10. Defective phages	j. lysogenic conversion
______	11. One gene, one enzyme	k. transduction
______	12. Evolution	l. restriction endonucleases
______	13. Needed for transformation	m. genetic engineering
______	14. Antibiotics	

Completion Questions

Provide the correct term or phrase for the following. Spelling counts.

15. Genetic information is contained in molecules of (a) ________________. This was shown by the experiments of (b) ________________ with mice and *Streptococcus pneumoniae* and by (c) ________________ and (d) ________________ with a bacterial virus.
16. The (a) ________________ of an organism is comprised of its genes, which are found on structures called (b) ________________.
17. The (a) ________________ of an organism may change in response to different environments by selective expression of different (b) ________________.
18. The three forms of RNA are called (a) ________________, (b) ________________, and (c) ________________.
19. The smallest of the RNA molecules is (a) ________________, each of which transports a single type of (b) ________________ for protein synthesis.
20. When DNA molecules are replicated, each strand serves as a(an) ________________ for the synthesis of the other.
21. (a) ________________ molecules combine with specific amino acids and bring them to the (b) ________________ during translation.
22. The complementary pairing of (a) ________________ and (b) ________________ molecules is essential during translation.
23. The genes of procaryotes do not contain (a) ________________, while those of eucaryotes almost always do. When these noncoding regions are removed from the (b) ________________, the remaining (c) ________________ form the mature mRNA.
24. Natural (a) ________________ occurs without human involvement. (b) ________________ selection is used by scientists to obtain pure cultures of new mutant strains.
25. Recombination requires DNA from two organisms, the (a) ________________ and the (b) ________________.
26. Auxotrophs are derived from (a) ________________ by (b) ________________.
27. Transformation requires cells that are ________________.
28. The U-tube was first used to demonstrate ________________.
29. The transfer of plasmids occurs through ________________.
30. All portions of a bacterial chromosome can be transferred by phages in ________________.
31. The specific enzymes needed for genetic engineering work are called ________________.
32. These enzymes are used to insert new genes into ________________.
33. The new DNA molecules are then put into bacteria via the process of ________________.

Essay Questions

Answer the following questions on a separate sheet of paper.

34. Describe one medically important aspect of bacterial recombination.

35. Genetic engineering often involves the use of bacteria to express human genes. Why is this desirable?

Answers

1. c, h
2. j
3. e
4. d
5. f
6. h
7. m
8. e
9. i
10. c
11. c
12. b
13. a
14. g

15a. DNA
15b. Griffith
15c. Hershey
15d. Chase
16a. genome
16b. chromosomes
17a. phenotype
17b. genes
18a. transfer
18b. ribosomal
18c. messenger
19a. transfer (or tRNA)
19b. amino acid
20. template
21a. tRNA
21b. ribosomes
22a. tRNA
22b. mRNA
23a. introns
23b. precursor mRNA
23c. exons
24a. selection
24b. Artificial
25a. donor
25b. recipient
26a. prototrophs
26b. mutation
27. competent
28. conjugation
29. conjugation
30. generalized transduction
31. restriction endonucleases
32. plasmids
33. transformation

34. The existence of resistance transfer factors (RTF) poses a serious problem since RTFs are genetic elements (plasmids) that are transferred quickly from one bacterium to another and allow the recipient bacterium to become resistant to several antibiotics at one time.

35. Bacteria are easier and cheaper to grow; therefore, their expression of human genes (such as insulin) allows commercial production of large quantities cheaply.

VI. ENRICHMENT

Use of Mutant Organisms

This chapter has emphasized the basic scientific facts related to microbial genetics, facts that have had important applications in the area of industrial microbiology. Microorganisms are very useful in many industrial, medical, and agricultural syntheses. The principles of genetics have been used to increase the efficiency of many of these processes. Consider the process of antibiotic production. Most antibiotics are produced by microorganisms through their specific biochemical capabilities. These organisms are grown in huge culture vessels (perhaps 50,000 gallons) called fermentors. Samples of fungi or bacteria have been treated to cause mutations and the mutant organisms tested for their ability to perform the desired process. In this way, the efficiency and rate of production of some antibiotics have been increased by as much as 1,000-fold. Similar benefits have been obtained in other industrial processes that use microorganisms.

PART II
MICROBIOLOGY TRIVIA PURSUIT
(Chapters 4 through 6)

This section examines your attention to detail, and your fact-gathering ability. While there are *four* challenging Parts to this section, a certain number of questions must be answered in each Part before you proceed up the MICROBIOLOGY TRIVIA PURSUIT trail. The answers and directions for continuing are given at the end of each Part. Try Part 1.

PART 1

Completion Questions

Provide the correct term or phrase for the following. Spelling counts.

1. The temperature at which an organism grows fastest is called its ______________________ temperature.
2. If a bacterial culture remains in the stationary phase too long, it will enter the ______________ phase.
3. Enzymes are ______________________ which catalyze biological reactions.
4. CO_2 is released by ______________________ as they grow.
5. The two types of nucleic acid are DNA and ______________________.

Directions

Check your responses with the *Answers* section, and add up your score. Enter the number of correct answers in the space provided.

Total Correct Answers for Part 1: __________.

If your score was 4 or higher, proceed to Part 2.

Answers

1. optimum; 2. death; 3. proteins; 4. heterotrophs; 5. RNA.

PART 2

Good show! Congratulations on the successful completion of Part 1. Now try your hand with Part 2.

Completion Questions

Provide the correct term or phrase for the following. Spelling counts.

6. Bacteria generally grow faster in a ______________________ medium.
7. The only type of ATP synthesis which occurs during fermentation is called ______________________.

8,9. The (8) ______________________ cycle, which is an essential part of respiration, is also very important in providing precursors for (9) ______________________.

10. The transfer of information from DNA to mRNA is called ______________________.

Directions

Check your responses with the *Answers* section, and add up your score. Enter the number of correct answers in the space provided.

Total Correct Answers for Part 2: ____________.

If your score was 4 or higher, proceed to Part 3.

Answers

6. complex; 7. substrate phosphorylation; 8. Krebs; 9. anabolism (or biosynthesis); 10. transcription.

PART 3

Great performance! You are doing very well. Here is a challenging Part 3.

Completion Questions

Provide the correct term or phrase for the following. Spelling counts.

11,12. An organism which requires a small amount of oxygen for growth is called (11) ______________. Such organisms are also often found to be (12) ______________________.

13. The term for the complete oxidation of a substrate linked to the reduction of an inorganic compound other than O_2 is ______________ ______________.

14. Operons for biosynthetic pathways are usually ______________________.

15. F factors and RTF factors are both types of ______________________.

Directions

Check your responses with the *Answers* section, and add up your score. Enter the number of correct answers in the space provided.

Total Correct Answers for Part 3: ____________.

If your score was 5, proceed to Part 4.

Answers

11. microaerophilic; 12. capneic; 13. anaerobic respiration; 14. repressible; 15. plasmids.

PART 4

You have come a long way. Now for the last and final challenge.

Completion Questions

Provide the correct term or phrase for the following. Spelling counts.

16. Cells in hypertonic environments may undergo ______________________.
17. An allosteric effector is a type of ______________________ which is important in feedback inhibition.
18. ______________________ is the name given to oxidative phosphorylation which involves the formation of a proton gradient.

19,20. Genes in eucaryotes usually contain (19) ______________________ which must be removed from precursor mRNA to form the (20) ______________________ which is used in translation.

Directions

Check your responses with the *Answers* section, and add up your score. Enter the number of correct answers in the space provided. Then add all your scores for Parts 1 through 4 together and find your score on the *Performance Score Scale.*

Total Correct Answers for Part 4: ____________.

Total Correct Answers for Parts 1 through 4: ____________.

Answers

16. plasmolysis; 17. noncompetitive inhibitor; 18. Chemiosmosis; 19. introns; 20. mature mRNA.

PERFORMANCE SCORE SCALE

Number Correct	*Ranking*
12	You should have done better
13	Better
14	Good
16	Excellent
18	Outstanding! Keep it up

Note that each major section of this Study Guide has a *MICROBIOLOGY TRIVIA PURSUIT* section. If you did not do as well as you expected, there are many more opportunities ahead.

7

Procaryotes: Their Structure and Organization

I. INTRODUCTION

The staining of bacteria with reagents found in a laboratory usually shows little detail beyond the shape (morphology), size, and arrangement of cells. This chapter describes the distinctive organization, ultrastructure, functions, and chemical composition of procaryotic cells.

II. PREPARATION

Read Chapters 2, 3, and 7 before continuing. In addition, knowing the following terms will be helpful to understanding the subject matter of this chapter. Refer to the glossary and the appropriate chapters if you are uncertain of any of them. A pronunciation guide for selected terms is provided as a learning aid.

1. adenosine triphosphate
2. adkinete
3. atrichous (a-TRIK-us)
4. axial (AK-see-al) filament
5. bacillus (bah-SIL-us)
6. bacteriophage (bak-TEE-ree-oh-faj)
7. basal granule
8. Brownian movement
9. capsule (KAP-sull)
10. cell
11. cell membrane
12. cell wall
13. chemotaxis (key-moh-TAKS-is)
14. chloroplast
15. coccus (KOK-us)
16. conidium (koh-NID-ee-um)
17. cyanobacterium
18. cyst (SIST)
19. cytoplasm
20. deoxyribonucleic acid
21. diplococcus (dip-low-KOK-us)
22. dipicolinic (dye-pick-oh-LIN-ik) acid
23. encapsulation
24. endospore
25. eucaryote (u-kare-EE-oat)
26. exoenzyme

27. exospore
28. flagellum (flah-JEL-lum)
29. gas vesicle
30. genera (JEN-er-ah)
31. germination
32. glycocalyx (gly-koh-KAY-lix)
33. gram-negative
34. gram-positive
35. lipid
36. lophotrichous (low-FOE-trik-us)
37. mesosome
38. metachromatic granules
39. micrometer
40. mitochondrion
41. morphology
42. mucoid (MEW-koyd)
43. murein sacculus (MEW-rein SACK-u-lus)
44. mycoplasma (my-koh-PLAZ-mah)
45. nucleoid (NEW-klee-oyd)
46. nucleus
47. peptidoglycan (pep-tid-oh-GLI-kan)
48. periplasmic (pair-ee-PLAZ-mik)
49. periplasmic space
50. peritrichous (pair-ee-TRIK-us)
51. pilus (PIE-lus)
52. plasmid (PLAZ-mid)
53. porin (PORE-in)
54. procaryote (pro-kare-EE-oat)
55. protein
56. protoplast (PRO-toe-plast)
57. ribosome
58. sarcinae
59. S-layer
60. spheroplast
61. spinae
62. spirae
63. spirillum (spy-RILL-um)
64. staphylococcus (staff-ee-loh-KOK-us)
65. sterol (STER-ol)
66. streptococcus (strep-toh-KOK-us)
67. teichoic (tie-KOH-ik) acid
68. tetrad
69. vibrio (VIB-ree-oh)

III. PRETEST

Correct answers to all questions can be found at the end of this section. Write your responses in the appropriate space provided.

Completion Questions

Provide the correct term or phrase for the following. Spelling counts.

1. Individual bacteria can exhibit one of these four steps: (a) ______________, (b) ______________, (c) ______________, and (d) ______________.
2. Cocci in a boxlike arrangement form a(an) ______________.
3. An irregular, grapelike cluster of cocci is called a(an) ______________ arrangement.
4. Small, rounded rods that are difficult to distinguish clearly from cocci are called ______________.

5. Mechanical means of cell disintegration include (a) ________________, (b) ________________, and (c) ________________.
6. Bacterial surface-associated structures include (a) ________________, (b) ________________, (c) ________________, and (d) ________________.
7. A quivering to-and-fro motion of bacterial cells is called ________________.
8. Bacteria lacking flagella are referred to as being ________________.
9. The origin of a flagellum is a(an) (a) ________________, which is located just beneath the (b) ________________.
10. Pili are ________________ than flagella.
11. Chemical analyses of pili show them to consist of protein subunits called ________________.
12. Capsule-producing colonies usually are ________________ in appearance.
13. Bacterial cell walls contain the two simple sugars related to glucose, (a) ________________ and (b) ________________.
14. A unique compound in cell walls lacking lysine is ________________.
15. An L-form does not have a(an) ________________.
16. The genetic information of a procaryote is contained within a ________________.
17. Extra chromosomal circular DNA molecules that are not essential to an organism are called ________________.
18. ________________ are nonunit membrane organelles that contain the enzyme needed for CO_2 fixation in photosynthesis.
19. The heat resistance of bacterial spores is associated with (a) ________________ and (b) ________________.
20. Bacterial sporulation can be considered to be a mechanism for an organism's ________________.
21. Physical factors needed for sporulation to occur include (a) ________________, (b) ________________, and (c) ________________.
22. The overall process of spore germination consists of (a) ________________, (b) ________________, and (c) ________________.
23. Cyanobacteria can reproduce by forming differentiated cells called (a) ________________ and (b) ________________.
24. The glycocalyx includes (a) ________________ and (b) ________________.
25. The movement of bacterial cells toward chemicals is known as ________________.
26. ________________ are enzymes that function as carriers of substrates across bacterial membranes.
27. In gram-negative cells, the ________________ separates the cell wall from the plasma membrane.

Photographic Questions

28. What are the refractile bodies (arrows) shown in Figure 7-1? ______________________

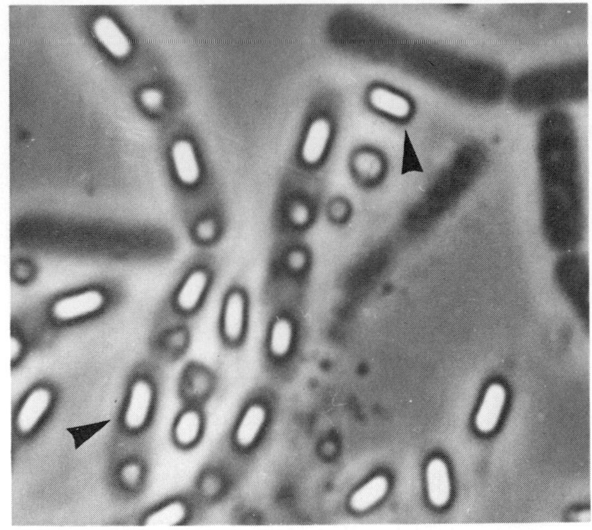

(Courtesy of Dr. P. Fitz-James, The University of Western Ontario.)

Figure 7-1

29. Identify the labeled structures shown in Figure 7-2. (a) ______________ (b) ______________

(c) ______________ (d) ______________

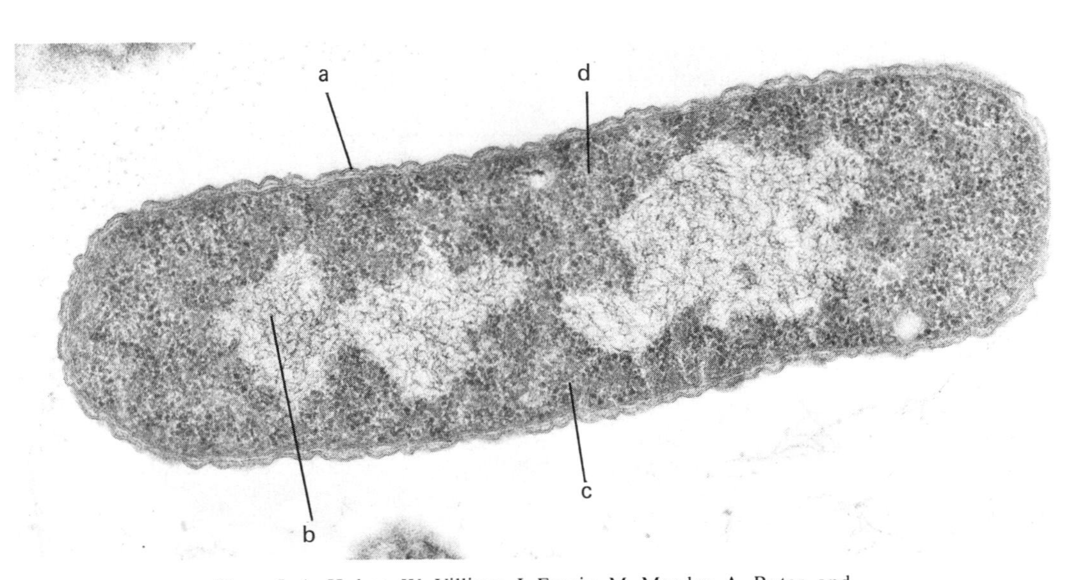

(From J. A. Hobot, W. Villiger, J. Escaig, M. Maeder, A. Ryter, and E. Kellenberger, *J. Bacteriol.* **162**:960-971 [1985]).

Figure 7-2

Answers

1a. rods
1b. cocci
1c. spirals
1d. squares
2. tetrad
3. staphylococcus
4. coccobacilli
5a. grinding with abrasives
5b. pressure cell disintegration
5c. sonic and ultrasonic disintegration
6a. flagella
6b. pili
6c. spines
6d. spirae
7. Brownian movement
8. atrichous
9a. basal granule or body
9b. cytoplasmic membrane
10. thinner
11. pilin
12. mucoid
13a. N-acetyl-glucosamine
13b. N-acetyl-muramic acid
14. diaminopimelic acid
15. cell wall
16. single chromosome
17. plasmids
18. carboxysomes
19a. calcium ions
19b. dipicolinic acid
20. survival
21a. narrow temperature range
21b. suitable pH
21c. increased oxygen for certain cells
22a. activation
22b. germination
22c. outgrowth
23a. heterocysts
23b. akinetes
24a. capsules
24b. S-layers
25. positive chemotaxis
26. permeases
27. periplasmic space
28. bacterial spores
29a. cell wall
29b. nucleoid
29c. cell or plasma membrane
29d. ribosomes

IV. CONCEPTS AND TERMINOLOGY

Bacterial cells are distinguished by morphological features, including size, shape, patterns of cell arrangement, and ultrastructure. Many of these properties are important for the identification of a particular species, uncovering specific structures and their related functions, and correlating the various intracellular structure with the overall functioning of the organisms.

Bacterial Size

Procaryotes such as bacteria border on the limits of the resolution of the bright-field microscope. Many of these organisms range in size from 0.2 to 1.2 μm in diameter and 0.4 to 14 μm in length.

Shapes and Patterns of Arrangement

Individual bacteria have one of four shapes: spherical, rodlike, spiral, or square.

Spherical bacteria, called *cocci* (singular *coccus*), can assume a variety of arrangements. Among the most common are:

1. Diplococci—pairs of cells
2. Streptococci—chains of four or more cells
3. Tetrad—four cocci in a boxlike or square arrangement
4. Sarcinae—cubical packet consisting of eight cells
5. Staphylococci—irregular grapelike clusters of cocci

Rodlike bacteria may occur singularly, in pairs (diplococci), or in chains (streptococci). Small, rounded cell rods that are difficult to distinguish clearly from cocci are referred to as coccobacilli. Spiral organisms, known as *spirilla* (singular *spirillum*), exhibit significant differences as to the number and fullness of spiral turns or coils. *Vibrios* are bacteria that consist of only a portion of a spiral. Square cells appear in flat, rectangular boxes with perfectly straight edges.

Structures and Functions

Most of the structures of procaryotes, together with their functions, activities, and chemical composition, are summarized in Table 7-1.

TABLE 7-1 Procaryotic Structures—Their Properties, Functions, Activities, and Chemical Composition

Structure	*Major Chemical Components*	*Properties, Functions, and Activities*
Akinete	General components of a cyano-bacterial cell	1. Limited protection 2. Resting cell (spore) 3. Nitrogen fixation
Axial filament	Protein	Movement in spiral types of organisms
Carboxysome	Protein	Utilization of carbon dioxide
Cell membrane	Protein, fatty acids; no sterols	1. Selective barrier between the cell's interior and exterior 2. Biosynthesis 3. Chromosome separation
Cell wall	Amino sugar (*N*-acetylglucosamine and *N*-acetylmuramic acid), protein lipopolysaccharides, and porins in gram-negatives	1. Encloses procaryotic cell 2. Provides shape and mechanical protection 3. Contains bacterial virus receptor sites
Chlorobium vesicle	Protein, lipid, photosynthetic pigment	Photosynthesis
Cyst	General components of procaryotic cell	1. Limited protection 2. Resting stage
Endospore[a]	General components of a procaryotic cell plus calcium and dipicolinic acid (DPA)	1. Protection against physical heat, pH changes, and drying 2. Cellular differentiation 3. Reproduction for some cyanobacteria
Exospore	General components of a procaryotic cell; lacks dipicolinic acid	1. Protection against physical heat and drying 2. Cellular differentiation
Flagellum	Protein	Movement
Gas vesicle	Protein and common gases	1. Regulates buoyancy 2. Light shielding
Genome (nuclear region or nucleoplasm)	Deoxyribonucleic acid	Contains all of the genetic information of the procaryote
Glycocalyx (includes capsules, and S layers)	Glycoprotein, polysaccharide	1. Protection against antibiotics, bacteriocins, bacterial viruses, immunoglobulins, and phagocytosis 2. Increases virulence 3. Enables bacteria to stick to other cells and inert substances
Heterocyst	Protein and lipid	Nitrogen fixation
Mesosome	Protein and lipid	1. Nucleoplasm division 2. Sporulation 3. Biosynthesis 4. Cell wall formation
Metachromatic granules	Nucleic acids, lipid, protein, and phosphate	Storage of reserve nutrients
Periplasmic space	Not applicable	Found in gram-negatives; provides environment for enzymatic activities involving preprocessing of nutrients and postprocessing of cellular waste products

TABLE 7-1 Procaryotic Structures—Their Properties, Functions, Activities, and Chemical Composition *(continued)*

Structure	*Major Chemical Components*	*Properties, Functions, and Activities*
Plasmid	Extrachromosomal DNA	Carries genetic factors associated with drug resistance and certain metabolic enzymes
Pilus	Protein	1. Attachment 2. Transfer of genetic material 3. Receptor sites for viruses
Ribosome	Protein and ribosomal RNA	Protein synthesis
Spine	Protein	Unknown
Spira	Protein	Unknown
Thylakoids	Protein, lipid, and photosynthetic pigment	Photosynthesis

[a]Endospores of cyanobacteria differ in both chemical composition and function from those of other procaryotes.

Specialized Cells

The cellular events in the cycles of certain procaryotes may change and lead to the formation of new cell types. This type of activity is *differentiation* at a primitive level. In bacteria, *dormant,* or resting, structures of four kinds can be produced: heat *endospores* and *exospores, cysts,* and heat-susceptible *conidia.* Endospores, exospores, cysts, and conidia are formed asexually.

Table 7-2 lists selected characteristics of these specialized cells.

TABLE 7-2 Selected Characteristics of Bacterial Dormant Structure

	Structures			
Property	*Endospores*	*Exospores*	*Cysts*	*Conidia (Heat-Susceptible Spores)*
Heat resistance	Characteristically present	Absent	Absent	To a limited degree
Cortex	Present	Absent	Absent	Absent
Dipicolinic acid (DPA)	Present	Absent	Absent	Absent
Number formed per cell	1	1-4	1	Formed in chains

V. CHAPTER SELF-TEST

Continue with this section only after you have read and have studied the tables and figures in Chapters 2, 3, and 7. A score of 80 percent or better is good. If your score is less than 65 percent, reread the chapters.

Correct answers to all questions can be found at the end of this section. Write your responses in the appropriate space provided.

Completion Questions

Provide the correct term or phrase for the following. Spelling counts.

1. Bacterial cells are about the size of the eucaryotic organelle, ______________________.
2. Four or more cocci in a chain exhibit a(an) ______________________ arrangement.

3. The function of bacterial spines is ________________.
4. The to-and-fro motion known as Brownian movement is caused by ________________.
5. Flagella all around a bacterial cell is called ________________.
6. Organized accumulations of gelatinous material on bacterial cell walls are called ________________.
7. Three examples of bacteria whose virulence is associated with encapsulation are (a) ________________, (b) ________________, and (c) ________________.
8. The ________________ separates the outer membrane from the plasma membrane.
9. The subunits of bacterial ribosomes are (a) ________________ and (b) ________________.
10. The movement of flagellated bacteria is associated with mechanical changes in the (a) ________________ and (b) ________________.
11. The glycocalyx provides protection against antibacterial agents such as (a) ________________, (b) ________________, (c) ________________, and (d) ________________.

Matching

Select the answer from the right-hand side that corresponds to the term or phrase on the left-hand side of the question sheet. An answer may be used more than once. In some cases, more than one answer may be required.

Topic: Structure and Function

______	12. Exospore	a.	locomotor organelle
______	13. Nucleoid	b.	regulation of substances into and out of cells
______	14. Ribosome	c.	protection against heat and drying
______	15. Spine	d.	pellicle formation
______	16. Sex pili	e.	protection against phagocytosis
______	17. Cell wall	f.	contains receptor sites for viruses
______	18. Axial filament	g.	extrachromosomal pieces of DNA
______	19. Capsule	h.	contains genetic information of cell
______	20. Cytoplasmic membrane	i.	involved with protein synthesis
______	21. Slime	j.	photosynthesis
______	22. Protoplast	k.	nitrogen fixation
______	23. Murein sacculus	l.	unknown function
______	24. Plasmid	m.	none of these
______	25. Thylakoids		
______	26. Akinetes		
______	27. Gas vacuoles		
______	28. Endospores		
______	29. Cysts		

Topic: Major Chemical Composition

______ 30.	Amino sugars, lipopolysaccharides, and protein	a.	cell wall
		b.	capsule
______ 31.	Polysaccharides, proteins, or both	c.	endospore
______ 32.	Calcium ions and dipicolinic acid	d.	spine
______ 33.	Primarily protein	e.	flagellum
______ 34.	DNA	f.	pilus
		g.	genome

Essay Questions

Answer the following questions on a separate sheet of paper.

35. Of what benefit are flagella to bacterial cells?

36. You are given the assignment to construct an efficient procaryotic cell system by a genetic engineering firm. However, you can only use four organelles. Indicate which four organelles you would choose and the reasons for your choices.

Photographic Question

37. Identify and give the function(s) of the labelled parts of the cells shown in Figure 7-3.

(a) ______________________ (b) ______________________ (c) ______________________

(d) ______________________ (e) ______________________

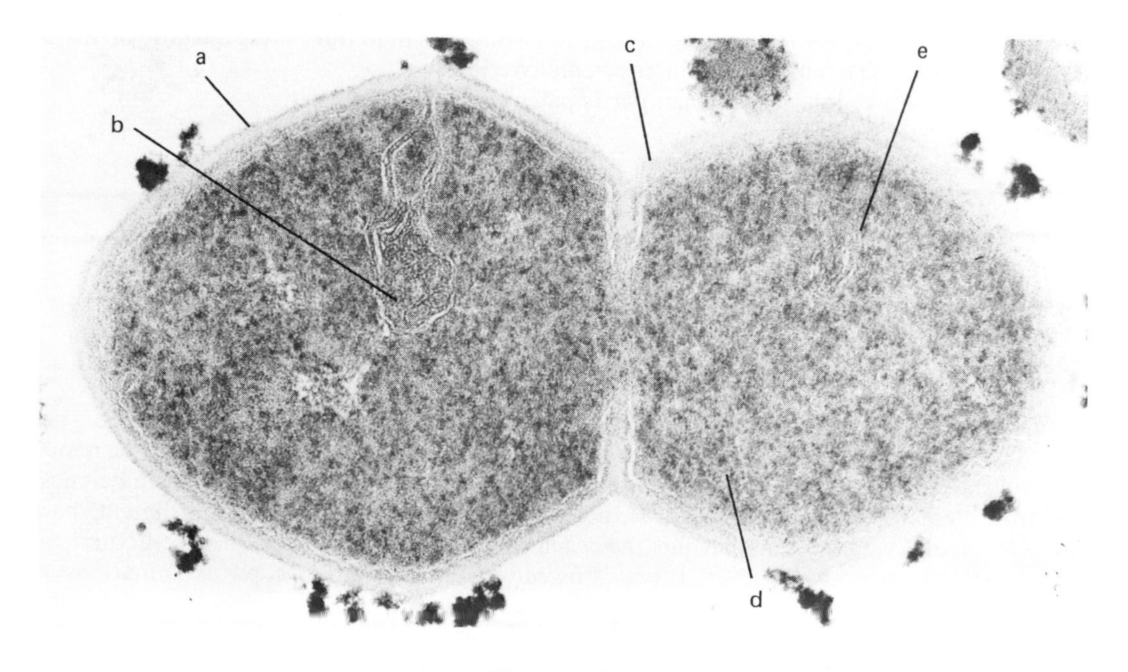

Figure 7-3

Answers

1. mitochondrion
2. streptococcus
3. unknown
4. the bombardment of cells by molecules in surrounding fluid environments
5. peritrichous
6. capsule

7a. *Bacillus anthracis*
7b. *Clostridium perfringens*
7c. *Streptococcus pneumoniae*
8. periplasmic space
9a. 50S
9b. 30S
10a. basal body
10b. ATP generation
11a. antibiotics
11b. bacterial viruses
11c. immunoglobulins
11d. phagocytes
12. c
13. h
14. i
15. m
16. m
17. f
18. a
19. e
20. b
21. m
22. m
23. m
24. g
25. j
26. k
27. m
28. c
29. m
30. a
31. b
32. c
33. d, e, f
34. g

35. Bacterial cells benefit from flagella in the following ways: (a) they can migrate toward favorable environments and away from unfavorable or harmful ones; (b) they can increase the concentration of nutrients or decrease the concentration of poisonous materials near their surfaces by causing a change in the flow rate of environmental fluids; and (c) they can migrate to uninhabited areas where colony formation can be achieved. It has also been suggested that flagellated pathogens may more easily penetrate certain host defense barriers, such as mucous secretions.

36. The four organelles that would be used for the construction of an efficient procaryotic cell and the reasons for their inclusion are as follows:
    a. *Ribosomes.* These organelles are essential for protein synthesis.
    b. *Nucleoid.* This organelle is important to the regulation of cellular activities such as metabolism and reproductive capabilities.
    c. *Mesosomes.* These organelles are needed for the generation of ATP to drive cellular activities.
    d. *Cell membrane.* This organelle is necessary to contain the other three organelles and to efficiently regulate the various substances entering and leaving the proposed cell.

37. a. *Cell wall.* The structure provides cellular form or shape.
    b. *Mesosome.* This structure is used to generate ATP for cellular activities.
    c. *Cell membrane.* This structure regulates the passage of materials into and out of the cell; it is a selective barrier between the cell's interior and exterior.
    d. *Ribosomes.* This cellular component participates in protein synthesis.
    e. *Mesosome.* See 37b answer.

VI. ENRICHMENT

Eucaryotes, Procaryotes, and Viruses

Three basic levels of organizational complexity are found among living systems—the eucaryotic, procaryotic, and viral levels. The differences among these three groups are considerable, as this chapter and the remaining ones in the division will show. Generally, there is no doubt as to which group a particular organism belongs.

The development of the electron microscope, a remarkable technological accomplishment, has provided scientists with a view of living systems beyond the reach of the light microscope. The structural distinctions among eucaryotic and procaryotic cells were revealed by careful electron microscopic examinations of different bacterial animal and plant cells. Other studies demonstrated the noncellular nature of viruses. The following table compares these living systems as to their structural properties and emphasizes the most important differences among eucaryotes, procaryotes, and viruses. Refer to this table after you have completed the other chapters in this division. The table will serve as a good summary.

TABLE 7-3 Structural Comparison of Eucaryotes, Procaryotes, and Viruses

	Life Form				
	Eucaryote				
Organelle or Related Structure	***Animal***	***Plant***	***Protist***	***Procaryote***	***Virus***
9 + 2 cilia or flagella	Often present	Generally absent	Present with many protozoa and some fungi	Absent (primitive flagella present)	Absent
Pili (fimbriae)	Absent	Absent	Present with some fungi	Present	Absent
Capsule	Absent	Absent	Present with some fungi	Present	Absent
Cell membrane	Present	Present	Present	Present	Absent
Cell wall	Absent	Cellulose-containing	Chitin, cellulose, or other polysaccharide-containing fungi and algae	Amino sugar-containing	Absent
Chromosome	Multiple containing DNA and protein	Multiple containing DNA and protein	Multiple containing DNA and protein	Single DNA circular molecule	Single DNA circular molecule
Centrioles	Present	Not observed	Absent	Absent	Absent
Endoplasmic reticulum (both smooth and rough)	Present	Present	Present	Absent	Absent
Golgi bodies	Present	Present	Present	Absent	Absent
Lysosomes	Often present	Usually absent	Usually absent	Absent	Absent
Mitochondria	Present	Present	Present	Absent	Absent
Mesosome	Absent	Absent	Absent	Present	Absent
Well-defined nucleus (nuclear membrane present)	Present	Present	Present	Absent	Absent
Pili	Absent	Absent	Absent	Present	Absent
Plastids	Absent	Present (chloroplasts and related structures)	Present in photosynthesizers	Primitive chloroplasts present in photosynthesizers	Absent
Ribosomes	Present	Present	Present	Present (though smaller)	Absent
S-layer	Absent	Absent	Absent	Present	Absent
Heat-resistant spores	Absent	Absent	Absent	Present	Absent
Vacuoles	Small or absent	Present in various sizes	Present in various sizes	Present in certain forms	Absent

8

Fungi

I. INTRODUCTION

The kingdom of Fungi includes a wide variety of organisms such as the molds, mushrooms, puffballs, rusts, smuts, and yeasts. This chapter is concerned with the structure, organization, distribution, and activities of this most interesting group.

II. PREPARATION

Chapters 2, 3, 4, and 8 should be read before continuing. The following terms are important for you to know. Refer to the glossary and the appropriate chapters of the text if you are uncertain of any of them. A pronunciation guide for selected terms is provided as a learning aid.

1. aflatoxin
2. *Amanita* (ah-man-EE-tah)
3. anamorph
4. antibiotic
5. arthroconidium (ar-throw-koh-NID-ee-um)
6. arthrospore (ar-THROW-spor)
7. ascocarp (AS-koh-carp)
8. Ascomycotina
9. ascospore (AS-koh-spore)
10. ascus (AS-kus)
11. asexual
12. basidiocarp (bah-SID-ee-oh-carp)
13. Basidiomycotina
14. basidiospore (bah-SID-ee-oh-spore)
15. blastoconidium
16. cell wall
17. chitin (KI-tin)
18. conidium
19. conjugation (kon-ju-GAY-shun)
20. dimorphic
21. Deuteromycotina
22. germinate
23. heterotroph
24. hypha (HI-fah)
25. medium
26. microfibril
27. mold
28. multinucleate
29. mushroom
30. mycelium (my-SEE-lee-um)
31. mycotoxin
32. *Neurospora* (new-RAH-spor-ah)
33. parasitic
34. pathogen

35. plasmodia (plaz-MOH-dee-ah)
36. pseudomycelium
37. rhizoid (RYE-zoid)
38. Sabouraud's agar
39. saprophyte
40. slime mold
41. sporangiospore
42. spore
43. sporocarp
44. synthetic medium
45. telomorph
46. thermophilic
47. toxin
48. truffle
49. yeast
50. zoospore
51. Zygomycotina
52. zygospore

III. PRETEST

Correct answers to all questions can be found at the end of this section. Write your responses in the appropriate space provided.

Completion Questions

Provide the correct term or phrase for the following. Spelling counts.

1. Yeast cells are ______________________ than bacteria.
2. Fungi exhibit a(an) (a) ______________________ type of cellular organization and do not have (b) ______________________ pigments.
3. The form of nutrition found with fungi is ______________________.
4. Molds show differentiation even though they do not have (a) ______________________, (b) ______________________, or (c) ______________________.
5. The basic structural unit of a mold is the ______________________.
6. The fungal colony is called a(an) ______________________.
7. A(An) ______________________ mycelium extends below the surface of the medium on which a mold grows.
8. The portion of the mold growth that is above the substrate material is called the ______________________.
9. Rhizoids function to (a) ______________________ and to (b) ______________________.
10. Several yeasts reproduce asexually by ______________________.
11. Exhibiting different forms under different environmental conditions is known as ______________________.
12. Fungal spores can be used for species (a) ______________________ and (b) ______________________.
13. Examples of asexual spores of fungi include (a) ______________________, (b) ______________________, (c) ______________________, (d) ______________________, (e) ______________________, and (f) ______________________.

14. Fungal spores germinate under suitable conditions of (a) ________________, (b) ________________, (c) ________________, and (d) ________________.
15. Fungal sexual spores are produced by ________________.
16. Two examples of fungal sexual spores are (a) ________________ and (b) ________________.
17. The membranes of fungi contain ________________, compounds not found with the procaryotes.
18. The fungal cell wall is ________________ to the cell membrane.
19. Fungi, but not the majority of bacteria, grow well on media with a pH of (a) ________________ to (b) ________________.
20. In baking with yeast, these organisms produce (a) ________________, the actions of which bring about the formation of (b) ________________ and (c) ________________.
21. Two examples of edible ascomycetes are (a) ________________ and (b) ________________.
22. Certain yeasts are important sources of ________________ vitamins.
23. Mushroom poisonings, when not associated with bacteria, are most often caused by species of ________________.
24. The unique flavor and appearance of Roquefort cheese is due to a species of ________________.
25. Poisons produced by fungi are generally called (a) ________________. The poisons produced by *Aspergillus flavus* growing on cereal products and nuts are (b) ________________.
26. Hyphae that enable certain fungi to spread horizontally are called ________________.
27. Slime molds can be divided into two groups: (a) ________________ and (b) ________________.
28. In their mode of nutrition, slime molds can be either (a) ________________ or (b) ________________.

Answers

1. larger
2a. eucaryotic
2b. photosynthetic
3. heterotrophic
4a. true roots
4b. stems
4c. leaves
5. hypha
6. mycelium
7. vegetative
8. aerial mycelium
9a. obtain food
9b. anchor the mycelial growth
10. budding
11. dimorphism
12a. identification
12b. classification
13a. arthroconidium
13b. blastoconidium
13c. chlamydoconidium
13d. conidia
13e. sporangiospores
13f. zoospores
14a. nutrition
14b. pH
14c. moisture
14d. temperature
15. nuclear fusion
16a. ascospores or basidiospores
16b. oospores or zygospores
17. sterols
18. external
19a. 5, 6
19b. 6
20a. sugar-fermenting enzymes
20b. alcohol
20c. carbon dioxide
21a. morel
21b. truffle
22. B complex
23. *Amanita*
24. *Penicillium*
25a. mycotoxins
25b. aflatoxins
26. stolons
27a. cellular
27b. acellular
28a. saprophytic
28b. parasitic

IV. CONCEPTS AND TERMINOLOGY

The fungi, a unique group of eucaryotic, nonphotosynthetic microorganisms, are larger and more complex than are bacteria.

Distribution and Activities

Fungi are widespread in nature and include a large number of different forms such as mushrooms, puffballs, woody bracket fungi, and the smaller molds and yeasts. Fungi are found in terrestrial and aquatic environments.

Fungi are heterotrophic in their mode of nutrition. Such organisms, called *saprophytes,* utilize the organic products of other organisms, either living or dead, as sources of energy. Fungi produce digestive enzymes that have beneficial effects such as the recycling of chemical elements to the soil for plant growth and development. Enzymes of certain fungi are responsible for diseases of animals and plants, as well as food spoilage.

Structure and Function

Fungi differ from bacteria in several ways, including size, structural development, cellular organization, and methods of reproduction. The designation *fungus* is a general term to include the two different forms—*molds* (which appear as cottony growths on surfaces) and *yeasts* (which appear as smooth, glistening colonies).

Molds do not have true roots, stems, or leaves. They show some differentiation in their general organization. Most molds consist of tubular, branching cells called *hyphae* (singular *hypha*) which, depending on the species, may or may not have *cross walls* or *septa.* Septa in several fungi are perforated by one or more pores.

As molds grow and hyphae branch and intermingle, accumulation of fungal cells form visible masses called *mycelia* (singular *mycelium*). Mycelia, which are analogous to bacterial colonies, may be dry and powdery. The portions of mycelia appearing above surfaces of material on which molds grow (substrate) are the reproductive portions (aerial mycelia). Most other parts in direct contact with substrates obtain nutrients and occasionally anchor the mycelium.

Yeasts, which are oval to spherical in shape, form glistening colonies. Many commonly reproduce by the asexual process of budding. Pseudomycelia are formed when newly formed cells (*buds*) do not separate and remain connected.

Several fungi exhibit two distinct forms under different environmental conditions. This property is called *dimorphism.*

Reproduction and Spores

Spores are reproductive units of fungi. They can be formed both asexually and sexually, and serve as a basis for identification and classification. Fungal spores do not exhibit the heat resistance associated with bacterial spores. Fungal spores are not more resistant than any other type of eucaryotic cell. The properties of both sexual and asexual spores are summarized in Table 8-1.

Spores may be produced on or within hyphae or isolated cells. In several complex fungi multicellular spore-forming structures, *sporocarps* are produced. The mushroom is an example of such a structure.

Fungal spores give rise to new cells (hyphae) by germinating under suitable conditions of moisture, nutrients, pH, and temperature. The general relationships of fungal parts are shown in Figure 8-1.

The Ultrastructure of Fungi

Both yeasts and molds exhibit a eucaryotic cellular organization. Membranes of these organisms contain sterols such as cholesterol-like compounds. The cell walls of filamentous fungi consist of threadlike structures called *microfibrils.* Microfibrils consist chemically of the complex polysaccharide *chitin.* Cellulose, another complex polysaccharide, also can be found in fungal cell walls.

Yeast cell walls contain a variety of polysaccharides such as glucan and mannan, lipids, proteins, and the amino sugar glucosamine. Some yeasts also have been found to have protein surface structures known as *pili.* Such structures may have a role in conjugation, a form of sexual reproduction.

Cultivation of Fungi

The cultural methods used for molds and yeasts are similar to those used for bacteria. Most fungi are aerobic but grow more slowly than bacteria. Various substances can be added to media to discourage the growth of bacteria. Maintaining the pH of a medium for molds and yeasts at 5.6 is very effective in such cases.

The appearance of mycelia (e.g., color, texture) can be useful for identification purposes.

TABLE 8-1 Properties of Fungal Spores

Spore Type	*Site and/or Type of Formation*	*Single or Multicellular*	*Shape*	*Resistance to Environment*	*Examples of Genera That Form Spore Type*[a]
Asexual					
Arthro-conidium (arthrospore)	Fragmentation of hyphae	Single	Cylindrical to round	Usually none	*Coccidioides, Geotrichum, Trichosporon*
Blasto-conidium (blastospore) (buds)	Formed on main cell	Single	Round to oval	Usually none	*Candida, Saccharomyces*
Chlamydo-conidium (chlamydo-spore)	Enlargement of terminal hyphal cells	Single	Considerable variation but usually round	These thick-walled cells exhibit unusual resistance to drying and heat	*Candida, Mucor*
Conidium (conidio-spore)	Borne on specialized hyphal branches, conidiophores	Single (microconidia)	Round to oval	Usually none	*Aspergillus, Cephalosporium, Penicillium*
		Multicellular (macroconidia)	Long and tapering	None	*Alternaria, Microsporum, Trichophyton*
Philalospore (modified conidium)	Borne on specialized hyphal branches, conidiophores, philalides	Single	Round to oval	Usually none	*Philalophora*
Sporangio-spore	Formed within sacs, sporangia, at end of hyphal cells	Single	Round	None	*Absidia, Mucor, Rhizopus*
Zoospore	Formed within sacs, sporangia, at end of hyphal branches	Single, flagellated	Round	None	*Saprolegnia*
Sexual					
Ascospore	Formed with sac-like cells, asci, after cellular and nuclear union	Single (usually 8 per ascus)	Round to oval	None	*Allescheria, Neurospora*
Basidio-spore	Formed at end of club-shaped structures, *basidia*	Usually single (usually 4 in number)	Round to oval	None	*Amanita, Agaricus, Coprinus*
Oospore	Developed within a fertilized egg cell, *oogonia*	Single (usually 1 to 20 per oogonium)	Round	More resistant than most asexual spores	*Saprolegnia*
Zygospore	Formed after cellular and nuclear fusion	Large, thick-walled, single structure	Round to oval	None	*Rhizopus*

[a]Certain fungi can form more than one type of spore.

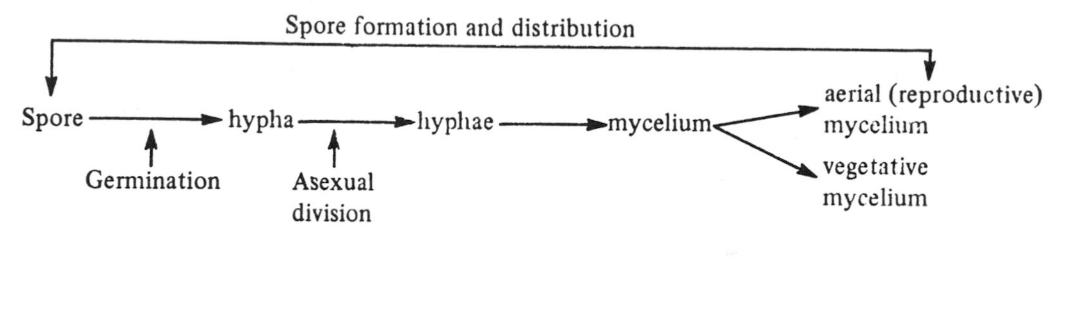

Yeast

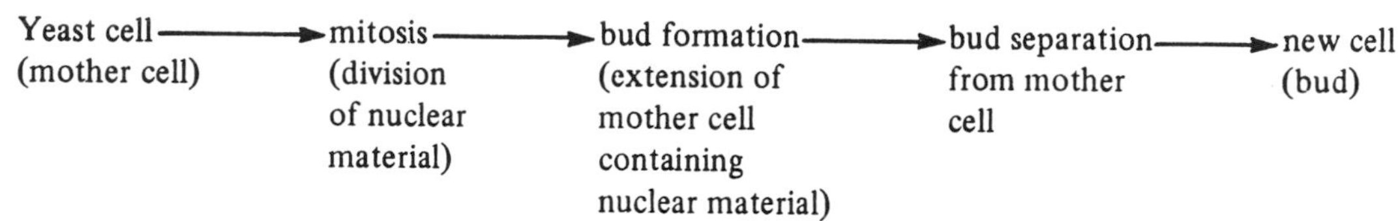

Figure 8-1 Diagrammatic representations of general life fungal asexual reproduction cycles: molds and yeast.

Classification

Several properties of fungi are important to their classification and identification: method of reproduction, mycelial formation, and cellular structure and formation. Additional approaches based on biochemistry, physiology, and mathematical analyses are used to determine relationships, compare properties, and analyze the degrees of similarities and differences that exist among fungi. Table 8-2 summarizes several properties of five groups of fungi.

Gymnomyxa, the Slime Molds and Associates

Slime molds, so called because of the slimy appearance of their vegetative forms (*slugs* or *plasmodia*), have properties resembling those of both fungi and protozoa. Most slime molds are saprophytic; however, some are parasitic and destructive to plants.

Two groups of slime molds are recognized: *cellular* and *acellular*. The vegetative forms of the former consist of amoebalike cells, whereas those of the latter exhibit an indeterminate size and shape. One outstanding feature of these organisms is their unique life cycle, which involve forms of the organism resembling both protozoan and fungal cells.

V. CHAPTER SELF-TEST

Continue with this section only after you have read and studied the tables and figures in Chapter 8. A score of 80 percent or better is good. If your score is less than 65 percent, reread the chapter.

Correct answers to all questions can be found at the end of this section. Write your responses in the appropriate space provided.

Completion Questions

Provide the correct term or phrase for the following. Spelling counts.

1. With respect to size, molds are ____________________________ than procaryotes.
2. Yeast cells exhibit an ____________________________ type of cellular organization.

TABLE 8-2 Selected Characteristics of the Major Subdivisions of the Kingdom of Fungi

		Sites of Spore Formation			
Subdivision	*Type of Mycelium*	*Asexual*	*Sexual*	*Representative Groups*	*Distinctive Activities*
Ascomycotina (Ascomycetes)	Septate	At the tops of hyphae	Within sacs; some species form multicellular fruiting bodies, *ascocarps*[a]	Morels, truffles, *Claviceps purpurea,* and yeasts	Bread and alcoholic beverage production; animal and plant pathogens; edible forms (e.g., truffles and morels)
Basidiomycotina (Basidiomycetes)	Septate	At the tips of hyphae	On a surface of a basidium	Poisonous mushrooms (*Amanita* spp.), mushrooms, rusts, smuts	Edible as well as poisonous forms
Deuteromycotina (Deuteromycetes, or Fungi Imperfecti)	Septate	At the tips of hyphae	None present	Most human pathogens	Cheese and antibiotic production; animal and plant pathogens; mycotoxin production including aflatoxins
Oomycetes[b]	Aseptate	In sacs	Within a unicellular female sex organ (oogonium)	Some aquatic forms, pathogens responsible for powdery mildew, blights of plants, and fish infections	Plant pathogens
Zygomycotina (Zygomycetes)	Almost completely aseptate (coenocytic)	In sacs	In mycelium	Bread mold (*Rhizopus nigricans*), aquatic species	Spoilage organisms and animal pathogens

[a]Four major types are known: apothecium, cleistothecium, locule, and perithecium.
[b]This is a class belonging to the division of Mastigomycote.

3. One way in which the cell membranes of fungal cells differ from those of procaryotes is that they contain chemicals such as ______________________.
4. The basic reproductive structure of a mold is the ______________________.
5. A mold colony is called a(an) ______________________.
6. Several yeasts reproduce asexually by ______________________.
7. The basic structural unit of molds is the ______________________.
8. Poisons produced by fungi are generally called ______________________.
9. In using yeasts for beer production, the fermentative actions of these organisms result in the formation of the gas ______________________.
10. A yeast used in the production of beer and in the making of bread is ______________________.
11. The mode of nutrition found among the fungi is ______________________.
12. The microfibrils of fungi are composed of ______________________.
13. The saclike structure enclosing ascospores is called a(an) ______________________.

True or False

Indicate all true statements with a "T" and all false statements with an "F".

_______ 14. Fungal spores germinate only under conditions of extreme heat.

_______ 15. Molds can grow well on media having a pH range of 5.6 to 6.

_______ 16. A truffle is an example of an edible basidiomycete.

_______ 17. Aflatoxins are generally produced by several yeast species.

_______ 18. A yeast is a special type of mold.

_______ 19. A zygospore is an example of an asexual spore.

_______ 20. The surfaces of fungi are important in sexual reproduction.

_______ 21. A septated mycelium contains cross walls.

_______ 22. Lysergic acid diethylamide (LSD) is a secretion product of certain ascomycetes.

_______ 23. *Amanita* species are noted for their characteristic flavoring of cheeses.

_______ 24. The vegetative forms of slime molds are called hyphae.

_______ 25. Aflatoxins primarily poison plants.

Essay Questions

Answer the following questions on a separate sheet of paper.

26. What is an ascocarp? How many types of ascocarps are known?
27. What properties of slime molds are unlike those of true fungi?

Photographic Quiz

28. Identify the fungus shown in Figure 8-2. ______________________________
29. Name the labeled parts of the fungus shown in Figure 8-2. (a) ______________________________,
    (b) ______________________________, and (c) ______________________________

Answers

1. larger
2. eucaryotic
3. sterols
4. spore
5. mycelium
6. budding
7. hypha
8. mycotoxins
9. carbon dioxide
10. *Saccharomyces cerevisiae*
11. heterotrophic
12. chitin
13. ascus
14. F
15. T
16. F
17. F
18. F
19. F
20. T
21. T
22. T
23. F
24. F
25. F
26. An ascocarp is a multicellular fruiting body found among the ascomycotina. Asci may or may not be produced within or upon such structures.

    Four major types of ascocarps are known: *apothecium, cleistothecium, locule,* and *perithecium.* The organization and morphological differences among ascocarps are of value in fungal classification.

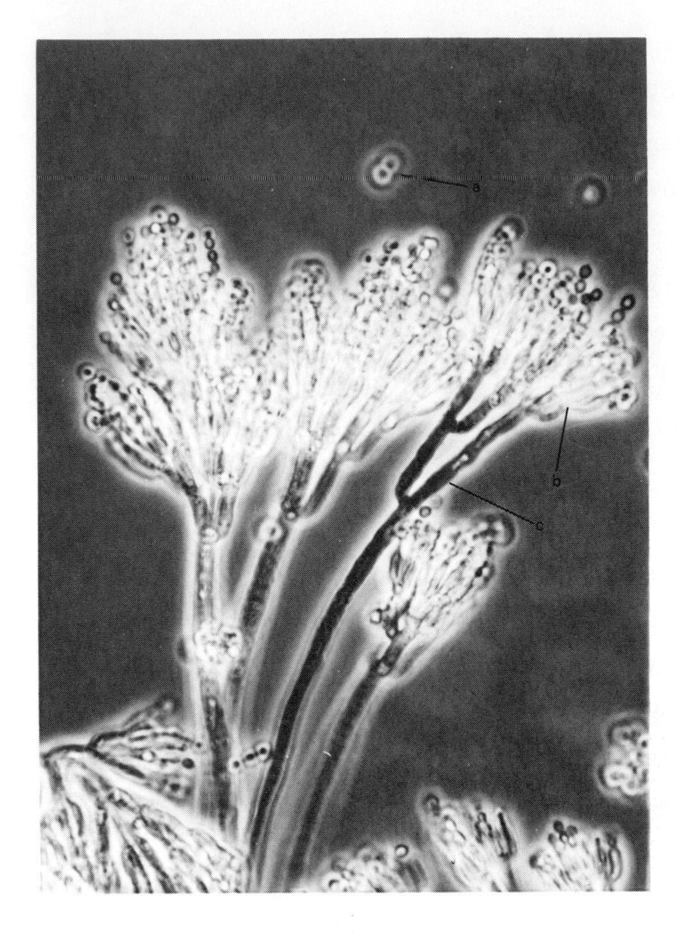

(Courtesy of A. Ciegler, D. I. Fennel, G. A. Sansing, R. W. Detroy, and G. A. Bennett, *Appl. Microbiol.* **26**:271-278 [1973]).

Figure 8-2

27. Slime mold have several properties that are unlike those of true fungi. For example, these forms of life have a naked cell mass of protoplasm, called *slugs* or *plasmodia* (singular *plasmodium*), that exhibit animal-like movement and may creep about. In addition, their actively feeding vegetative structures are composed of masses of amoebalike cells without cell walls.

28. *Penicillium* species
29a. conidiospore
29b. sterigma
29c. conidiophore

VI. ENRICHMENT

Key Terms

commensalism (koh-MEN-sal-izm): a situation in which two organisms live together and neither is harmed, but one is benefitted by the association

estrogen (ES-troh-jen): the female sex hormones

Candida albicans Pathogenicity and Predisposing Causes of Infection

Candida albicans is a fairly widespread commensal microorganism. It can be recovered from various body locations and up to about 40 percent in the normal population. Hospital patients show a somewhat higher percentage of recovery. In the majority of cases, the yeast is present in the absence of clinical symptoms. Some controversy exists, however, with regard to the presence of *C. albicans* in the vagina and its involvement in diseases. This appears to be especially the finding during pregnancy. The incidence of vaginal candidosis increases markedly during pregnancy, and abruptly decreases in the immediate post-partum (after delivery) period.

The question that needs to be asked and, of course, answered is "Why is *C. albicans* pathogenic in the vagina?" Understanding why this situation occurs might lead to an effective preventative approach in dealing with the disease state, and provide some insight into the properties of *C. albicans* and the factors that may trigger or predispose an individual to infection.

In general, *C. albicans* is considered to be an opportunist, and the trigger for infection is almost always some local or systemic (entire body) increase in host susceptibility. There are a number of such triggers or predisposing factors associated with vaginal candidosis; these include pregnancy, the use of oral contraceptives, immune defects, the use of clothing which increases the warmth and humidity of the vaginal environment, and antibiotic therapy.

Pregnancy and the use of oral contraceptives have been reported to cause an increased incidence of vaginal candidosis. It is believed that the estrogen content of a particular contraceptive preparation is probably the key factor as to whether *Candida* infection occurs or not. Hormones have been found to be bound by receptors on the organisms. The exact rolc played by such hormone binding on causing or predisposing an individual to vaginal infection remains to be determined.

Systemic fungal infections frequently occur as a result of a breakdown in a host's immune system. A number of findings suggest that vaginal candidosis could have a similar predisposing cause.

An interesting observation has been reported concerning the environmental influence of certain types of clothing on the incidence of candidosis. Documented evidence has been gathered that shows an increased prevalence of clinical vaginitis in women wearing tight-fitting clothing such as nylon underwear and tights. Women with loose-fitting and/or cotton clothing were not similarly affected. Apparently, the warmth and humidity generated by the tight-fitting undergarments contributed in some way to an environment favoring the transition of the commensal type of *Candida* to a parasitic one.

Although topical treatment with various antibiotic preparations or oral treatment with the antimycotic drug ketoconazole provide quick relief of symptoms in the majority of cases of vaginal thrush, such treatment does nothing to correct the underlying situation which allowed the infection to develop in the first place. The apparent failures of treatment, and relapses which occur within a matter of days or weeks following completion of treatment, probably are caused by the widespread occurrence of *C. albicans* and the continuing existence of one or more predisposing causes of infection. (Refer to Chapters 24 and 28 for additional features of *Candida* infections.)

9

The Protists

I. INTRODUCTION

Of the five kingdoms of organisms, the protists are far and away the most diverse in structure and life cycles. They include some of the simplest as well as many of the most complex forms of life. This chapter surveys the protozoa and algae, two forms of protists. The kingdom of Protista includes a variety of microbial groups whose members are predominantly unicellular. Some organisms included in this kingdom tend to have animallike features, others have plantlike qualities, and still others have features befitting fungi. However, they also share many properties. The activities, along with the structure and organization, of protists are presented in this chapter. Several portions of this chapter will serve as an important reference for Chapters 22, and 25 through 29.

II. PREPARATION

Read Chapters 2, 3, 5, and 9 before continuing. Knowing the following terms will help you to understand the subject matter of this chapter. Refer to the glossary and the appropriate chapters of your text if you are uncertain of any of them. A pronunciation guide for selected terms is provided as a learning aid.

1. African sleeping sickness
2. amebic dysentery (ah-ME-bik DIS-en-ter-ee)
3. *Amoeba* (ah-ME-ba)
4. aplanospore (a-plan-OH-spore)
5. autogamy (AW-toh-gah-me)
6. autotroph (AW-toh-troff)
7. budding
8. cellulose
9. cell wall
10. centric (SEN-trick)
11. chlorophyll
12. cilium (SIL-ee-um)
13. commensalism (koh-MEN-sal-izm)
14. conjugation (kon-ju-GAY-shun)
15. contractile vacuole (kon-TRAK-til VAK-yule)
16. cyanobacterium
17. cyst (sist)
18. cytostome (si-TOE-stom)
19. diatom (die-AH-tom)
20. diatomaceous (die-AH-tom-a-shus) earth
21. *Didinium* (dye-DIN-ee-um)
22. DNA
23. endozoic (en-doh-ZOH-ik)
24. *Entamoeba histolytica* (en-tah-ME-bah hiss-TOL-it-ee-kah)
25. epivalve
26. euglenoid (u-GLEN-oid)
27. filamentous (fill-ah-MEN-tus)

28. filopodium (fi-low-POH-dee-um)
29. flagellum (flah-JEL-um)
30. food chain
31. frustule (FRUS-tool)
32. gamete (GAM-eat)
33. heterotroph (HET-er-oh-troff)
34. holozoic (hole-oh-ZOH-ik)
35. hypertonic
36. hypovalve
37. karyogamy (kar-ee-oh-GAM-ee)
38. kinetoplast (ki-neh-TOE-plast)
39. lichen (LIE-ken)
40. lobopodium (low-boh-POH-dee-um)
41. loculi (LOK-u-lie)
42. macronucleus
43. malaria (mah-LARE-ee-ah)
44. meiosis (my-OH-sis)
45. mutualism (MEW-chu-al-izm)
46. organelle (OR-gah-nell)
47. parasite
48. pellicle (PELL-ee-kel)
49. pennate (PEN-ate)
50. phototaxis
51. plankton
52. plasmogamy (plaz-MOH-gam-ee)
53. plasmotomy
54. protista
55. pseudopodium (see-doh-POH-dee-um)
56. puncta (PUNK-tah)
57. pyrenoid
58. saprobic
59. saxitoxin (sak-see-TOK-sin)
60. sporozoa (spore-oh-ZOH-ah)
61. striae (STRY-ee)
62. symbiosis
63. symmetry (SIM-eh-tree)
64. test
65. toxoplasma (tok-soh-PLAZ-mah)
66. trichocyst (trick-OH-sist)
67. trophozoite (trof-oh-ZOH-ite)
68. zoospore (ZOH-oh-spore)
69. zygote (ZIE-goat)

III. PRETEST

Correct answers to all questions can be found at the end of this section. Write your responses in the appropriate space provided.

Completion Questions

Provide the correct term or phrase for the following. Spelling counts.

1. The organelles of locomotion found among protozoa are (a) ______________________________ , (b) ______________________, and (c) ______________________ .
2. The 9 + 2 microtubule arrangement is a characteristic of (a) ______________________________ and (b) ______
3. The three methods of obtaining food used by protozoa are (a) ______________________________ , (b) ______________________, and (c) ______________________ .

4. The form of asexual reproduction exhibited by multinucleated protozoa resulting in the formation of two or more multinucleated cells is called ______________________________ .

5. The ______________________________ of lost or damaged parts is a characteristic property of most protozoa.

Matching

Select the answer from the right-hand side that corresponds to the term or phrase on the left-hand side of the question sheet. An answer may be used more than once. In some cases, more than one answer may be required.

Topic: Structure and Function

______	6. Axopodia	a. excretion
______	7. Trichocyst	b. regulation of reproductive processes
______	8. Basal granule	c. osmoregulation
______	9. Micronucleus	d. locomotion
______	10. Cytopyge	e. protection
______	11. Pellicle	f. regulation of metabolism and development
______	12. Macronucleus	g. none of these
______	13. Cirri	
______	14. Cyst	
______	15. Lobopodia	
______	16. Contractile vacuole	
______	17. Test	

Topic: General Properties and Representatives of the Protozoa

______	18. Amebic dysentery	a. ciliophora
______	19. Leishmaniasis	b. sporozoa
______	20. Conjugation	c. flagellates
______	21. Tentacles	d. Sarcodina
______	22. Foraminifera	e. suctorians
______	23. *Didinium*	f. none of these
______	24. *Paramecium*	
______	25. Malaria	
______	26. Multiple fission	
______	27. African sleeping sickness	

Topic: Algal Protist Divisions

______ 28. Pellicle
______ 29. Saxitoxin
______ 30. Diatoms
______ 31. Valves
______ 32. Asexual reproduction only
______ 33. Eucaryotic organization
______ 34. Cellulose-containing cell walls
______ 35. Silicon in cell walls
______ 36. Diatomaceous earth
______ 37. Theca

a. golden algae
b. euglenoids
c. dinoflagellated
d. all of these
e. none of these

Topic: Structure and Function

______ 38. Raphe
______ 39. Diatomaceous earth
______ 40. Pellicle
______ 41. Frustule
______ 42. Striae
______ 43. Valve
______ 44. Auxospore
______ 45. Zoospore

a. provides protection to euglenoids
b. toxin gland
c. locomotor organelle
d. enables organism to float near water surface
e. used in filtration processes
f. sexual spore
g. motile spore
h. starch formation
i. cell wall component
j. none of these

Answers

1a. cilia
1.b flagella
1c. pseudopodia
2a. cilia
2b. flagella
3a. autotrophic
3b. saprobic
3c. holozoic
4. plasmotomy
5 regeneration
6. d
7. e
8. d
9. b
10. a
11. e
12. f
13. d
14. e
15. d
16. a and c
17. e
18 d
19. c
20. a
21. e
22. d
23. a
24. a
25. b
26. b
27. c
28. b
29. c
30. a
31. a
32. a
33. d
34. d
35. a
36. a
37. c
38. i
39. e
40. a
41. i
42. i
43. i
44. f
45. g

IV. CONCEPTS AND TERMINOLOGY

Protozoa belong to the kingdom of Protista and are unicellular, eucaryotic microorganisms. Many are motile and require organic nutrients, which they obtain from their environments. These organisms vary widely in shape, size, structure, and physiological properties.

Distribution and Activities

Protozoa are widely distributed in nature. Many have beneficial relationships with other organisms, whereas others are noted for their harmful, parasitic activities. Some protozoa are autotrophs; others are heterotrophs. Some heterotrophic protozoa absorb nutrients through the cell surface (*saprobic*); others ingest entire organisms or particles (*holozoic*).

Structure and Function

The organelles of protozoa are involved with *movement, nutrient gathering, excretion, osmoregulation, reproduction,* and *protection.* The properties of these structures are summarized in Table 9-1.

Trophozoites and Cysts

Trophozoites are active, normal feeding forms of several parasitic and free-living protozoa. These forms are sensitive to various unfavorable environmental factors, including chemicals, pH or temperature changes, and food deficiencies.

Cysts are resting cells that have a thick resistant covering. For some protozoa, the cyst serves as a means for reproduction.

Methods of Reproduction

Both asexual and sexual reproduction occur among protozoa. Table 9-2 describes the general methods of reproduction.

Cultivation

Not all protozoa can be cultured under laboratory conditions. Some must be grown on artificial media or in tissue culture systems. Others require the addition of other living protozoa or bacteria as food sources.

Classification

Several properties of protozoa are used for classification purposes: method of obtaining nutrients; method of reproduction; cellular organization, structure, and function; and organelles of locomotion. Table 9-3 summarizes several properties of the Ciliophora, Opalinata, Sarcodina, Mastigophora, and Sporozoae.

Algae: "The Grass of the Waters"

Algae serve as a basic food for numerous forms of life. They are members of the various floating populations of life in most bodies of water called *plankton.* Algae are *phytoplankton,* while animal-like organisms are *zooplankton.* In addition, algae fix or capture more carbon by photosynthesis than do all land plants combined.

These photosynthetic aquatic protists are generally free floating and free living. Some have a beneficial symbiotic relationship (mutualistic) with fungi in forming *lichens.* The presence of others can give water a disagreeable taste or, because of occasional spurts of growth, can produce *algal blooms,* which result in increased levels of algal waste products and disrupt aquatic communities. Table 9-4 lists the major divisions of algae and summarizes several of their features.

TABLE 9-1 Properties of Protozoan Organelles

Structure	*Function*	*General Description*
Pseudopodia	Locomotion Capturing food	Temporary protoplasmic extensions of cells. Types include *lobopodia* (finger-shaped, round-tipped), *filopodia* (thin-pointed), *rhizopodia* (branching, slender, pointed), and *axopodia* (slender structure with several fibers forming an axial filament).
Flagella	Locomotion Respond to chemicals and touch	Delicate whiplike structures whose beating action propels the organism. Individual flagella usually consist of two central microtubules surrounded by nine double tubules (9 + 2). Each flagellum terminates in a cytoplasmic DNA containing basal granule (kinetoplast).
Cilia	Locomotion	Miniature (shorter) flagella with a similar organization. Ciliary structures are coordinated by a network of fibers.
Cirri	Locomotion	Stiffened tufts of cilia.
Food cups	Food capture	Extensions of cells that surround food.
Cytosome (mouth)	Food capture	Direct means of ingestion.
Tentacles	Food capture and protection	Specific permanent structures used to spear prey.
Cytopyge	Elimination of indigestible material	Temporary or permanent cellular opening.
Contractile vacuole	Excretion and osmoregulation	Found in several freshwater ciliates and some amoebae. It separates a dilute solution of water and electrolytes from the cytoplasm for eventual elimination.
Pellicle	Protection against chemicals, drying, and mechanical injury	Surface covering of several flagellates, amoebae, and all ciliates.
Tests (shells)	Protection	Coverings found with several stationary and some free-living forms. Structures can consist of sand grains cemented together, or secretions of calcium carbonate or silica.
Trichocysts	Protection Food capture Anchoring to surfaces	Threadlike structures discharged in response to various stimuli.
Macronucleus	Regulation of metabolism and development Maintenance of visible features	A large cellular structure varying in size. Found in ciliates.
Micronucleus	Overall control of cell Regulation of reproduction	Small cellular structure found in ciliate.

Structure and Organization

Algae range in size from microscopic dimensions to lengths greater than sixty meters. Some are unicellular; others are highly differentiated and multicellular. Multicellular algae can be filamentous (threadlike) or exhibit colonial arrangements. Some may even have rootlike organs, stems, and leaves. Table 9-5 lists and describes several of the structures, cell forms, and arrangements found among the major algal protist divisions.

Algae exhibit a typical eucaryotic cellular organization, which includes a cell wall, nucleus, nucleolus, mitochondria, cell membrane, vacuoles, granules, and chloroplasts. Flagellated cells of most algal groups have directional light wave antennae. These specialized structures are used for phototaxis.

TABLE 9-2 Methods of Protozoan Reproduction

General Category of Reproduction	*Specific Methods*	*General Features*
Asexual	Binary fission	Division into two parts of original organisms. Most common form of reproduction. May be nuclear, cytoplasmic, or both.
	Budding	Formation of a new individual by an extension of the cell containing a full complement of genetic material. The extension eventually separates from its parent cell.
	Multiple fission (schizogony)	A multinucleated organism undergoes division, producing cells with a single nucleus.
	Plasmotomy	Division of multinucleated cells into two or more multinucleated daughter cells.
	Regeneration	Replacement of lost or damaged parts.
Sexual	Autogamy	A modified version of conjugation. The micronucleus of a cell divides into two parts that in turn reunite to form a zygote nucleus. Two cells eventually form, each with the full complement of nuclear structures.
	Conjugation	Involves the temporary union of two organisms for the transfer of one *micronucleus*. The donor nucleus unites with the recipient to form a zygote. Characteristically found with ciliates.
	Syngamy	The union of two different sex cells (gametes) resulting in the formation of a fertilized cell or *zygote*.

TABLE 9-3 Description of the Protozoa

			Selected Differentiating Properties—Method of Reproduction				
Phylum	*Subphylum or Class*	*Means of Movement*	*Asexual*	*Sexual*	*Repre-sentative*	*Habitats*	*Activities*
Ciliophora	Kinetofrag-minophorea	Cilia	Trans-verse fission	Conjugation	*Balantidium coli*,[a] *Euplotes*, *Paramecium* spp., *Stentor*, *Tetrahymena*, *Vorticella*	Fresh and salt water, animal intestinal tracts	Free living; some are parasitic.
Sarco-mastigophora	Opalinata	Cilia	Binary fission	Syngamy	*Opalina*, *Protoopalina*	Large intestines of frogs and toads	Generally commensals.
	Sarcodina	Pseudo-podia (false feet)	Binary fission	When present, involves flagellated sex cells	*Amoeba* spp., *Difflugia*, *Entamoeba histolytica*[a]	All bodies of water, soil, certain animal tissues	Free living; some are parasitic.

TABLE 9-3 Description of the Protozoa *(continued)*

Phylum	Subphylum or Class	Means of Movement	Selected Differentiating Properties—Method of Reproduction: Asexual	Selected Differentiating Properties—Method of Reproduction: Sexual	Representative	Habitats	Activities
	Mastigophora	Flagella	Binary fission	None	*Chlamydomonas, Cryptobia, Giardia lamblia,*[a] *Leishmania, Trichomonas* spp., *Trichonympha, Trypanosoma brucei gambiense, T. cruzi*[a]	Fresh and salt water, soil, intestinal tract of some animals	Free living; some are parasitic.
Apicomplexa	Sporozoa[b]	Generally nonmotile except for certain sex cells	Multiple fission	Involves flagellated sex cells	*Eimeria,*[c] *Plasmodium* spp.,[a] *Toxoplasma gondii*[a]	Intracellular in animals	Generally parasitic.

[a]Human parasite.
[b]Class.
[c]Small animal pathogen.

Methods of Reproduction

Algae can reproduce asexually or sexually. At the cellular level, the following mechanisms associated with reproduction may be involved:

1. Union of cells—*plasmogamy*
2. Union of nuclei—*karyogamy*
3. Chromosome number reduction—*meiosis*

See Table 9-4 for the general features of algal reproduction.

Cultivation

Various types of liquid and solid media are used for algal cultivation in the laboratory. Exact amounts of inorganic and organic substances required for growth must be provided. Depending on the type of algae, fresh or marine water is used to support growth. Suitable light and temperature must also be provided.

Lichens

Lichens are formed through a mutually beneficial (mutualistic) relationship between certain blue-green bacteria or green algae and fungi. The resulting body is called a *thallus.* The bacteria or algae provide nutrients through photosynthesis, whereas the fungi provide moisture and minerals and a structural framework for the association.

TABLE 9-4 Major Divisions of Algal Protists

	Divisions		
Property	***Chrysophycophyta***	***Euglenophycophyta***	***Pyrrophycophyta***
Common Name	Golden algae (includes diatoms)	Euglenoids[a]	Dinoflagellates[b]
Habitat	Fresh and salt water	Fresh water	Fresh and salt water
General Structural Arrangement	Mainly unicellular	Unicellular	Unicellular
Pigments Contained	Chlorophylls *a* and *c*, special carotenoids, xanthophylls	Chlorophylls *a* and *b*, carotenes, xanthophylls	Chlorophylls *a* and *c*, carotenes, xanthophylls
Selected Reserve Materials	Oils, leucosin, chrysolaminarin, oils	Fats, paramylum	Starch, oils
Motility	Unique movement with diatoms; others utilize flagella	Motile by means of flagella	Motile
Method of Reproduction	Asexual and sexual	Asexual only by binary fission	Asexual; sexual rare
Economic and Ecological Importance	1. The abrasive property of diatomaceous earth is used in various types of polishes and in the manufacture of insulating materials and dynamite sticks. 2. Production of reserve food substances such as oils, which have been significant sources of petroleum.	Used for laboratory study and teaching situations.	1. Important members of food webs in aquatic environments by serving as a food source. 2. Some discolor and produce offensive odors and taste in water. 3. Associated with parolytic shellfish poisoning (PSP) and red tides. 4. Form algal blooms.

[a]These microorganisms possess characteristics of both animals and plants. Euglenoids seem intermediate between algae and protozoa.

[b]One genus, *Gonyaulax,* occurs in algal blooms referred to as the "red tide."

TABLE 9-5 Structure and Organization of Algal Protists

Structure or Stage	***Description and Function***	***Algal Division***
Aplanospore	Nonmotile spore used for reproduction	Dinoflagellates
Zoospore	Motile flagellated spore used for reproduction	Dinoflagellates
Auxospore	Sexual reproductive structure	Diatoms
Pyrenoid	Center for starch formation during photosynthesis	Euglenoids
Frustules	Cell wall halves that fit together like a pillbox, giving diatoms a characteristic appearance.[a] Holes (*puncta*) arranged in rows (*striae*) or ridges of diatom. Frustules distinguish them from green algae. Some halves or valves have a raphe.	Diatoms
Pellicle	An outer thickened cell membrane with spiral ridges, protective in function	Euglenoids
Eyespot	Pigmented structure that enables organism to respond to light stimuli	Euglenoids
Contractile vacuole	Osmoregulating structure that excretes water and waste, similar to type found with certain protozoa	Euglenoids
Cyst	Rounded form appearing in certain stages of the life cycle; provides protection against unfavorable environments	Euglenoids
Theca	Heavy cell wall; protective in function	Dinoflagellates

[a]Silicates are major components of these cell walls.

Three major types of lichens are recognized:

1. *Crustose*—irregular crustlike patches found on rocks or tree bark
2. *Foliose*—curled, leafy, usually greenish gray growths on tree bark
3. *Fruticose*—highly branched growths that either hang from tree parts or originate in the soil

Lichens serve as sources of food, chemicals such as litmus, and antibiotics. They also aid in the decomposition of rock and soil formation.

V. CHAPTER SELF-TEST

Continue with this section only after you have read Chapters 8 and 9. A score of 80 percent or better is good. If your score is less than 65 percent, reread the chapter.

Correct answers to all questions can be found at the end of this section. Write your responses in the appropriate space provided.

Completion Questions

Provide the correct term or phrase for the following. Spelling counts.

1. Most algae are ______________ eucaryotic cells.
2. The occasional accumulations of algae known as (a) ______________ may disrupt aquatic communities by increased concentration of (b) ______________.
3. Algae may be either (a) ______________ or (b) ______________ in structural arrangement.
4. Algae can reproduce either (a) ______________ or (b) ______________.
5. Algal protist divisions are distinguished by several properties, including (a) ______________, (b) ______________, (c) ______________, (d) ______________, (e) ______________, and (f) ______________.
6. Algae found in animal life are called ______________.
7. The walls of diatoms consist of two halves, referred to as (a) ______________, with the upper half or (b) ______________ fitting together with the lower half or (c) ______________.
8. A long narrow opening found with some diatom walls is called ______________.
9. Deposits of fossil diatoms known as (a) ______________ are used in many commercial applications, including (b) ______________, (c) ______________, and (d) ______________.
10. The outer thickened cell membrane of euglenoids is a ______________.
11. The toxin of *Gonyaulax catenella,* known as (a) ______________, affects the human (b) ______________.
12. Lichens represent a(an) (a) ______________ symbiosis between (b) ______________ and (c) ______________.
13. Highly branched lichens that usually hang from tree parts are called ______________.

Matching

Select the answer from the right-hand side that corresponds to the term or phrase on the left-hand side of the question sheet. An answer may be used more than once. In some cases, more than one answer may be required.

Topic: Structure and Function

14. Contractile vacuole
15. Trophozoite
16. Micronucleus
17. Filopodia
18. Trichocysts
19. Cilium
20. Tests
21. Pellicle
22. Cyst
23. Cytosome
24. Gametes
25. Macronucleus
26. Tentacles

a. organelle of locomotion
b. provides protection
c. involved in food capture
d. excretion
e. reproduction
f. protects against excessive water intake
g. regulation of metabolic functions
h. none of these

Topic: Protozoan Activities and Associations

27. Foraminifera
28. Conjugation, a major means of reproduction
29. Malaria
30. Amebic dysentery
31. Utilize pseudopodia for locomotion
32. Some members have cirri
33. Multiple fission
34. African sleeping sickness

a. Ciliophora
b. Sarcodina
c. Mastigophora
d. Sporozoa
e. all of these
f. none of these

Topic: Properties of Algal Protists

______ 35. Sex cell
______ 36. Movement
______ 37. Cell wall
______ 38. Nonmotile spore
______ 39. Starch formation centers

a. pyrenoid
b. aplanospore
c. zoospore
d. gamete
e. flagella

_______ 40. Motile spore

_______ 41. Photosynthesis

_______ 42. Puncta

f. frustule

g. chloroplast

h. none of these

Essay Questions

Answer the following questions on a separate sheet of paper.

43. What special protective devices do protozoa have to survive in life-threatening conditions?
44. Distinguish between a trophozoite and a cyst.
45. What purpose do pyrenoids serve the alga that contains them?
46. Of what economic value is diatomaceous earth?

Photographic Quiz 1

47. What type of cellular organization is evident in the organism shown in Figure 9-1? _______________
48. Identify the labeled structures: (a) _______________, (b) _______________, (c) _______________, and (d) _______________

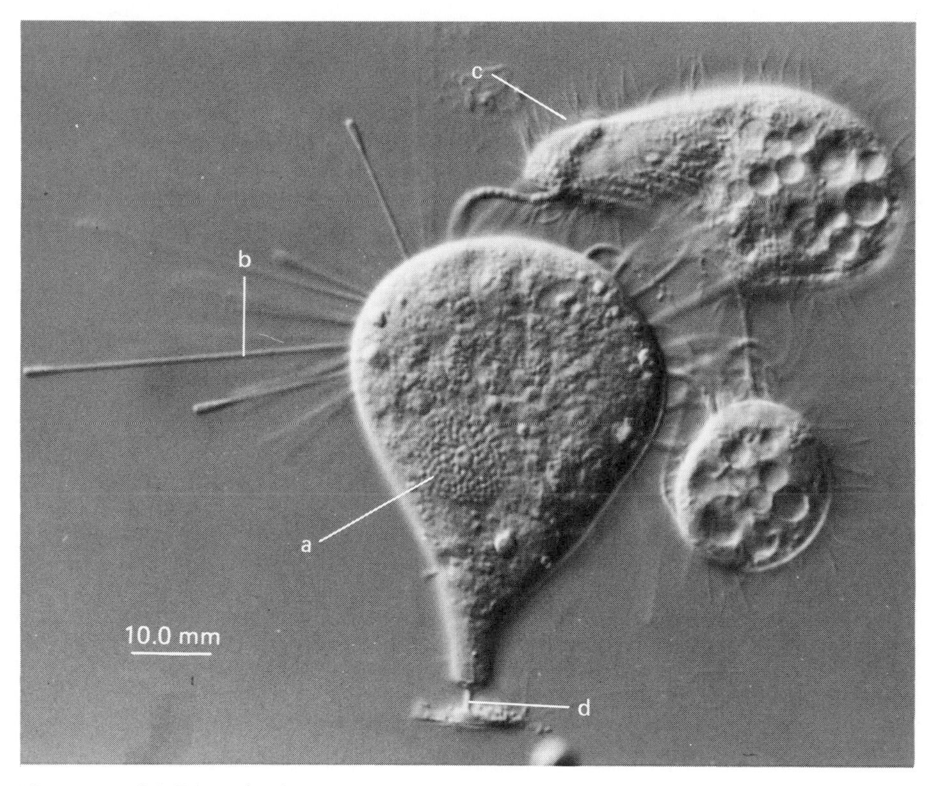

(Courtesy of J. B. Tucker.)

Figure 9-1

Photographic Quiz 2

49. Identify the algal cell shown in Figure 9-2. ____________________

50. Identify the labeled parts of the algal cell in Figure 9-2: (a) ____________________, (b) ____________________, and (c) ____________________

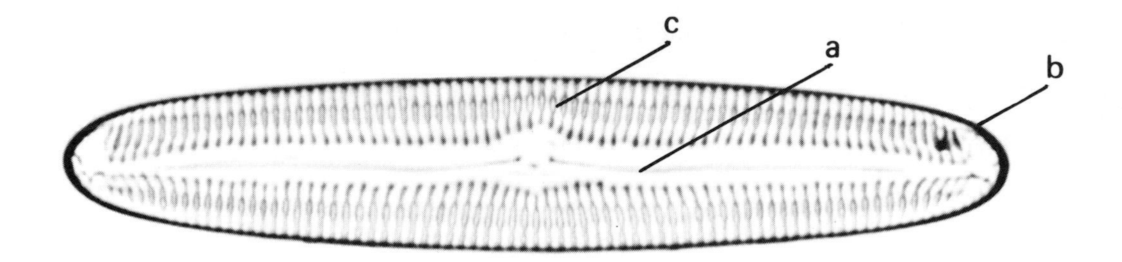

Figure 9-2

Answers

1. photosynthetic
2a. algal blooms
2b. waste product
3a. unicellular
3b. multicellular
4a. asexual
4b. sexual
5a. cellular organization
5b. cell wall chemistry
5c. flagellation or its absence
5d. pigmentation
5e. reserve storage products
5f. reproduction
6. endozoic
7a. valves
7b. epivalve
7c. hypovalve
8. raphe
9a. diatomaceous earth
9b. polishing agents
9c. insulation
9d. dynamite sticks
10. pellicle
11a. saxitoxin
11b. nervous system
12a. mutualistic
12b. algae or cyanobacteria
12c. fungi
13. fruticose
14. d
15. c
16. e
17. a
18. b
19. a
20. b
21. b
22. b, e
23. c
24. e
25. g
26. c
27. b
28. a
29. d
30. b
31. b
32. a
33. d
34. c
35. d
36. e
37. f
38. b
39. a
40. c
41. i
42. f

43. Most of the protective structures of protozoa prevent mechanical injury or protect against drying, excessive water intake, and predators. They include several types of surface coverings and trichocysts. Other protozoa resort to encystment.

44. In several parasitic species of free-living protozoa found in temporary bodies of water, the normal, active feeding form is known as a *trophozoite.* This form often cannot withstand the effects of various chemicals, food deficiencies, temperature or pH changes, and other harsh factors in the environment. To overcome such conditions, many protozoa can secrete a thick, resistant covering and develop into a resting stage called a *cyst.*

45. Pyrenoids produce a special type of starch called paramylum, which is stored as a reserve food and is used for the resynthesis of chloroplast constituents.

46. Diatomaceous earth has economic value. For example, because of its abrasive quality, it is used as a polishing agent in toothpastes and metal polishes. It is also used in the manufacture of insulating materials and dynamite sticks. Other industries use diatomaceous earth for the filtration of beer, oil, and other fluids of economic importance.

47. eucaryotic
48a. macronucleus
48b. tentacle
48c. cilia
48d. stalk
49. pennate diatom
50a. raphe
50b. frustule
50c. striae

VI. ENRICHMENT

Cryptosporidiosis, At One Time an Unrecognized Human Protozoan Infection

Cryptosporidium is a protozoan parasite belonging to the class of Sporozoa and the subclass of Coccidia. The microorganism completes its life cycle on the intestinal and respiratory surfaces of mammals, birds, and reptiles.

Until recently, cryptosporidiosis was thought to be an uncommon infection, and *Cryptosporidium* was considered to be an opportunistic microorganism with a definite specificity for only certain animal hosts. Since its discovery in a laboratory mouse by E. E. Tyzzer in 1907, *Cryptosporidium* infections have been described in 16 different animal species. Interestingly, some animals, although readily infected, appear to have an inborn resistance to the effects of the protozoon. Other animal species, such as rats, mice, and guinea pigs, show no apparent symptoms on infection. Still others, such as cows and sheep, are susceptible and can become quite ill if infected by *Cryptosporidium* at an early age. Cryptosporidiosis causes more severe illness in lambs and calves and some older animals than do certain toxin-producing strains of *Escherichia coli* and rotavirus gastrointestinal infections. Toxin-producing *E. coli* causes scours, a fatal disease for certain farm animals. Humans also are susceptible to infection. Moreover, there is little doubt that cryptosporidiosis can have serious consequences for individuals with a lowered resistance. Infections in victims of AIDS (acquired immune deficiency syndrome) have proved to be fatal. Evidence also exists to suggest that *Cryptosporidium* may be a cause of a temporary, mild, rapidly occurring diarrheal illness in immunologically normal persons. Humans apparently acquire the protozoan pathogen from diseased lower animals.

Cryptosporidium infects the entire intestinal tract, but most commonly the lower portions of the small intestine are attacked more severely. The protozoon causes partial destruction, fusion, and distortion of intestinal villi. Such injuries result in poor digestion and malabsorption. Death may result from dehydration.

Diagnosis of the disease is generally based on the demonstration of the organism in fecal matter or in gastrointestinal epithelial tissue. Giemsa, a modified acid-fast staining technique, generally is used for identification and diagnosis.

Cryptosporidium appears to be resistant to a great variety of disinfectants and drugs that have been used successfully against other protozoan pathogens. Such preparations include broad-spectrum antibiotics and drugs for worm infections (antihelminthics).

Refer to Chapter 26 for additional details of this disease.

10

Viruses, Viroids, and Prions

I. INTRODUCTION

A virus is a submicroscopic microorganism consisting of a nucleic acid core surrounded by some form of protective coating. These forms do not have metabolic systems and therefore must use the machinery of the cells they invade to replicate and produce more viruses. This chapter surveys the structure and organization of viruses together with their replication cycles, relationships with higher forms of life, and methods for their cultivation. Attention also is given to smaller forms such as viroids, other disease-associated small RNAs, and prions.

II. PREPARATION

Chapters 2, 3, 7, and 10 should be read before continuing. The following terms are important to know. Refer to the glossary and the appropriate chapters of the text if you are uncertain of any of them. A pronunciation guide for selected terms is provided as a learning aid.

1. antibiotic (an-tie-by-OT-ik)
2. artificial media
3. autoclave
4. bacteriophage (back-TEE-ree-oh-fayj)
5. binal (BY-nal)
6. capsid (KAP-sid)
7. capsomere (KAP-so-mere)
8. cell fusion
9. computer imaging
10. cyanophage (sye-ann-OH-fayj)
11. cytopathic effect
12. defective interfering particle
13. DNA polymerase
14. embryonated egg
15. envelope
16. genome (GEE-nom)
17. granulosis (gran-u-LOW-sis)
18. helical (HE-lee-kal)
19. hemagglutination (he-mah-gloo-teh-NAY-shun)
20. hemagglutinin (he-mah-GLOO-teh-nin)
21. hypersensitivity
22. icosahedron (eye-CO-sa-he-dron)
23. interferon (in-ter-FEER-on)
24. intracellular
25. *in vitro* (in VEE-troh)
26. latent (LAY-tent)
27. lysogenic conversion
28. lysogeny

29. lytic (LIT-ik) cycle
30. minus strand
31. nucleic acid
32. nucleocapsid (noo-klee-oh-KAP-sid)
33. occlusion (oh-KLOO-shun)
34. one-step growth curve
35. organelle
36. phage typing
37. plaque (plack)
38. plasmoptysis (plaz-MOP-teh-sis)
39. plus strand
40. polyhedral (pol-ee-HE-dral)
41. prion (PRE-on)
42. prophage
43. protein
44. replication
45. ribosome
46. RNA
47. RNA-dependent-RNA polymerase
48. reverse transcriptase
49. satellite virus
50. spike
51. syncitia (sin-SEH-sha)
52. synthesis
53. tissue culture
54. transcriptase
55. uncoating
56. virino (VYE-ree-no)
57. virion (VEER-ree-on)
58. viroid (VYE-royd)
59. virusoid (VYE-roo-soyd)

III. PRETEST

Correct answers to all questions can be found at the end of this section. Write your responses in the appropriate space provided.

Completion Questions

Provide the correct term or phrase for the following. Spelling counts.

1. A mature virus particle, or (a) ______________________, has at least five specific properties: they are (b) ______________________, (c) ______________________, (d) ______________________, (e) ______________________, and (f) ______________________.
2. The inner nucleic acid core of a virus particle is surrounded by a protein outer coat called a(an) ________.
3. Viral nucleic acid cores can contain either (a) ______________________ or (b) ______________________, but never both.
4. Viral nucleic acid can be either (a) ______________________ or (b) ______________________ stranded.
5. Subunits of capsids are called ______________________.
6. The viral capsid, together with its nucleic core, form the ______________________.
7. Basic shapes of virus particles include (a) ______________________, (b) ______________________, (c) ______________________, (d) ______________________, and (e) ______________________.

8. The outer covering of a virus obtained from a host membrane is called a(an) ______________ .
9. Projections on the surfaces of some virus particles enable them to ______________ red blood cells.
10. A 20-sided viral particle is called a(an) ______________ .
11. The properties of capsids used in their classification include (a) ______________ , (b) ______________ , (c) ______________ , and (d) ______________ .
12. The properties of viral nucleic acids used in their classification include (a) ______________ , (b) ______________ , (c) ______________ , (d) ______________ , and (e) ______________ .
13. Bacterial viruses, or (a) ______________ , can only be cultured in (b) ______________ .
14. A bacterial cell containing a prophage is ______________ .
15. A presence of bacterial viruses cultivated on seeded agar plates is indicated by the appearance of ______ .
16. Phages that ultimately cause cell destruction in bacterial hosts are referred to as being ______________ .
17. The specific steps in a *Escherichia coli,* T-even lytic cycle are (a) ______________ , (b) ______________ , (c) ______________ , (d) ______________ , and (e) ______________ .
18. In the absorption stage, phages attach to ______________ .
19. The enzyme ______________ in a phage's tail aids the viral penetration into the host cell.
20. In a one-step growth curve experiment, the burst size is the approximate number of ______________ produced per host cell.
21. Cyanophages are similar to other bacterial viruses in (a) ______________ and (b) ______________ .
22. Procedures for the cultivation of cyanophages ______________ those used for bacteriophages.
23. The ability of viruses to produce malignant tumors in chickens was discovered by ______________ .
24. Viral insect diseases may be either (a) ______________ or (b) ______________ .
25. Polyhedral inclusions may be formed in either the (a) ______________ or (b) ______________ of infected cells.
26. Most enveloped viruses gain entrance into a host cell by (a) ______________ of the envelope with the (b) ______________ .
27. The cellular removal of an envelope and capsid from the viral particle is called ______________ .
28. Enveloped viruses exit a host cell by a(an) ______________ process.
29. Chicken embryo inoculations have several important uses in virology: (a) ______________ , (b) ______________ , (c) ______________ , (d) ______________ , (e) ______________ , and (f) ______________ .
30. Local lesions produced by certain viruses in chick embryos are called ______________ .

31. Other signs of viral infection involving chick embryos include (a) ______ and (b) ______.
32. The destructive action of viruses in tissue culture systems is known as ______.
33. Sites of destruction and large concentrations of viruses in animal tissue culture systems are called ______.
34. Certain cancer-producing viruses can destroy some cultured cells but also can ______ others into malignant systems.
35. A specific group of proteins known for their ability to inhibit viral growth are known as the ______.
36. Most plant viruses contain ______.
37. External signs of plant virus infection include (a) ______, (b) ______, and (c) ______.
38. Four small RNA molecules associated with disease states are (a) ______, (b) ______, (c) ______, and (d) ______.
39. Scrapie is caused by a submicroscopic particle called a(an) ______.

Photographic Quiz

40. What type of virus particle is shown in Figure 10-1? ______
41. What component generally is contained in the structure labeled "a"? ______
42. What function does the structure labeled "b" have? ______

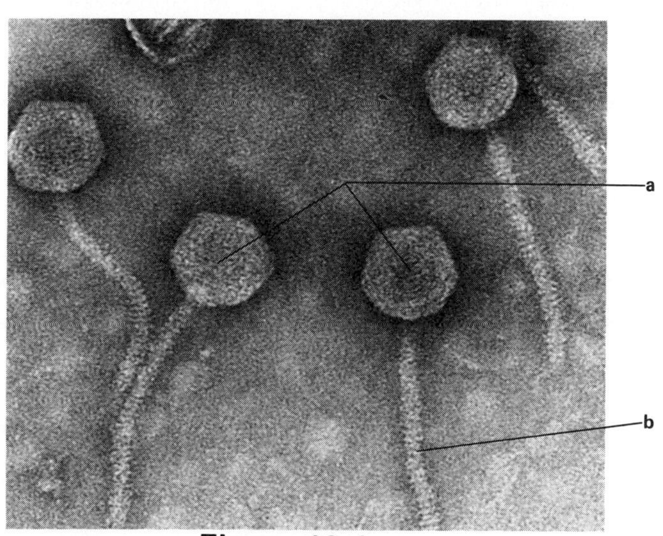

Figure 10-1

Answers

1a. virion
1b. possession of either RNA or DNA
1c. ability to use and direct host cell's enzyme systems for virus replication
1d. absence of binary fission
1e. absence of energy-producing metabolic cycle
1f. dependence on host ribosomes for synthesis of viral proteins
2. capsid
3a. DNA
3b. RNA
4a. single
4b. double
5. capsomeres
6. nucleocapsid
7a. cubic
7b. helical
7c. binal
7d. bullet
7e. filamentous
8. envelope
9. clump
10. icosahedral virion
11a. shape and size
11b. numbers of capsomeres
11c. presence or absence of an envelope
11d. nucleocapsid symmetry
12a. type of nucleic acid
12b. number of nucleic acid strands
12c. molecular weight of nucleic acid
12d. genetic direction of protein synthesis
12e. presence of a transcriptase
13a. bacteriophage
13b. bacteria
14. lysogenic
15. plaques
16. virulent or lytic
17a. absorption
17b. penetration
17c. new viral DNA replication
17d. maturation
17e. mature virus release
18. specific receptor sites on bacterial host cell walls
19. lysozyme
20. virus particles
21a. structure
21b. infection cycle
22. differ from
23. P. Rous
24a. granuloses
24b. polyhedroses
25a. cytoplasm
25b. nuclear region
26a. fusion
26b. plasma membrane
27. uncoating
28. budding
29a. vaccine production
29b. virus isolation
29c. investigating mechanisms of virus infection
29d. determining effectiveness of antiviral drugs
29e. studying viral morphology
30. pocks
31a. hemagglutination with infected fluids of embryo
31b. death of the embryo
32. cytopathic effect
33. plaques
34. transform
35. interferons
36. RNA
37a. discoloration
37b. misshapen fruits
37c. abnormal growths
38a. viroids
38b. satellite viruses
38c. satellite RNA molecules
38d. defective interfering particles
39. prion
40. bacteriophage
41. nucleic acid molecules
42. viral attachment to a host cell

IV. CONCEPTS AND TERMINOLOGY

What Is a Virus?

Viruses are unlike any other form of microorganism. This is obvious not only from their *submicroscopic* size but from other differences related to the way in which they function. For example, the basic processes of a virus are only active inside a living cell. A virus can be defined as an infectious agent having the following specific properties:

1. The possession of only one type of nucleic acid, DNA or RNA, never both;
2. The ability to transfer viral nucleic acid from one host to another;
3. The absence of binary fission;
4. The lack of any energy-harnessing metabolic cycles such as those found in other life forms;
5. The ability to direct a host cell's enzyme system for viral replication.

The general properties of viruses compared with other microbial groups, viroids, and prions, are presented in Table 10-1. Viroids and other disease-associated RNA molecules are discussed later.

TABLE 10-1 Properties of Viruses, Other Microorganisms, and Viroids

Microbial Group	Microbial Components: Cell Wall	Internal Membrane Parts (for example, mitochondria)	Ribosomes	Both DNA and RNA	Growth Requirements: Cultivation in or on Artifical Media	Require Living Cells	Sensitivity to Antibiotics
Algae	Present	Present	Present	Present	Yes	No	Variable
Bacteria	Present	Absent	Present	Present	Yes	Some	Present
Fungi	Present	Present	Present	Present	Yes	No	Variable
Protozoa	Absent	Present	Present	Present	Yes	Some required	Variable
Viruses	Absent	Absent	Absent[a]	Absent[b]	No	Yes	Absent
Viroids	Absent	Absent	Absent	RNA only	No	Yes	Absent
Prions	Absent	Absent	Absent	None	No	?	Absent

[a]One group of viruses, the *Arena* viruses, which cause natural infections in rodents, contain host cell ribosomes.
[b]Individual virus particles contain either DNA or RNA, never both.

Basic Structure of Extracellular Viruses

In a typical viral life cycle, both intracellular (active) and extracellular (inactive) phases occur. The intracellular phase involves the replication of new viral parts leading to the formation of new mature virus particles (virions). In the extracellular phase, virus particles are free to invade other susceptible cells.

Individual virions may consist of a nucleic acid (either DNA or RNA) inner core surrounded by a protein outer coat or shell called a *capsid,* which is made up of protein subunits called *capsomeres.* The nucleic acid core and capsid together form the *nucleocapsid.* As a result of their intracellular development pattern, some virus particles acquire a small portion of a host cell's cytoplasmic and nuclear membranes as they complete their cycle and are extruded. The outer coat of the virus is called an *envelope.* Some envelopes may be covered with projections used by virus particles for attachment purposes. These projections are helpful in virus identification.

The nucleic acids of viruses contain the viral genetic information or *genome.* The nucleic acids may be single stranded (*ss*) or double stranded (*ds*). These compounds also may assume a long, filamentous form, either folded or coiled. The genomes of certain RNA viruses contain *plus strands* that, upon infection, act as messenger RNA molecules for new viral proteins. Several viruses have enzymes associated with the formation of their nucleic acids and, in certain situations, an enzyme that helps viruses attach to host cells. In the case of influenza, envelope spikes contain *neuraminidases,* which bind these agents to cells of the respiratory tract.

Virions differ in size as well as shape. The capsomeres of different viruses are arranged in definite geometric patterns. The capsids of some are long and helical, others are many sided or *polyhedral,* and still others show a combination of polyhedral and helical known as *binal.* Several viruses exhibit icosahedral virions or particles having 20 triangular faces.

Classification

Virus classification is still in its infancy when compared with the systems used for other forms of life. Current classification systems include the following properties of virions:

1. Capsid organization:

    a. Shape and general size of virus particle
    b. Number of capsomeres
    c. Presence or absence of an envelope
    d. General symmetry of nucleocapsid

2. Nucleic acid chemistry:
   a. Type of nucleic acid (whether DNA or RNA)
   b. Number of strands (whether single or double)
   c. Molecular weight of nucleic acid
   d. Manner in which genetic information is translated into proteins
   e. Presence of a *transcriptase* (an enzyme involved in the formation of nucleic acids)
   f. Nucleic acid hybridization information

Table 10-2 shows how the properties can be used.

Bacteriophages (Bacterial Viruses)

Structure

Almost all readily cultivatable bacterial species can serve as hosts for bacteriophages or bacterial viruses. The host range of a bacteriophage (or phage) may be limited to one bacterial species or several genera.

Bacterial viruses contain either DNA or RNA. The DNA is double stranded in most phages. In RNA phages, the nucleic acid is either single or double stranded. The nucleic acid in most bacterial viruses is contained in a polyhedral capsid also known as the "head." In several viruses, this capsid is attached to a helical protein structure called a "tail." The tail and its additional parts are used for attachment to a susceptible bacterial host cell.

Filamentous viruses, which are among the smallest forms known to date, are long deoxyribonucleoproteins measuring about 5.5 nm in diameter. Several gram-negative bacteria, such as *Escherichia coli, Pseudomonas aeruginosa,* and *Vibrio parahaemolyticus,* act as hosts for these viruses. Filamentous viruses may be assembled during their extrusion from a host cell. Various types of bacteriophages can be isolated from natural sources.

Cultivation and Detection

Bacteriophage cultivation is relatively uncomplicated and requires appropriate actively growing, young bacterial cells. Either broth or agar cultures can be used. The presence of phages and their destructive activities in liquid preparations are evident from the clearing of turbid (cloudy) cultures. On agar plate cultures, transparent clear areas or *plaques* develop in the dense background of bacterial growth.

Several methods are used to estimate accurately the number of virus particles in a sample. Procedures employing specific bacteriophages also are used to identify specific bacterial species tracing sources of epidemics and identifying strains of pathogenic bacterial species. Examples of such methods include:

1. *The plaque count:* the procedure involves the mixing of different dilutions of the phage suspension to be assayed with a standard concentration of bacterial host cells on specially prepared agar plates. The number of viral particles in the original sample is determined by multiplying the number of plaques formed by the dilution factor and volume of phage sample used. The result is expressed as the number of plaque-forming units (PFU) per milliliter of the original sample. Quantitative determinations of virus particles in a sample also can be made using certain electron microscopy methods.

2. *Bacteriophage (phage) typing:* the procedure is performed by growing the unknown bacterial culture—isolated from a patient, for example—on an agar plate. Known phages are systematically spotted onto the "lawn" of bacteria. After a suitable incubation period, zones of lysis (plaques) appear if the appropriate phage was applied. From these results, the bacterial host can be identified.

Replication Cycle of Bacteriophages

Virulent or *lytic phages* regularly infect, replicate, and complete their life cycles in bacterial hosts, ultimately causing the host's destruction. Other bacteriophages are known that multiply and the release of their progeny occurs without the destruction of the host cell. The DNA of *temperate phages* is generally incorporated into that of the host cell's chromosome. The integrated form is known as *prophage,* and the bacterial cell containing the prophage is *lysogenic.* In some cases, temperate phages may be converted into virulent forms.

TABLE 10-2 Representative Characteristics Used in the Classification of Viruses

Nucleic Acid (NA)	*Double or Single Stranded*	*Symmetry of Capsid*	*Number of Capsomeres*	*Size of Capsid in Angstroms*	*Enveloped or Naked*	*Host Range of Viruses*			
						Animal	*Bacterial*	*Insect*	*Plant*
DNA	Single	Filamentous		50 x 8,000	Naked		Coliphage fd		
	Double	Helical		90-100	Enveloped	Poxviruses			
	Single	Polyhedral	12	220	Naked		Coliphage ϕX-174		
	Double	Polyhedral	72	450-550	Naked	Polyoma and Papilloma			
			252	600-900	Naked	Adeno-viruses			
			812	1,400	Naked			Tipula iridescent virus	
			162	1,000	Enveloped	Herpes-viruses			
		Binal (a combination of polyhedral and helical components)		Polyhedral head 950 x 650 Helical tail 170 x 1,150	Naked		Coliphages $T_2T_4T_6$		
RNA	Single	Helical		175 x 3,000	Naked				Tobacco mosaic virus
				90	Enveloped	Myxo-viruses			
				180	Enveloped	Paramyxo-viruses			
		Polyhedral		200-250	Naked		Coliphage f_2		
			32	280	Naked	Picorna-viruses			Turnip yellow mosaic virus and tomato bushy stunt virus
	Double	Polyhedral	92	700	Naked	Reo-viruses			Wound tumor virus

The lytic cycle for certain phages in which *Escherichia coli* and other bacteria serve as hosts has been studied quite extensively. The steps of a cycle such as the T-even phages ($T_2T_4T_6$) include:

1. *Adsorption:* the phage adsorbs onto specific receptor sites of the host cell's wall. Phage tail fibers serve as adsorption sites. The enzyme *lysozyme* is released from the phage tail to digest a small portion of the host's cell wall and to aid in the injection.
2. *Penetration:* phage DNA is introduced by the phage's tail, through the host's cell wall, and across the cell membrane. The exact mechanism for the process is unknown.
3. *Replication of new viral DNA:* following the injection of phage DNA, a latent period occurs during which no infective virus can be observed. Once infection begins, the total synthetic machinery of the host is directed toward the production of phage parts.
4. *Maturation:* viral DNA is used to form other essential viral parts, including the capsomeres of the capsid. New phage particles are assembled under the control of certain viral genes.
5. *Liberation of viruses:* as the maturation phase comes to an end, the enzyme bacteriophage lysozyme accumulates. It chemically disrupts the host cell wall, causing an osmotic imbalance that leads to a release of newly formed virus particles (*virions*).

The general cycle is shown in Figure 10-2.

From events of the lytic bacteriophage cycle, it is possible to experimentally determine the time lapse between phage nucleic acid injection and the release of mature virus particles, and to estimate the approximate number of these phages produced per cell (*burst size*). The experimental procedure is called a *one-step growth curve.* It is carried out by periodically removing samples of phage-infected bacterial suspension and determining the number of free virus particles and infected bacteria as shown by plaque count.

Cyanophages

Cyanophages are very similar to other bacterial viruses in structure and in infection cycles. Conditions, media, and procedures for the preparation of suitable hosts differ significantly from those used for bacteriophages.

Animal Viruses

Several major illnesses of humans and other animals are viral in nature. The properties of representative animal viruses are shown in Table 10-3.

The sequence of events in animal virus replication include the following steps:

1. Virion attachment to specific host cell receptor sites;
2. Penetration of entire virus particles by phagocytosis or pinocytosis;
3. Viral envelope and capsid in a phagocytic vacuole (if present) are removed from the nucleic acid core by a systematic uncoating process;
4. Viral replication, which may involve viral capsids being enclosed by host cell membrane material (envelope);
5. Viral release, which differs according to the virus involved (refer to the text for specific details);
6. Possible host cell death following release of virus particles.

Living embryonated chicken eggs are inexpensive and easy to handle, making suitable systems for animal virus cultivation. They yield large quantities of viruses for vaccines and furnish diagnostic reagents to study the mechanisms of viral infection, isolate viruses, and determine the effectiveness of antiviral drugs. Several portions of embryonated eggs are used for viral cultivation, including the allantoic and amniotic cavities, the yolk sac, the chorioallantoic membrane, and the embryo itself. The particular region used depends on the virus to be cultured.

Signs of virus infection in chick embryos may be detected in several ways, including (1) the formation of local swellings or lesions called *pocks,* (2) the demonstration of a blood clumping reaction or *hemagglutination* by allantoic or amniotic fluids, and (3) the detection of virus particles by electron microscopy. With certain viruses, no obvious signs of infection appear.

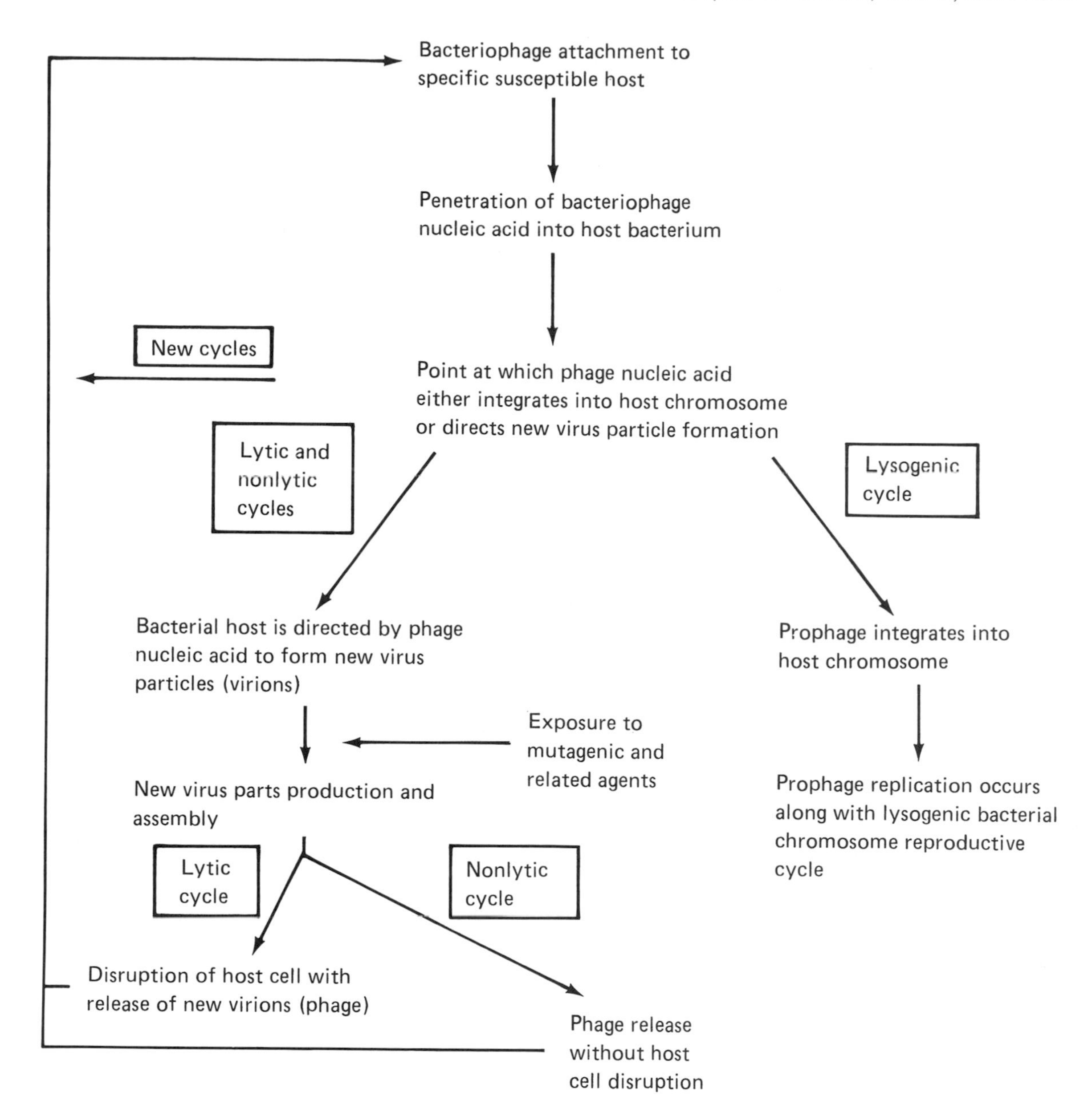

Figure 10-2
Replication cycles of bacteriophages. Three types of bacteriophage-host cell interactions are shown. The lysogenic cycle is shown on the right. Both lytic and nonlytic cycles are shown on the left.

The cultivation of animal tissues (tissue culture) also is used for isolation, cultivation, and the study of viruses. Three specific advances made it possible to use animal tissues for the cultivation of microbes requiring living cells: (1) the introduction of antibiotics to prevent contamination by bacteria and fungi, (2) the development of a functional, defined growth medium for the cells in a tissue culture system, and (3) the incorporation of the enzyme trypsin to free cells from tissue fragments so that single cell layer systems could be grown. The tissues of plants and cold- and warm-blooded animals can be cultured.

The presence of viruses in mammalian tissue culture systems can be detected by various morphologic changes and destruction of cells, known as *cytopathic effects,* and by certain alterations in cellular metabolic reactions. Occasionally, the effects produced in a culture are caused by unsuspected and unknown contaminating viruses.

TABLE 10-3 Properties of Major Animal Virus Groups

Virus Group[a]	*Type of Nucleic Acid*	*Single or Double Stranded*	*Enveloped*	*General Properties*
Adenovirus	DNA	Double, linear	No	Found in several animal species, including humans; associated with respiratory infections and with tumors in laboratory animals.
Baculovirus	DNA	Double, circular	Yes	Group contains insect viruses that are in a protein-crystalline substance (matrix).
Hepadnavirus	DNA	Double, circular	Yes	Causative agent of hepatitis B infection.
Herpesvirus	DNA	Double, linear	Yes	Important causative agents of human disease such as chickenpox, infectious mononucleosis, and infections of the skin and mucous membranes.
Iridovirus	DNA	Double, segmented	No	Group includes viruses having the widest range of insect hosts.
Papovavirus	DNA	Double, circular	No	Causes warts; used in the study of tumor development.
Parvovirus	DNA	Single, linear	No	Group includes satellite viruses that are incapable of replication except in the presence of a helper virus and insect viruses.
Poxvirus	DNA	Double, linear	No	Found in several animal species, including humans; examples of infections include smallpox and molluscum contagiosum; all viruses affect the skin.
Arenavirus	RNA	Single, circular	Yes	Particles contain cellular ribosomes; viruses cause natural inapparent infections of rodents.
Bunyavirus	RNA	Single, circular	Yes	Viruses are spread by a variety of arthropods (arboviruses).
Calcivirus	RNA	Single, plus strand	No	Causative agents of gastrointestinal infections.
Coronavirus	RNA	Single, linear, plus strand	No	Includes several agents of the common cold.
Flavivirus	RNA	Single, circular	Yes	Includes causative agents of yellow fever and several central nervous system diseases.
Orthomyxo-virus	RNA	Single	Yes	Includes the influenza viruses.
Paramyxo-virus	RNA	Single	Yes	Many produce human childhood diseases such as measles and mumps and localized respiratory infections.
Picornavirus	RNA	Single	No	Includes several agents of diseases such as poliomyelitis, rashes, meningitis, and mild upper respiratory infections, and gastrointestinal infections.
Reovirus	RNA	Double	No	Most are spread by fecal-contaminated water and food; several have yet to be associated with specific diseases.
Retrovirus	RNA	Double, segmented	No	Includes tumor- and cancer-causing agents and human immunodeficiency virus (HIV).

TABLE 10-3 Properties of Major Animal Virus Groups *(continued)*

Virus Group[a]	*Type of Nucleic Acid*	*Single or Double Stranded*	*Enveloped*	*General Properties*
Rhabdovirus	RNA	Single, minus strand	Yes	Group includes large bullet-shaped viruses such as the agents for rabies and vesicular stomatitis as well as insect viruses.
Togavirus	RNA	Single, plus strand	Yes	Arthropod-spread diseases, including yellow fever and several nervous system diseases as well as insect viruses.

[a]Several new viruses, as well as some unclassified viruses, may require the formation of new groups.

Most cytopathic effects (CPE) can be observed using a standard light microscope. Typical effects include cytoplasmic granulation, vacuole formation, development of giant and multinucleated cells, and condensation of nucleic acid and protein in the nuclear membranes of infected cells. A particular CPE can be of diagnostic importance.

Viruses are cultivated in a monolayer tissue culture system overlaid with nutrient agar and inoculated at a desired temperature. Evidence of infection includes the formation of clearly defined areas of cellular destruction called *plaques,* which can be used for viral identification in a particular system. In addition, pure viral strains can be obtained from them by a procedure called *plaque purification,* which is based on the assumption that the offspring of a single virus produce each plaque.

Other plaquing techniques also are employed in the study of viruses. The *plaque assay* can determine the number of infectious particles in a specimen. Specific dilutions of virus-containing material are used. The number of infectious particles is computed from the number of plaques formed, and from the volume and dilution of the specimen used.

Transformation assays are used for the detection and measurement of viruses that do not cause cell destruction and death. Several viruses can transform normal cells into malignant ones whose uninhibited growth results in *tumors.*

Interferons

Interferons are a family of proteins that has several effects including antiviral activity, cell growth inhibition, and suppression of the immune response. Three different types of human interferons are known and originally were named according to their sources or origins: *fibroblast, immune,* and *leukocyte.*

Plant Viruses

Most plant virions consist of RNA surrounded by a protein capsid. Plant viruses also are known to contain single- and double-stranded DNA. The shapes of plant viruses are either polyhedral or helical.

The cultivation of plant viruses is carried out for many of the same reasons associated with animal and bacterial viruses. Three different procedures are used: *tissue culture, cell culture,* and *protoplast culture.* Protoplasts here refer to plant cells whose cell walls have been removed by enzymatic treatment.

Plant viruses do not have specific mechanisms to ensure their penetration into host cells. They can enter hosts through breaks and abrasions, or by the activities of insects, parasitic worms, or other plants such as dodder. The genetic makeup of the host plant contributes to the outcome of the particular virus infection. The general signs of infection are summarized in Table 10-4.

Viruses of Eucaryotic Microorganisms

Viruses are found in a variety of algae, fungi, and protozoa. Such association can take one of two forms. In one type, the eucaryotic microbe functions as a *vector* or transmitter of the virus. In the other, the virus uses the microbe as a host.

TABLE 10-4 Signs of Plant Virus Infection

External	*Internal*
Discoloration of plant parts	Degeneration of conductive tissues and vessels
Formation of blisterlike spots, misshapen fruits	Increase in cell numbers
Development of abnormal growths on roots and other plant parts	Presence of cell inclusions containing virus particles

Viroids and Other Disease-Associated Small RNAs

Several small RNA molecules associated with disease states are known and include *viroids, satellite viruses, satellite RNA molecules,* and *defective interfering particles.* Properties of these molecules are compared in Table 10-5.

TABLE 10-5 Comparison of Disease-Associated Small RNAs

Property	*Viroids*	*Satellite Virus*	*Satellite RNA*	*Defective Interfering (DI) Particles*
Causes infections	Yes	No	No	No
Possesses a specific capsid	No	Yes	No	No
Is contained in helper capsid	No	No	Yes	Yes
Replicates only in presence of helper	No	Yes	Yes	Yes
Interferes with helper replication	No	Yes	Yes	Yes
Exhibits *in vivo* and *in vitro* stability of RNA	No	High	High	Low

V. CHAPTER SELF-TEST

Continue with this section only after you have read Chapters 2, 3, 7, and 10 and have reviewed all illustrations in these chapters. A score of 80 percent or better is good. If your score is less than 65 percent, reread the chapters.

Correct answers to all questions can be found at the end of this section. Write your responses in the appropriate space provided.

Matching

Select the answer from the right-hand side that corresponds to the term or phrase on the left-hand side of the question sheet. An answer may be used more than once. In some cases, more than one answer may be required.

Topic: General Properties of Viruses and Other Microorganisms

______ 1. Virion	a.	clearly sensitive to antibiotics
______ 2. Molds	b.	cell wall present
______ 3. Protozoa	c.	contains both RNA and DNA

_____ 4. Virus particle
_____ 5. Algae
_____ 6. Eucaryotic plant cell
_____ 7. Bacteria
_____ 8. Yeast

d. generally can be grown on artificial media
e. always requires living cells for replication
f. contains ribosomes as a general property
g. all of these
h. none of these

Topic: Structure and Function of Viruses

_____ 9. Capsid
_____ 10. Nucleocapsid
_____ 11. Virion
_____ 12. Genome
_____ 13. Envelope
_____ 14. Spike
_____ 15. RNA
_____ 16. Icosahedron

a. can be either single or double stranded
b. a subunit of the capsid
c. contains the viral genetic material
d. a mature virus particle
e. viral outer covering obtained from host membrane material
f. used for attachment to host cells
g. a 20-sided figure
h. composed of capsomeres

Topic: Replication Cycle of Bacteriophages

_____ 17. Initial release of lysozyme
_____ 18. Dissolving of viral capsid
_____ 19. Capsomere production
_____ 20. Attachment to bacterial cell wall receptor sites
_____ 21. Injection of viral nucleic acid
_____ 22. Stepwise assembly of virus parts

a. adsorption
b. penetration
c. new viral DNA replication
d. burst size
e. mature virus release
f. all of these
g. none of these

Topic: Diseases Caused by Viruses

_____ 23. Rous sarcoma
_____ 24. Granuloses
_____ 25. Tulip break
_____ 26. Rabies
_____ 27. Chrysanthemum stunt disease
_____ 28. Warts
_____ 29. Food-and-mouth disease
_____ 30. Cauliflower mosaic disease
_____ 31. Polyhedroses
_____ 32. Potato spindle tuber disease
_____ 33. Influenza
_____ 34. Mumps

a. human virus
b. plant virus
c. bacteriophage
d. viroid
e. cancer-causing virus
f. insect virus
g. choices *a* and *c* only
h. virusoids
i. prions
j. none of these

True or False

Indicate all true statements with a "T" and all false statements with an "F".

______ 35. A bacterial cell containing a prophage is referred to as being virulent.

______ 36. Bacteriophage lysozyme controls the stepwise assembly of new virus particles.

______ 37. Plaques are local lesions that appear in virus-infected chicken embryos.

______ 38. Plant viruses enter their hosts by pinocytosis.

______ 39. Parasitic worms can introduce viruses in susceptible plants.

______ 40. Viroids are larger than viruses.

______ 41. Capsids surround envelopes.

______ 42. A temperate phage ultimately causes host cell destruction.

______ 43. The shapes of most plant viruses are either binal or helical.

______ 44. Viruses may be found in association with all categories of microorganisms.

Completion Questions

Provide the correct term or phrase for the following. Spelling counts.

45. Viruses ______________ be cultivated outside of a living system.
46. A mature virus particle is called a(an) ______________.
47. Capsids are constructed from a definite number of protein subunits known as ______________.
48. ______________ are viral nucleic acid components that act as messenger RNA for the new viral proteins.
49. The clumping of red blood cells by certain viruses such as influenza involves protein molecules called ______________.
50. Capsids consisting of 20 triangular faces exhibit a(an) ______________ symmetry.
51. The specific effects of interferons are (a) ______________, (b) ______________, and (c) ______________.
52. Individual plant virions may consist of (a) ______________, (b) ______________, or (c) ______________.
53. Small RNA molecules associated with disease states include (a) ______________, (b) ______________, (c) ______________, and (d) ______________.
54. Satellite viruses are dependent on ______________ for their replication.
55. Prions cause the fatal disease of sheep and goats known as ______________.
56. Two related human diseases believed to be caused by prions are (a) ______________ and (b) ______________.

Essay Questions

Answer the following questions on a separate sheet of paper.

57. When are the basic processes of a virus active?

58. What purpose do viral hemagglutinins serve?

59. What are satellite RNAs?

Photographic Quiz

60. In the transmission electron micrograph shown in Figure 10-3, identify the labeled component of the virus particle. ______________________________

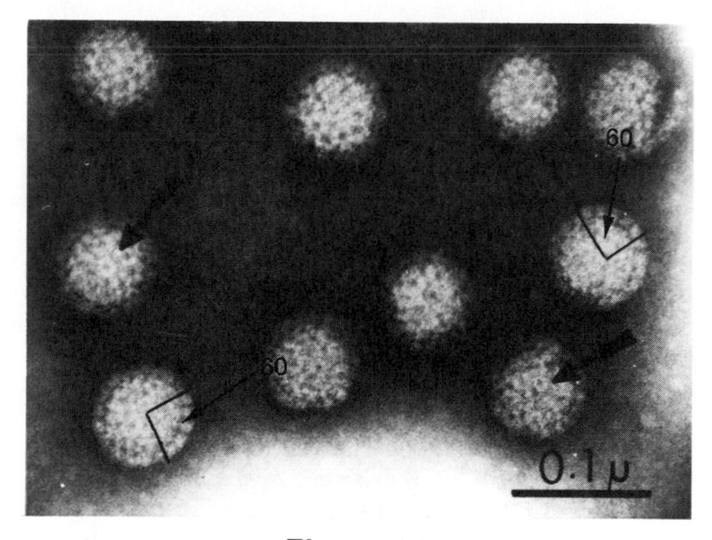

Figure 10-3

Answers

1. e
2. b, c, d, f
3. b, c, d, f
4. e
5. b, c, d, f
6. b, c, d, f
7. a, b, c, d, f
8. b, c, d, f
9. h
10. c
11. d
12. c
13. e
14. f
15. a, c
16. g
17. a
18. g
19. g
20. a
21. b
22. g
23. e
24. f
25. b
26. a
27. d
28. a
29. h
30. b
31. f
32. d
33. a
34. a
35. F
36. F
37. F
38. F
39. T
40. F
41. F
42. F
43. F
44. T
45. cannot
46. virion
47. capsomeres
48. plus strands
49. hemagglutinins
50. icosahedron

51a. antiviral activity
51b. cell growth inhibition
51c. suppression of the immune response
52a. RNA
52b. single-stranded DNA
52c. double-stranded DNA
53a. viroids
53b. satellite viruses
53c. satellite RNA molecules
53d. defective interfering particles
54. helper viruses
55. scrapie
56a. kuru
56b. Creutzfeldt-Jakob disease

57. The basic processes of a virus are active only when the viral nucleic acid, upon entering a cell, causes the formation of new virus parts. In a typical life cycle, viruses are alive when they are intracellular, inducing the formation of parts for mature virus particles. Outside cells, viruses are metabolically inactive and are no more alive than pieces of nucleic acids.
58. Viral hemagglutinins are contained in spike-shaped structures or the envelopes of viruses such as influenza. The hemagglutinins enable those viruses to attach themselves to cells and to clump red blood cells in diagnostically related tests.
59. *Satellite RNAs* are present in varying numbers in the protein coats of certain helper or satellite viruses. These molecules are similar in size to viroids, replicate only in the presence of a specific virus, and may or may not contribute to devastating effects on infected plants. The mechanism of action and the activities of these molecules are not fully understood at this time.
60. capsomere

VI. ENRICHMENT

Potato Spindle Disease

The potato provides, with greater efficiency, a more nutritionally well-balanced food than does any other major crop. This crop has great potential for countries with high populations and low incomes. This vegetable, however, is susceptible to a disease caused by the potato spindle tuber viroid (PSTV). The disease agent is easily spread by touching (contaminating) plant leaves with infected hands or tools and through seed potatoes, true seed, and pollen. The disease becomes more severe as successive generations of infected potatoes are planted. Eradication of the disease, once it is established, is difficult and no cure is known.

In severely infected crops, PSTV disease can result in losses of over 50 percent. Potatoes that do develop are cracked and spindle shaped, a feature that gave the disease its name. By detecting and eliminating diseased potatoes from the breeding stock, yields would be greatly increased.

Tests generally used to detect PSTV are laborious and expensive to perform with a large number of samples. In 1981, with the benefit of recombinant DNA technology, a form of genetic engineering, Robert Owens and Theodor O. Diener devised a new method for screening large numbers of potatoes for the potato spindle tuber viroid.

To perform the test, a small piece of material from the potato is homogenized, and a portion of the homogenate is fixed onto a thin membrane filter. These filters are then incubated with laboratory-prepared radioactive DNA. Those spots containing PSTV combine with the radioactive DNA and expose a photographic film placed against them.

This screening method can be easily automated to certify viroid-free seed potatoes. In addition, a wide variety of specimens can be used including sprouts, buds, or eyes, or the skin of a potato.

A number of informative articles describing the properties of various viruses have been written in recent years.

PART III
MICROBIOLOGY TRIVIA PURSUIT
(Chapters 7 through 10)

This section examines your attention to detail, and your fact-gathering ability. While there are *four* challenging Parts to this section, a certain number of questions must be answered in each Part before you proceed up the MICROBIOLOGY TRIVIA PURSUIT trail. The answers and directions for continuing are given at the end of each Part. Try Part 1.

PART 1

Completion Questions

Provide the correct term or phrase for the following. Spelling counts.

1 and 2. The two simple amino sugars found in bacterial cell walls are (1) ________________ and (2) ________________.

3. Identify the labeled structures in the light micrograph shown in Figure III-1. ________________

4 and 5. What two chemicals account for the heat resistance of the labeled structure shown in Figure III-1?
(4) ________________ (5) ________________

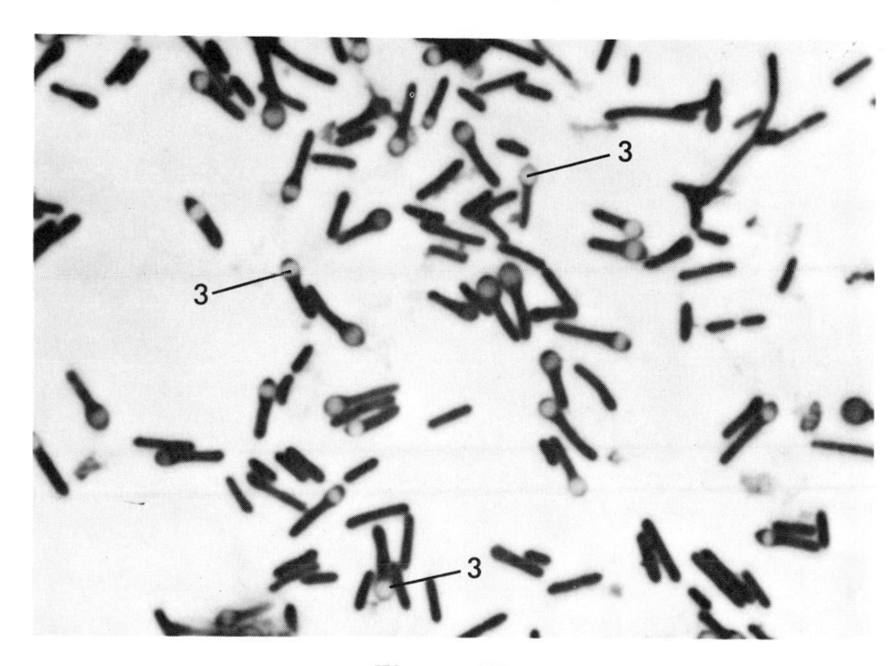

Figure III-1

Directions

Check your responses with the *Answers* section, and add up your score. Enter the number of correct answers in the space provided.

Total Correct Answers for Part 1: ____________.

If your score was 4 or higher, proceed to Part 2.

Answers

1. N-acetylglucosamine; 2. N-acetylmuramic acid; 3. bacterial endospore; 4. calcium; 5. dipicolinic acid.

PART 2

Very good! Congratulations on the successful completion of Part 1. Now try your hand with Part 2.

Completion Questions

Provide the correct term or phrase for the following. Spelling counts.

6. What is the morphology of the cells shown in Figure III-2? ______________________

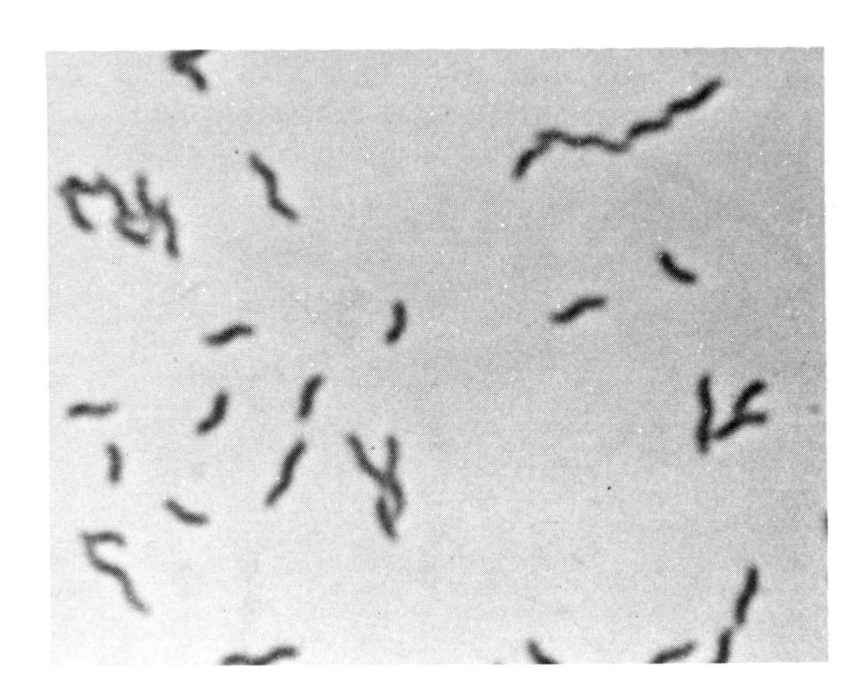

Figure III-2

7. The origin of the word root *phyco* comes from the Greek *phykos* and has the meaning ______________________.

8 and 9. The two means of movement available to euglenoids are (8) ______________________ and (9) ______________________.

10. Identify the morphology of the procaryote shown in Figure III-3. ______________________

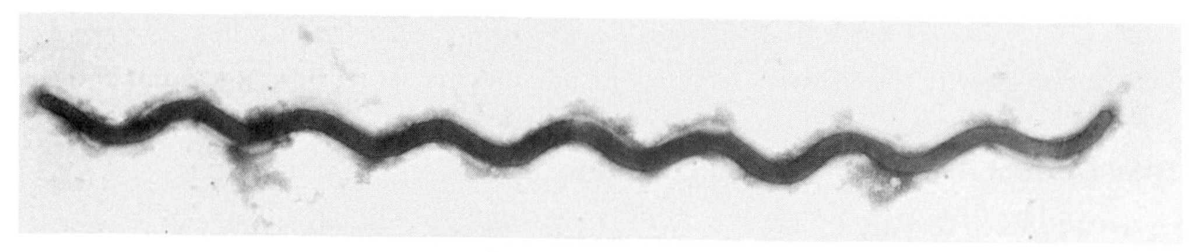

Figure III-3

Directions

Check your responses with the *Answers* section, and add up your score. Enter the number of correct answers in the space provided.

Total Correct Answers for Part 2: ________________.

If your score was 4 or higher, proceed to Part 3.

> ***Answers***
>
> 6. vibrios; 7. seaweed; 8. euglenoid movement; 9. flagella; 10. spiral.

PART 3

Excellent! You are doing very well. Here is a challenging Part 3.

Completion Questions

Provide the correct term or phrase for the following. Spelling counts.

11. The term ________________________ is derived from proteinaceous infectious particle.

12 and 13. The nucleic acid of specific virus particles may be either a (12) ________________________ strand or a (13) ________________________ strand.

14. What type of nucleic acid is found in virusoids? ________________________

15. Identify the type of microorganism indicated by the arrows in the transmission micrograph shown in Figure III-4. ________________________

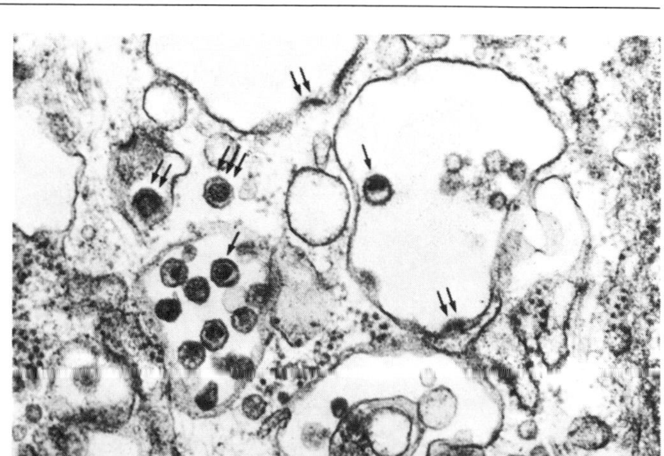

(From C.-Y. Kang, *et al.*, *J. Virol.* **16**:1027-1039 [1975].)

Figure III-4

Directions

Check your responses with the *Answers* section, and add up your score. Enter the number of correct answers in the space provided.

Total Correct Answers for Part 3: ______________.

If your score was 5, proceed to Part 4.

Answers

11. prion; 12. plus; 13. minus; 14. RNA; 15. virus particles.

PART 4

You have come a long way. Now for the last and final challenge.

Completion Questions

Provide the correct term or phrase for the following. Spelling counts.

16. The upper valve of a diatom's cell wall is called the ______________________________.
17. Identify the types of microorganisms shown in Figure III-5. ______________________________

(Courtesy of C. M. Pringle.)

Figure III-5

18. Identify the labeled structures shown in Figure III-6. ______________________________
19. Give the function of the labeled structures in Figure III-6. ______________________________

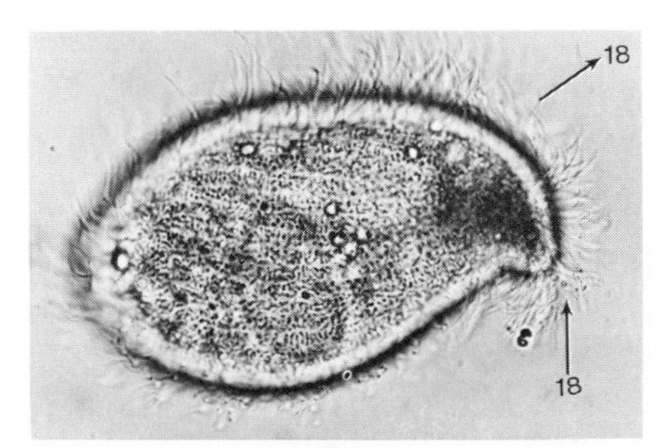

(Courtesy of Dr. J. J. A. van Bruggen.)

Figure III-6

20. What is the specific shape of the virus particles (arrows) shown in Figure III-7? ______________________

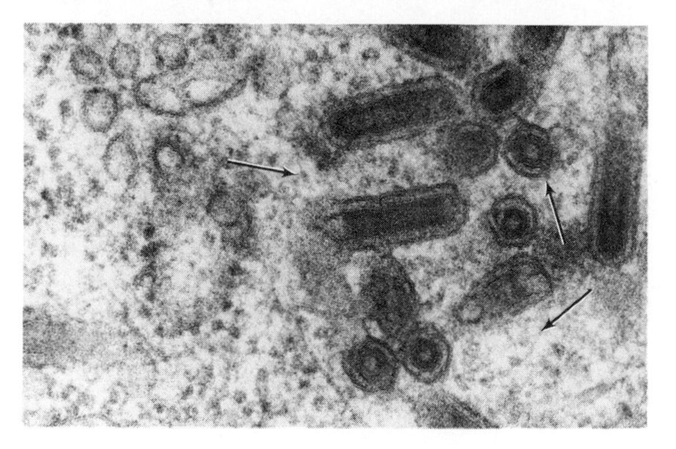

(Courtesy of Dr. Gholamreza Darai.)

Figure III-7

Directions

Check your responses with the *Answers* section, and add up your score. Enter the number of correct answers in the space provided. Then add all your scores for Parts 1 through 4 together and find your score on the *Performance Score Scale.*

Total Correct Answers for Part 4: ______________.

Total Correct Answers for Parts 1 through 4: ______________.

Answers

16. epivalve; 17. diatoms; 18. cilia; 19. movement; 20. bullet.

PERFORMANCE SCORE SCALE

Number Correct	*Ranking*
12	You should have done better
13	Better
14	Good
16	Excellent
18	Outstanding! Keep it up

Note that each major section of this Study Guide has a *MICROBIOLOGY TRIVIA PURSUIT* section. If you did not do as well as you expected, there are many more opportunities ahead.

11

Chemicals and Physical Methods in Microbial Control

I. INTRODUCTION

Microorganisms are found in every natural and human-made environment on earth, including our bodies and the food we eat. In many cases, these microorganisms perform activities which are beneficial to humans. However, a number of microbial activities are clearly undesirable from a human point of view. For example, several microorganisms can cause disease in humans or domesticated animals, others can damage or decompose materials such as leather and wood, and still others may alter foods, making them unfit to eat or even making them sources of toxins or infectious agents. It is therefore sometimes necessary to control microbial activities, usually by killing the microorganisms. This chapter discusses physical and chemical procedures which have been developed to control undesired microorganisms in a variety of environments. The treatment of infected patients (chemotherapy) will be covered in Chapter 12.

II. PREPARATION

Chapter 11 should be read before continuing. The following terms are important for you to know. Refer to the glossary and the appropriate chapters if you are uncertain of any of them. This chapter will serve as an important reference for later chapters. A pronunciation guide for selected terms is provided as a learning aid.

1. acriflavine (ak-rih-FLAY-vin)
2. alcohol
3. aldehyde (AL-dee-hide)
4. antiseptic (an-tee-SEP-tick)
5. autoclave
6. bactericidal (back-teer-ee-SYE-dal)
7. bacteriostatic (back-teer-ee-oh-STAT-ik)
8. biological indicator
9. crystal violet
10. D-value
11. dark reactivation
12. detergent
13. direct spray
14. disinfect
15. ethylene oxide
16. filtration
17. flash pasteurization
18. germicide
19. halogen (HAL-oh-jen)
20. heavy metal

21. hydrophilic (hi-drow-FIL-ik)
22. hydrophobic (hi-drow-FOH-bik)
23. hypochlorite (hi-poh-KLORE-ite)
24. ionizing radiation
25. laminar air flow
26. pasteurization
27. peroxide
28. phenol (FEE-noll)
29. phenol coefficient
30. photoreactivation
31. sanitization
32. sporicide (SPOR-ee-side)
33. sterility monitoring
34. sterilize
35. target theory
36. thermal death point
37. thermal death time
38. tissue toxicity
39. ultraviolet radiation
40. use-dilution

II. PREPARATION

Correct answers to all questions can be found at the end of this section. Write your responses in the appropriate space provided.

Completion Questions

Provide the correct term or phrase for the following. Spelling counts.

1. A(An) ______________________ is used to kill pathogens on inanimate surfaces.
2. A(An) ______________________ is used to kill pathogens on living tissues.
3. ______________________ is the killing of all microorganisms in an area.
4. Depending on the organisms it attacks, a germicide may be a(an) (a) ______________________ , (b) ______________________ , or (c) ______________________ .
5. An antimicrobial agent that acts against bacteria by temporarily stopping growth rather than killing is said to be ______________________ .
6. The halogens (a) ______________________ and (b) ______________________ form one class of disinfectant.
7. Halogens apparently function by oxidizing (a) ______________________ and disrupting (b) ______________________ .
8. The group of disinfectants that function primarily by dehydration are the (a) ______________________ , (b) ______________________ , (c) ______________________ , and (d) ______________________ .
9. The first disinfectant was (a) ______________________ , which now has many biologically active derivatives such as (b) ______________________ , a common deodorant in soaps.
10. (a) ______________________ is an unstable molecule that acts as an oxidant and releases O_2 gas, therefore being valuable as a(an) (b) ______________________ agent in infections caused by anaerobes.

11. Two dyes that have antibacterial activity are (a) ______________ and (b) ______________ .
12. Detergents have (a) ______________ and (b) ______________ portions and appear to work by disrupting (c) ______________ .
13. Metal ions usually exert their antimicrobial effect by precipitating (a) ______________ . Three metals commonly used are (b) ______________ , (c) ______________ , and (d) ______________ .
14. The most effective gaseous sterilant is (a) ______________ , which is a(an) (b) ______________ agent.
15. The two basic types of heat used in physical methods of microbial control are (a) ______________ heat, created in a(an) (b) ______________ , and (c) ______________ heat, created in a(an) (d) ______________ .
16. Both types act by denaturing (a) ______________ and disrupting (b) ______________ .
17. ______________ heat is more effective for sterilization because it penetrates materials better.
18. The ______________ of a microorganism is determined by experiments at a constant temperature and different times.
19. The ______________ of a microorganism is determined by experiments at a constant time and different temperatures.
20. The (a) ______________ value is the amount of (b) ______________ necessary to kill 90 percent of a population at a given (c) ______________ .
21. The two basic methods of monitoring heat sterilization use (a) ______________ or (b) ______________ indicators.
22. Spores from bacteria of the genus ______________ are used as indicators.
23. Boiling and (a) ______________ are heat methods that disinfect but do not sterilize. Of these two, (b) ______________ is preferred when high temperatures might damage the material being treated.
24. An important concern in this procedure is uniform ______________ distribution.
25. There are two types of sterilization by radiation: (a) ______________ radiation and (b) ______________ radiation.
26. Organisms treated with (a) ______________ radiation can repair themselves by the light-mediated process of (b) ______________ or by (c) ______________ .
27. The more powerful (a) ______________ radiation appears to work on the basis of the (b) ______________ theory.
28. The first sterilization filters were made of (a) ______________ . The newest, extremely thin filters in wide commercial use are (b) ______________ filters.

Answers

1. disinfectant
2. antiseptic
3. sterilization
4a. bactericide
4b. fungicide
4c. viricide
5. bacteriostatic
6a. iodine
6b. chlorine
7a. proteins
7b. membranes
8a. alcohols
8b. methanol
8c. ethanol
8d. isopropanol
9a. phenol
9b. hexachlorophene
10a. peroxide
10b. antiseptic
11a. acriflavine
11b. crystal violet
12a. hydrophobic
12b. hydrophilic
12c. membranes
13a. proteins
13b. arsenic
13c. mercury
13d. silver
14a. ethylene oxide
14b. alkylating
15a. moist
15b. autoclave
15c. dry
15d. oven
16a. proteins
16b. membranes
17. moist
18. thermal death time
19. thermal death point
20a. D
20b. time
20c. temperature
21a. chemical
21b. biological
22. *Bacillus*
23a. pasteurization
23b. pasteurization
24. heat
25a. ultraviolet
25b. ionizing
26a. ultraviolet
26b. photoreactivation
26c. dark reactivation
27a. ionizing
27b. target
28a. porcelain
28b. membrane (cellulose acetate)

IV. CONCEPTS AND TERMINOLOGY

Definitions and Goals

There are three major categories of antimicrobial action: antisepsis, disinfection, and sterilization. The distinctions among these different activities are as follows. *Antisepsis* is the killing of pathogens in living tissue rather than inanimate materials. *Disinfection* is the killing of all pathogenic microorganisms on an inanimate surface; living organisms will remain. *Sterilization* is the killing of all microorganisms in an area on a surface. The term *sanitization* refers to disinfection in which the total number of microorganisms is reduced but without necessarily killing all the pathogens. Fortunately, pathogens are generally much easier to kill than other microorganisms so that the relatively difficult process of sterilization is not required in all cases.

The key factors of chemical procedures is to kill microorganisms without damaging the material on which the microorganisms are living. This concern is especially important when the material in question is a person, though it also applies to inanimate surfaces such as surgical instruments, where corrosion may be a problem.

Incomplete Killing

The goal of antimicrobial procedures is to kill the pathogenic microorganisms. The agents used are therefore referred to as *germicides.* Those that kill bacteria are bactericidal; fungi, fungicidal; viruses, viricidal. Unfortunately, the procedures are sometimes only partly effective and merely delay the growth of the microorganisms without killing them. This result is called *bacteriostatic.* It results from many different causes. Sometimes the disinfectant was made at the wrong concentration or pH, or perhaps the exposure time was not long enough. Bacteriostasis may be dangerous, since later dilution or other change of conditions may allow the pathogens to grow.

The greatest problems in disinfection are with bacterial spores and some viruses. These structures are extremely resistant to most treatments because of the chemical nature of their outer coverings.

Mechanisms of Disinfection

The textbook discusses each agent individually and describes their respective mechanisms of action. This chapter considers the antimicrobial activities by groups in terms of the biological inhibition that they cause. Under each of these groups, the individual agents will be discussed.

Protein Denaturation or Damage

This is the most common mode of action. Proteins are relatively easy to damage and are, of course, essential to the proper functioning of an organism. Many are enzymes, and the others are important structural components of ribosomes and membranes. Examples of agents that damage proteins are halogens, alcohols, phenols, peroxides, heavy metals, aldehydes, and ethylene oxide.

Membrane Disruption

The cell membrane must be intact for proper cell functioning. Agents that disrupt membranes are very effective antimicrobial compounds. Examples are halogens, alcohols, phenols, and detergents.

Miscellaneous

The remaining modes of action are rather different from each other but will be placed in one group here because there is usually only one example of each. The antiseptic dyes—acriflavine and crystal violet—function by interfering with DNA transcription and peptidoglycan synthesis, respectively. A potent alkylating agent, ethylene oxide, inhibits protein function, as mentioned, but also blocks the active functioning of reactive groups on many cellular compounds, not just proteins.

Ethylene oxide has become a very important sterilant in recent years. This gaseous material is very good at penetrating small areas of surfaces that liquid disinfectants cannot always reach. Ethylene oxide has found wide use in the sterilization of plastics and other materials that cannot withstand conventional heat sterilization methods.

Effectiveness Tests

Before an antimicrobial agent can be used commercially, the proper concentration and conditions for use must be clearly established. There are two major methods for this determination.

Phenol Coefficient

Phenol is the oldest disinfectant in common use. Therefore, the original test for an agent's effectiveness involves measurement of the activity relative to the activity of phenol. The resulting ratio is called the *phenol coefficient.* It is interesting that essentially all modern antimicrobial agents have phenol coefficients greater than one, meaning that at equivalent concentrations, they are more effective than phenol, the traditional standard.

Use-Dilution

The actual use of most disinfectants involves the treatment of surfaces of material rather than solutions. Therefore, the use-dilution procedure was devised as an improvement upon the phenol coefficient method so that tests of the effectiveness of an antimicrobial agent could be made under conditions which more closely simulated those expected in an actual application. A key feature of the procedure is the use of standard physical items such as surgical sutures or metal cylinders which are coated with the test organism. These objects are then exposed to different levels of the antimicrobial agent being tested to determine the concentration needed for effective action.

Direct Spray

This relatively new procedure was developed for two reasons. First, some antimicrobial agents mix poorly with water and therefore cannot be tested easily with the phenol coefficient or use-dilution methods. Second, other test procedures are done under conditions quite different from applications such as to skin; therefore, a more realistic method was desired. In the direct spray method, organisms are inoculated onto plates and are then exposed to a carefully controlled amount of the antimicrobial agent in an aerosol form.

Toxicity Tests

In addition to antimicrobial effectiveness, it is also necessary to determine the potential toxicity of an agent. This determination is important even if the agent will only be used as a disinfectant for surfaces, because of harmful residues. The toxicity tests employ either the tissue culture method described in the textbook or direct application of various dilutions of the agent to test animals.

Physical Control of Microorganisms

Clean versus Sterile

Before any antimicrobial agent can be effective, it must come in contact with the target microorganisms. This is not difficult to accomplish with chemical agents because the chemicals penetrate quite well. Physical agents, especially heat and radiation, have less penetrating power and therefore work much better on clean surfaces. Liquids (e.g., culture media) are an exception because heat is transferred quite readily through water. It is a simple point, but still important, that sterilization procedures cannot replace adequate cleaning.

Methods

The three classes of physical sterilization methods are heat, radiation, and filtration. The textbook explains the principles of these methods in detail. This chapter notes some of their special features. Be sure you read the textbook first to understand the mechanisms of action of these sterilizing agents.

Heat

Heat is the simplest procedure for sterilization and is therefore the most commonly used. There are two ways for heat to be applied: as wet heat or as dry heat. Wet heat is more efficient because of its greater penetration abilities. Sterilization times are much shorter at a given temperature for wet heat (steam) than for dry heat (oven). Nevertheless, it is sometimes necessary to use dry heat procedures because the objects to be sterilized may be harmed by the moisture (metal corrosion or fabric soaking, for example) or because the objects are needed in a dry condition after sterilization.

The *autoclave* is the common, convenient device used to create heat for sterilization. This device pressurizes steam to attain high temperatures. As the textbook emphasizes, the lethal action of an autoclave comes solely from the heat generated, not the excess pressure. An autoclave typically operates at two atmospheres of pressure, whereas most microorganisms are not harmed by pressure until it reaches hundreds of atmospheres.

Heat can, of course, be used as a disinfectant as well as a sterilant. The pasteurization of milk products and some alcoholic beverages is a common use of heat to disinfect. As discussed in Chapter 11, pathogens are usually easier to kill than are other microorganisms, and this relatively mild heat treatment serves to remove harmful organisms without damaging the commercial product. It is certainly possible to sterilize milk or wine by autoclaving, but the product would be substantially altered by the heat and thus rendered commercially useless.

The effect of a given heat treatment can be analyzed and some predictions made, as described in the text. The value of these calculations is that it is not necessary to try all possible combinations of time and temperature to determine the likely effect. Therefore, a few conditions can be tried and those results used to make further sterilizations more efficient. This procedure makes it clear that time and temperature interact in heat sterilization and that proper sterilization requires a knowledge of both parameters.

Ultraviolet Radiation

Ultraviolet radiation can be an effective sterilizing agent, but its use is limited. It penetrates poorly and therefore is useful only as a surface sterilant. Liquids can be sterilized this way only if they are spread very thinly first.

Many bacteria can repair the damage done by ultraviolet radiation, a fact important to consider when materials are to be stored for long periods after sterilization. Ultraviolet radiation is common and plentiful in our atmosphere. The bacterial repair mechanisms have presumably evolved as a result of selection applied by the radiation.

Ionizing Radiation

The high energy from ionizing radiations such as x-rays and gamma rays is a powerful sterilant. As the term implies, biological molecules are ionized, frequently through interaction with metal ions. The ionized molecules are often broken or otherwise damaged and lose their biological function. The cellular DNA appears to be the most sensitive component to the ionizing radiation. Once the genetic information is substantially damaged or altered, repair is impossible. This form of radiation, unlike ultraviolet, is relatively rare in our environment. It is therefore not surprising that organisms have evolved repair mechanisms for ultraviolet-induced damage but not for damage from ionizing radiations.

Ionizing radiation is gaining in popularity as a method for treating certain foods for the purpose of prolonging their shelf lives. Modern techniques have made the process more economical while minimizing the amount of change in the food product. There are unresolved concerns about the long-term safety of irradiated foods, and this area is currently the subject of active research.

Filtration

Air filtration has been used for many years with great success. Hospital operating rooms are protected by large filter systems that will not let organisms pass. Surgical masks also are air filters—typically wads of cotton and glass wool that function on the principle that microorganisms, although small, have physical dimensions and can therefore be trapped in small, winding pathways.

Liquid filtration is also based on the trapping of particles (organisms) by preventing them from passing through small holes. This procedure has gained very wide use since the end of World War II with the creation of extremely thin filters that have very little interaction with the liquid or air being filtered, permitting a solution to be made free of microorganisms without undergoing any chemical changes. One of the important applications of liquid filtration is the production of heat-sensitive compounds. Laboratory vitamins and sugar solutions can be sterilized by passage through a sterile filter with no damage, whereas heat sterilization would likely cause damage. An important industrial application is the sterilization of alcoholic beverages such as beer.

Monitoring Sterilization

It is necessary to be sure that an item has been sterilized before it is used. For heat sterilization procedures, this monitoring is relatively simple and is done either chemically or biologically. The chemical procedure involves chemicals that change color after exposure to enough heat to cause sterilization. These chemicals are commonly contained in tape or paper wrappers and are not visible before heating. After heating, the chemical is apparent (usually brown) and may spell out the word "sterile" for convenience. The biological procedure involves exposure of a test microorganism along with the materials to be sterilized. After the heating procedure, the biological test agent is incubated for 24 to 48 hours. If no growth occurs during that time, the culture is assumed to have been killed during the heat exposure. It is further assumed that all materials in the sterilizing chamber receive the same heat. This biological monitoring is more difficult and time consuming, but it also gives more definite results.

Monitoring the effectiveness of a radiation sterilization procedure is more difficult. There are no analogous chemicals that undergo simple color changes after lethal radiation doses. The normal biological test, most conveniently used in testing aqueous solutions, cannot be used because radiation (especially ultraviolet) penetrates water very poorly. One possible test procedure is to incubate an item in nutrient growth medium after radiation exposure and determine if growth occurs. This process involves extra handling and possible contamination and is rarely used.

V. CHAPTER SELF-TEST

Continue with this section only after you have read Chapter 11. A score of 80 percent or better is good. If your score is less than 65 percent, reread the chapter.

Correct answers to all questions can be found at the end of this section. Write your responses in the appropriate space provided.

Matching

Select the answer from the right-hand side that corresponds to the term or phrase on the left-hand side of the question sheet. An answer may be used more than once. In some cases, more than one answer may be required.

_____ 1. Active form of one type of halogen disinfectant

_____ 2. Method of testing new agents

_____ 3. Agent that does not completely kill eucaryotic microorganisms

_____ 4. Chlorine

_____ 5. Ethylene oxide

_____ 6. Complete killing

_____ 7. Heavy metal

_____ 8. Pathogen removal

_____ 9. Cationic or anionic

_____ 10. Cell wall synthesis inhibitor

a. disinfect
b. sterilize
c. crystal violet
d. mercury
e. detergent
f. phenol coefficient
g. fungistatic
h. halogen
i. tissue toxicity
j. hypochlorite

Matching

Select the answer from the right-hand side that corresponds to the term or phrase on the left-hand side of the question sheet. An answer may be used more than once. In some cases, more than one answer may be required.

_____ 11. Ultraviolet radiation

_____ 12. Relative killing measure

_____ 13. Disinfection

_____ 14. Cellulose acetate

_____ 15. Photoreactivation

_____ 16. Major heat target

_____ 17. Ionizing radiation

_____ 18. Sterilization monitoring

_____ 19. Surgical masks

_____ 20. High pressures

k. pasteurization
l. filtration
m. proteins
n. DNA
o. autoclave
p. *Bacillus*
q. D-value
r. moist heat

Completion Questions

Provide the correct term or phrase for the following. Spelling counts.

21. The major goal of antimicrobial chemical treatment is not sterilization but, rather, the elimination of ____________________.
22. If this is accomplished in living tissue, the procedure is said to be ____________________.
23. If bacteria in an area are only delayed in their growth rather than killed, the process is ____________.
24. The two microbial forms most difficult to control are (a) ____________ and (b) ____________.
25. ____________________ cause dehydration and protein denaturation.
26. Because it releases O_2, ____________________ is an effective antiseptic against anaerobes.
27. ____________________ are very effective membrane disrupters because of their chemical similarity to lipids.
28. New agents are tested for effectiveness by comparison to (a) ____________________ in the (b) ____________________ test.
29. Further testing involving surfaces is done with the ____________________ test.
30. A new procedure to test water-insoluble agents is the ____________________ method.
31. The ____________________ of antimicrobial agents to animals must also be determined.
32. The most common physical method used for sterilization is ____________________.
33. This agent is also used as a(an) ____________________ in pasteurization.
34. The most heat-resistant structures are bacterial ____________________.
35. These structures are also used to ____________________ sterilization procedures.
36. The damage done by (a) ____________________ radiation can often be repaired, even in the dark by (b) ____________________.
37. (a) ____________________ radiation sterilizes very well but (b) ____________________ most materials poorly.
38. Radiation affects cellular (a) ____________, while heat affects the (b) ____________ and (c) ____________________ of cells.
39. Air sterilization is routinely done by (a) ____________________ and is most effective in a(an) (b) ____________________ system.
40. Liquids that are heat sensitive can be sterilized by (a) ____________________ through (b) ____________________ without being changed.

Essay Questions

Answer the following questions on a separate sheet of paper.

41. Effective antisepsis is often more difficult to attain than is disinfection. Why?

42. List two parts of the cell membrane that are good targets for antimicrobial chemicals. Explain.
43. Why is liquid filtration sometimes used for sterilization?

Answers

1. j
2. f
3. g
4. h
5. b
6. b
7. d
8. a
9. e
10. c
11. n
12. q
13. k
14. l
15. n
16. m
17. n
18. p
19. l
20. o
21. pathogens
22. antiseptic
23. bacteriostatic

24a. spores
24b. viruses
25. alcohols
26. peroxide
27. detergents
28a. phenol
28b. phenol-coefficient
29. use-dilution
30. direct spray
31. toxicity
32. heat
33. disinfectant
34. spores
35. monitor
36a. ultraviolet
36b. dark reactivation
37a. ionizing
37b. penetrates
38a. DNA
38b. proteins
38c. membranes
39a. filtration
39b. laminar air flow
40a. filtration
40b. membrane filters

41. Antisepsis deals with living tissue, so the chemicals used must be mild enough not to damage the patient. Since disinfection is defined as the killing of pathogens on an inanimate surface, harsher chemicals can be used.
42. Lipids are the main structural components of the membrane; their disruption may lyse the cell. Proteins in the membrane may have specific transport activities essential to the cell's nutrition.
43. Many solutions contain heat-sensitive compounds that would be damaged by heating. The increased pressure produces higher temperatures, which are responsible for killing; the pressure does not kill.

VI. ENRICHMENT

Maintenance of Sterilization

We have examined several physical processes of sterilization and have seen the difficulties in using them. A related aspect that we did not discuss in detail is the maintenance of sterilization. That is, how is an object kept sterile until use? Let us consider the two simplest examples: dry supplies and solutions.

Dry supplies such as surgical instruments or dressings are usually sterilized by autoclaving. The easiest way to keep them sterile is to put them inside a container—say, a stainless steel pan—or wrap them in heavy paper before placing them in the autoclave. These coverings do not aid in the sterilization process, but they keep the objects from contact with the atmosphere when they are removed from the autoclave.

Solutions must be contained in flasks or bottles. These bottles must be at least partially open or vented during autoclaving or excess pressure might build up inside and burst the bottle. However, a lid or cap must be placed on the container after sterilization to prevent organisms from the atmosphere from entering. Sterile filtration procedures must be done with filters and receiving flasks that have been sterilized, usually by autoclaving. The filter must be put in place, filtration performed, and receiving flask closed afterward, all without external contamination. Since the air in most places, especially laboratories, contains many organisms, filtration processes are often carried out in laminar flow hoods to minimize airborne contamination.

12

Antimicrobial Chemotherapy

I. INTRODUCTION

Chemotherapeutic agents represent a wide range of compounds used in the treatment of disease. Most of these chemicals fall into two general categories: those produced biologically by microorganisms, such as fungi, and synthetic products prepared commercially. This chapter considers various types of chemotherapeutic agents used in the treatment of microbial diseases, modes of action, methods to determine antibiotic activity, and factors involved with drug resistance. This chapter will serve as an important reference for the treatment presentations in Chapters 23 through 31.

II. PREPARATION

Chapters 5, 6, 7, and 12 should be read before continuing. The following terms are important for you to know. Refer to the glossary and the appropriate chapters of the text if you are uncertain of any of them. A pronunciation guide for selected terms is provided as a learning aid.

1. additive effect
2. allergy (AL-err-gee)
3. aminoglycoside (ah-me-no-GLY-koh-side)
4. antagonistic (an-tag-oh-NIST-ik)
5. antibiotic (an-tie-by-OT-ik)
6. antibiotic abuse
7. AZT
8. bactericidal (back-teer-ee-SI-dal)
9. bacteriostatic (back-tee-re-oh-STAT-ik)
10. beta-lactamase (BAY-tah LAK-tam-ace)
11. broad spectrum
12. cell membrane
13. cell wall
14. cephalosporin
15. chemotherapy
16. coenzyme
17. dermatitis (der-mah-TIE-tis)
18. drug resistance
19. genital herpes
20. interferon (in-ter-FEAR-on)
21. *in vitro* (in VEE-troh)
22. *in vivo* (in VEE-voh)
23. malaise (MAL-aze)
24. minimal antibacterial concentration
25. minimal inhibitory concentration
26. monobactam (mon-oh-BACK-tam)

27. pathogen
28. penicillin
29. penicillin binding protein
30. plasmid (PLAZ-mid)
31. polypeptide
32. protein synthesis
33. substrate (SUB-straight)
34. synergistic (SIN-err-jist-ik)
35. tetracycline (teh-trah-SYE-clean)
36. therapeutic index
37. toxicity (tok-SIS-ee-tee)
38. transpeptidase
39. transposon
40. vaccine (VAK-seen)

III. PRETEST

Correct answers to all questions can be found at the end of this section. Write your responses in the appropriate space provided.

Completion Questions

Provide the correct term or phrase for the following. Spelling counts.

1. The groundwork for modern chemotherapy can be accredited to ______________________.
2. According to Ehrlich, a functional drug would (a) ______________ or (b) ______________ the growth of a parasite without ______________________ the host.
3. All chemotherapeutic agents are obtained from (a) ______________ or (b) ______________.
4. The term *antibiotic* is applied to a variety of antimicrobial drugs, which include (a) ______________ and (b) ______________________.
5. The use of an antibiotic for the treatment of human microbial infection depends on several factors, including (a) ______________, (b) ______________, (c) ______________, (d) ______________________, and (e) ______________________.
6. The lowest concentration of a drug that will prevent growth of a standardized suspension of an organism is called the ______________________.
7. To be effective, an antibiotic must (a) ______________ and (b) ______________.
8. A drug that kills bacteria is referred to as being (a) ______________________, whereas one that only inhibits growth is called (b) ______________________.
9. Drugs active against several gram-positive and gram-negative bacteria are considered to be ______________ in their mode of action.
10. Five specific mechanisms of action of antimicrobial drugs are (a) ______________, (b) ______________, (c) ______________________, (d) ______________________, and (e) ______________________.

Matching

Select the answer from the right-hand side that corresponds to the term or phrase on the left-hand side of the question sheet. An answer may be used more than once. In some cases, more than one answer may be required.

Topic: Specific Mechanisms of Antimicrobial Drugs

_____ 11. Cephalosporin
_____ 12. PABA
_____ 13. Sulfonamides
_____ 14. Penicillin
_____ 15. Ampicillin
_____ 16. Streptomycin
_____ 17. Chloramphenicol
_____ 18. Tetracycline
_____ 19. Polymyxin B
_____ 20. Adenine dinucleotide
_____ 21. Isonicotinic hydrazide
_____ 22. Rifampin

a. interferes with the synthesis of folic acid
b. interferes with cell wall formation
c. inactivated by beta-lactamase
d. causes misreading of the genetic code
e. blocks transfer of activated amino acid from transfer RNA to growing polypeptide chain
f. increases permeability of membranes
g. none of these

True or False

Indicate all true statements with a "T" and all false statements with an "F".

_____ 23. A mechanism by which an organism develops resistance to antimicrobial drugs is a change in the selective permeability of cell walls and membranes of organisms.

_____ 24. Antibiotics can be used as feed additives to protect plants and livestock against infections.

_____ 25. Antibiotic resistance can, in certain cases, be controlled by chemically changing the structure of the antibiotic involved.

_____ 26. Antibiotic resistance can be transferred between microorganisms by nucleoids.

_____ 27. In determining the antibiotic resistance of an organism using the filter-paper disk method, the type of medium used will not affect the size of the antibacterial activity zone.

_____ 28. Monitoring blood levels of antibiotics in an infected individual is important in providing proper treatment.

_____ 29. Polyene antibiotics interfere directly with fungal protein synthesis.

_____ 30. A major feature of the mechanisms of action of chemotherapeutic agents for protozoan diseases is the absence of any interference with protein and nucleic acid synthetic processes.

_____ 31. One drug interfering with the antimicrobial activity of another is referred to as a synergistic effect.

_____ 32. *Serratia marcescens* is the source for a group of antibiotics called monobactams.

_____ 33. The synthesis of beta-lactamases may be regulated by plasmids.

_____ 34. Methicillin, oxacillin, and nafcillin are effective against beta-lactamase-producing bacteria.

_____ 35. Imidazoles interfere with sterol formation.

______ 36. Interferons are not host specific.

______ 37. Alpha and beta interferons can be induced by most major virus groups.

______ 38. Current interferon therapy is not without side effects.

Identification Question

39. Select the letter that represents the beta-lactam ring (*a, b, c,* or *d*) in Figure 12-1. ________________

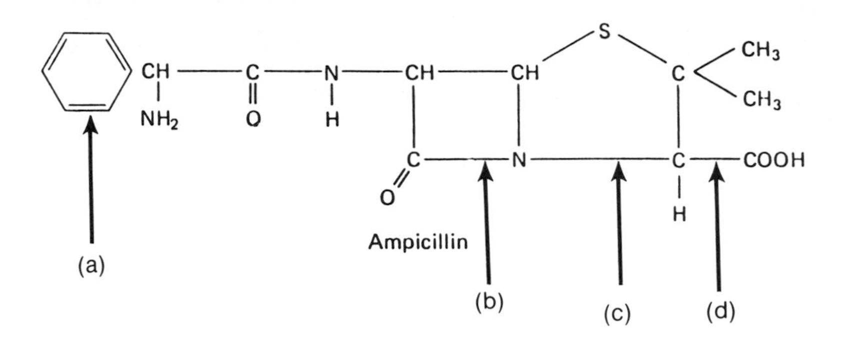

Figure 12-1

Answers

1. Paul Ehrlich
2a. kill
2b. inhibit
2c. injuring
3a. microbial
3b. laboratory-synthesized preparation
4a. totally synthesized laboratory products
4b. chemically modified or altered natural antibiotics
5a. patient's general physical condition
5b. existence of drug allergies
5c. the nature of the disease agent
5d. site or location of infection
5e. antibiotic dosage
6. minimal inhibitory concentration (MIC)
7a. be relatively non-toxic to the host
7b. exhibit anti-microbial activity in low concentrations in the host
8a. bactericidal
8b. bacteriostatic
9. broad spectrum
10a. inhibition of a specific product or metabolism
10b. inhibition of cell wall formation
10c. inhibition of protein synthesis
10d. irreversible damage to cell membranes
10e. inhibition of nucleic acid synthesis
11. b
12. g
13. a
14. b and c
15. b and c
16. d
17. e
18. e
19. f
20. g
21. g
22. g
23. T
24. T
25. T
26. F
27. F
28. T
29. F
30. F
31. F
32. F
33. T
34. T
35. T
36. F
37. T
38. T
39. b

IV. CONCEPTS AND TERMINOLOGY

Antibiotics are among the most frequently prescribed drugs for the treatment and control of microbial diseases. However, their use is not without problems, such as antibiotic-resistant microorganisms, the influence of the host (the individual being treated) on the drug's antimicrobial effectiveness, and drug toxicity.

Historical Background

The modern era of chemotherapy as a means of controlling infectious disease was launched by the discoveries of Salvarsan by Ehrlich in 1909, sulfa drugs by Domagk in 1935, and penicillin by Fleming in 1929.

The approach to modern drug development was initiated by Ehrlich, who stressed that a functional drug kills or inhibits the growth of a parasite without causing serious injury to the host.

Chemotherapeutic Agents

All chemotherapeutic agents originate in one of two ways: (1) as natural metabolic products of microorganisms or (2) as substances synthesized in the laboratory. The term *antibiotic* is applicable today to a variety of antimicrobial drugs, including compounds totally synthesized in the laboratory and the chemically altered (semisynthetic) forms of natural antibiotics.

The bacterial and fungal species used for the production of antimicrobial drugs are largely soil and water fungi. In natural environments, the production of antibiotics serves as a natural means of controlling microbe populations that constantly compete for food and space to live.

Antibiotics are used to treat diseases caused by microbes, to protect plants and livestock against infectious agents and accelerate their growth, and to maintain the freshness of various foods and related products.

Principles of Chemotherapy

The effective antibiotic treatment of infected individuals depends on the selection of the best antimicrobial drug. The administration of an antibiotic for treatment depends upon several factors, including the patient's general physical condition, existence of drug allergies, the pathogen, the site of infection, and the dosage needed.

Laboratories determine a patient's drug susceptibility and report the drug sensitivity and resistance pattern of a disease agent. With such results, physicians can select the appropriate drug for treatment. Unfortunately, because of factors such as certain tissues concentrating a drug or blocking its penetration, laboratory findings do not always correlate with the results of drug therapy in the body. Therefore, for effective treatment, it is important to know the achievable levels of antimicrobial drugs in the body and the relative sensitivities of the pathogens. The relative drug sensitivity is called the minimal inhibitory concentration (MIC), meaning the lowest drug concentration that will prevent the growth of a standardized microbial suspension.

Effective antimicrobial drugs must be relatively nontoxic to the host and exhibit antimicrobial activity at low concentrations in the body. Drugs that cause irreversible injury of susceptible microorganisms and ultimately cell death are called *cidal.* Drugs of this type that kill bacteria are referred to as *bactericidal.* Chemotherapeutic agents that do not cause death but inhibit microbial growth produce a *static* effect. The general properties of chemotherapeutic drugs are described in Table 12-1.

TABLE 12-1 General Categories of Antimicrobial Agents

General Category	*Description*
Cidal drugs	1. Cause irreversible injury and eventual death of susceptible microorganisms 2. Independent in their actions 3. Directly effects disease agents
Static drugs	1. Do not cause cell death but inhibit microbial growth 2. Dependent on the host's immune mechanisms for the eventual elimination of pathogenic microbes 3. Dilution or removal of drug enables microorganisms to resume normal activities
Broad spectrum	Drugs that are effective against a variety of gram-negative and gram-positive organisms
Limited (narrow) spectrum)	Drugs that are effective against only a limited number of species

Determining the usefulness of antibiotics for treatment takes into consideration factors such as the patient's history and clinical condition and the results of laboratory isolations and antibiotic sensitivity tests. Of particular importance is the *therapeutic index,* the ratio between the minimum toxic dose of the drug for the host and the minimum effective microbial lethal dose. Specific mechanisms of action exhibited by antimicrobial drugs include:

1. Inhibition of the formation of a specific product of metabolism (metabolite)
2. Inhibition of cell wall formation
3. Inhibition of protein synthesis
4. Irreversible damage to the cell membrane
5. Inhibition of nucleic acid synthesis (either RNA or DNA)

The effectiveness of a particular antibiotic can be increased when it is administered in combination with another antimicrobial agent. The resulting antimicrobial activity is greater than when each agent is given separately and is referred to as a *synergistic effect.* Some drug combinations may be *antagonistic;* that is, one drug may interfere with the antimicrobial activity of another.

Some Commonly Used Antimicrobial Drugs

Table 12-2 summarizes the mechanisms of action and range of activity of some commonly used antimicrobial drugs.

TABLE 12-2 Properties of Commonly Used Antimicrobial Drugs[a]

Antimicrobial Drug(s)	*Mechanisms of Action*	*Disease(s) Treated*
Sulfa drugs	Bacteriostatic; active against bacteria that form their own folic acid. A sulfonamide may replace paraminobenzoic acid (PABA) in such reactions, a form of competitive inhibition.	*Escherichia coli* urinary tract infections; meningococcal infections; chancroid; trachoma; *Nocardia* infections; certain protozoan diseases
Penicillins[b] (penicillin, ampicillin, carbenicillin, cephalosporin)	Bactericidal; interfere with bacterial cell wall synthesis.	Sensitive gram-negative and gram-positive strains; in high concentration, *E. coli* and *Proteus mirabilis* infections
Clindamycin	Bactericidal; inhibits protein synthesis.	Pneumococci; streptococci; most *Staphylococcus aureus* isolates
Chloramphenicol and tetracyclines (chlortetracycline, doxycycline, and minocycline)	Bacteriostatic; inhibit protein synthesis by blocking transfer of activated amino acids to a growing peptide chain.	Several gram-negative and gram-positive bacteria; rickettsia; chlamydia; certain *Salmonella* infections
Aminoglycosides (streptomycin, kanamycin, gentamicin)	Bactericidal; interfere with protein synthesis by combining with a subunit of a ribosome and causing a misreading of the genetic code.	A variety of gram-negative and gram-positive organisms. Streptomycin and kanamycin are active against *Mycobacterium tuberculosis*
Polymyxin B and colistin	Bacteriostatic or bactericidal (effect is dosage dependent); cause leakage of essential cytoplasmic components through all membranes.	Effective against most gram-negatives, with exceptions of *Proteus* species
Isonicotinic hydrazine (INH) and para-amino-sulicyclic acid (PAS)[c]	Generally bacteriostatic; blocks essential enzyme activities.	*Mycobacterium tuberculosis*
Ethambutol	Bacteriostatic; inteferes with RNA synthesis.	*Mycobacterium tuberculosis*
Rifampin	Bactericidal; interferes with nucleic acid synthesis.	Active against most gram-negatives, gram-positives, and *M. tuberculosis*

[a]Refer to Table 12-3 in the text for additional uses of chemotherapeutic drugs.
[b]Penicillin and ampicillin are inactivated by beta-lactamase, an enzyme produced by penicillin-resistant staphylococci and certain other organisms.
[c]Commonly used in combination with streptomycin.

Monobactams

Monobactams represent a group of monocyclic bacterially produced beta-lactam antibiotics. These compounds exhibit activity against aerobic gram-negative bacteria. Monobactams interfere with cell wall formation, and are produced by members of the bacterial genera such as *Acetobacter, Agrobacterium, Chromobacterium, Gluconobacter,* and *Pseudomonas.*

Mechanisms of Drug Resistance

Partly as a result of drug misuse, various antibiotic-resistant organisms in microbial populations have developed and reproduced. These organisms have become increasingly difficult to eliminate even with the development of more broad-spectrum and specific drugs.

The mechanisms by which organisms develop resistance to antimicrobial agents include (1) a change in the selective permeability of the cell walls and membranes of organisms, (2) a change in the sensitivity of affected enzymes, (3) an increased production of a competitive substrate, and (4) the inactivation or destruction of the drug by an enzyme such as beta-lactamase. Antibiotic resistance can be transferred between microorganisms by several means, including bacterial viruses. In many instances, most antibiotic resistance is determined and controlled by extrachromosomal particles called *plasmids.*

Antibiotic Sensitivity Testing Methods

The minimum inhibitory concentrations (MICs) of an antimicrobial agent can be determined using either liquid or solid media. Procedures for this purpose include the cylinder and well methods, the filter-paper disk method, the Kirby-Bauer (K-B) standardized single-disk method, and agar-overlay modification techniques. Representative procedures are briefly summarized in Table 12-3. The presence of clear zones around disks, wells, or cylinders containing antibiotics indicates the inhibition of a test organism's growth. Such organisms are termed susceptible. The extent of the zone indicates the degree of sensitivity. Resistant (R) organisms are not affected by the drugs used.

Several factors can affect the size of the zone of antibacterial activity: (1) the depth of the medium used, (2) the choice of medium, (3) the size of the inoculum, and (4) the diffusion rate of a particular antibiotic. The last factor, in particular, has resulted in unfortunate misinterpretations of results.

Blood Levels of Antimicrobial Agents

Proper treatment of life-threatening diseases often requires close monitoring of blood levels of antibiotics in patients. Several monitoring methods are available, including biologic and enzymatic assays. These procedures have not been standardized to the same extent as the better known antibiotic sensitivity tests.

Antimycotic Agents

Unlike treatment for bacterial infections, treatment of fungus infections or mycoses by chemical agents is still quite limited. The most important antifungal agents are those that affect cell membranes. The polyene antibiotics interact with the membranes of susceptible cells and distort their selective permeability. Other antimycotic agents are 5-fluorocytosine, which interferes with RNA synthesis, and the imidazoles, which interfere with sterol synthesis.

The Treatment of Protozoan Diseases

Chemotherapeutic agents used for treatment are not always readily available, can be highly toxic to the host, and may not be effective against all stages of a parasite. The mechanisms of action of antiprotozoan drugs include interference with energy metabolism; disruption of membrane functions; and interference with nucleic acid production, protein synthesis, and other biosynthetic reactions of a parasite.

TABLE 12-3 Antibiotic Sensitivity Testing Methods

Method	*Description*	*Interpretation of Results*
Broth (tube) dilution	Tubes of sterile broth containing decreasing concentrations of a specific antibiotic are inoculated with a standard concentration of the test organism. Tubes are incubated.	After 24 hours of incubation, the MIC is determined by the lowest drug concentration inhibiting growth.
Agar dilution	Different antibiotic concentrations are incorporated into an agar medium for both aerobes and anaerobes. A replicator device may be used to inoculate multiple specimens into a series of plates with varying concentrations of antibiotics. Plates are incubated.	After incubation, the minimum inhibitory concentration may be determined electronically.
Cylinder and well	Cylinders or wells (holes) containing antimicrobial agent are placed into agar plates seeded with test organism. Preparations are incubated.	Measurement of zones of growth inhibition (if present).
Filter-paper disk	Paper disks impregnated (antibiotic disks) with known concentrations of specific antibiotics are placed onto surfaces of seeded agar plates. Preparations are incubated.	Measurement of zones of inhibition.
Kirby-Bauer single disk	Paper disks impregnated with known concentrations of antibiotics are placed onto the surface of an agar plate containing a standardized concentration of test organisms. Preparations are incubated.	Measurement of zones of inhibition by calipers or a template (pattern) showing standard zone sizes.
Agar overlay	A standardized suspension of test organisms in melted agar are poured onto the surfaces of an agar plate. After solidification, antibiotic disks are applied and preparations are incubated.	Measurement of zones of inhibition and comparison to known drug sensitivities.

Antiviral Agents

Limited success has been achieved with chemotherapeutic drugs for viral diseases. The properties of some promising drugs are given in Table 12-4.

Interferons

Interferons (IFNs) are a family of proteins that serve as biological regulators of cell function. Human interferons are classified into three distinct antigenic types: α (alpha), β (beta), and γ (gamma). In addition to their antiviral actions, IFNs affect cell multiplication, cell motility, cell membrane rigidity, and various immunological processes, including immunoglobulin response, rejection of foreign tissue grafts, skin test reactions as in the tuberculosis skin test, the activation of macrophages, and the selection of natural killer T cells. Interferons induce enzymes that interfere with the translation of viral mRNA. The interferons appear to have great potential as therapeutic agents not only against viruses but also against certain cancer cells.

Antiviral Agent Sensitivity Testing

Determination of the clinical effectiveness of an antiviral agent involves evaluations for antiviral activity and toxicity both in cell culture in animal models and actual cases of infections.

TABLE 12-4 A Comparative Summary of Antiviral Drugs

Drug	*Representative Viral Disease(s) Affected*	*Possible Side Effects*	*Mechanism of Action*
Acyclovir	Genital herpes, fever blister, chicken pox	Local pain of short duration at site of application	Selectively inhibits viral DNA synthesis
Amantadine	Influenza A_2 (Asia)	General irritability, insomnia, confusion, hallucinations, inability to concentrate	Prevents penetration of certain viruses into host cells
3′-azido-3′-deoxythymidine (AZT)	AIDS	Inhibits normal blood cell formation; headache	Interferes with viral DNA synthesis
5-Iodo-2′-deoxyuridine	Severe herpes simplex virus type 1 infection	Nausea, vomiting, hair and fingernail loss, lowering of white blood cells (leukopenia) and platelets (thrombocyto-penia)	Blocks synthesis of nucleic acids
Cytosine arabinoside	Progressive varicella (chicken pox) and zoster (shingles) infections	Nausea, vomiting, loss of appetite, chromosomal changes, lowering of white blood cells and platelets, anemia	Inhibits DNA synthesis
Methisazone	Progressive vaccinia	Nausea, vomiting, loss of appetite, liver toxicity (hepatotoxicity)	Interferes with protein synthesis at the level of trans-lation
Ribavirin	Respiratory syncytial virus infections, AIDS	Difficulty in breathing, chest soreness, anemia, rash, and inflammation of eyelid lining (conjunctivitis)	Inhibits early replication step leading to viral nucleic acid synthesis
Vidarabine	Genital herpes, fever blister, cytomegalovirus infection, infectious mononucleosis	Nausea, vomiting, diarrhea; tremors, pain, and seizures in chronic hepatitis patients and kidney transplant recipients	Selective inhibition of viral DNA synthesis

Chemotherapy Side Effects

At times, unfavorable side effects or responses develop in patients that limit the usefulness of antibiotics in arresting or curing certain diseases. Examples of side effects and complications include an exaggerated allergic response (hypersensitivity), direct toxic effects on cells and/or organs, such as the kidney and liver, destruction of blood-forming tissues in the bone marrow, and changes in an individual's normal bacterial flora, which may contribute to the establishment and overgrowth by opportunistic microorganisms.

V. CHAPTER SELF-TEST

Continue with this section only after you have read Chapters 5, 6, 7, and 12. A score of 80 percent or better is good. If your score is less than 65 percent, reread the chapter.

Correct answers to all questions can be found at the end of this section. Write your responses in the appropriate space provided.

Matching

Select the answer from the right-hand side that corresponds to the term or phrase on the left-hand side of the question sheet. An answer may be used more than once. In some cases, more than one answer may be required.

Topic: Principles of Chemotherapy

_____ 1. Effective against several gram-negatives and gram-positives

_____ 2. Cidal drug

_____ 3. Effective against a single group of microbes

_____ 4. Static effect

a. inhibition of growth

b. causes death

c. broad spectrum

d. limited spectrum

e. none of these

Topic: Specific Mechanisms of Antimicrobial Drugs

_____ 5. Interferon

_____ 6. Mitomycin

_____ 7. Rifamycin

_____ 8. Penicillin

_____ 9. Bacitracin

_____ 10. Aminoglycosides

_____ 11. Tetracyclines

_____ 12. Cycloserine

_____ 13. Polyenes

_____ 14. Sulfa drugs

_____ 15. Polymyxin

_____ 16. 5-fluorocytosine

_____ 17. Methisazone

_____ 18. Vancomycin

_____ 19. AZT

a. inhibits some aspects of protein synthesis

b. inhibits cell wall formation

c. damages cell membrane

d. inhibits nucleic acid synthesis

e. none of these

True or False

Indicate all true statements with a "T" and all false statements with an "F".

_____ 20. Antibiotics can be totally synthesized in the laboratory.

_____ 21. In nature, antibiotic production serves as a means of controlling competing microbial populations.

_____ 22. The existence of drug allergies in an infected individual can affect the choice of antibiotic administered for treatment.

_____ 23. Laboratory drug-susceptibility tests correlate well with results in the body.

_____ 24. MIC refers to the maximum concentration of a drug that will prevent the growth of a standardized suspension of an organism.

______ 25. Interference with a host's natural defense mechanisms is one of several undesirable side effects caused by antibiotics.

______ 26. Static chemotherapeutic agents are independent in their actions.

______ 27. Cidal drugs are dependent on a host's immune mechanisms for the eventual elimination of a pathogen.

______ 28. Changes in the selective permeability of the cell membranes of drug-sensitive organisms can be a factor in the development of antibiotic resistance.

______ 29. Penicillinase acts on the beta-lactam ring of synthetically produced tetracyclines.

______ 30. Antibiotic resistance in organisms frequently is controlled by plasmids obtained from infected host cells.

______ 31. In the filter-paper disk antibiotic sensitivity test, the size of the inoculum is not an important factor in determining the effectiveness of an antibiotic.

______ 32. The Kirby-Bauer method is not applicable to determining the antibiotic sensitivities of rapidly growing pathogenic bacteria.

______ 33. The Kirby-Bauer and agar-overlay modification procedures are effective methods to monitor blood levels of antibiotics during treatment.

______ 34. In biological assays to determine the blood levels of antimicrobial agents, the minimal concentration of an antibiotic found to be effective against a specific pathogen by disk diffusion is the dosage administered.

______ 35. All microorganisms resistant to polyenes contain sterols.

______ 36. Some antiprotozoan preparations contain arsenic or antimony.

______ 37. An antagonistic effect occurs when the effectiveness of a particular antibiotic is increased by administering it in combination with another drug.

______ 38. Beta-lactamases are important enzymes responsible for resistance to many beta-lactam antibiotics such as imidazoles.

______ 39. Interferons (IFNs) are a family of proteins having antiviral and antitumor activities.

______ 40. Five distinct types of interferons are known: alpha, beta, gamma, F, and I.

______ 41. Interferon inhibits viral replication by interference with the host cell's protein translation process.

______ 42. One possible side effect associated with the use of certain chemotherapeutic drugs is an individual's normal bacterial flora.

Identification Question

43. Select the letter that represents the beta-lactam ring (*a, b, c,* or *d*) in Figure 12-2. ______________

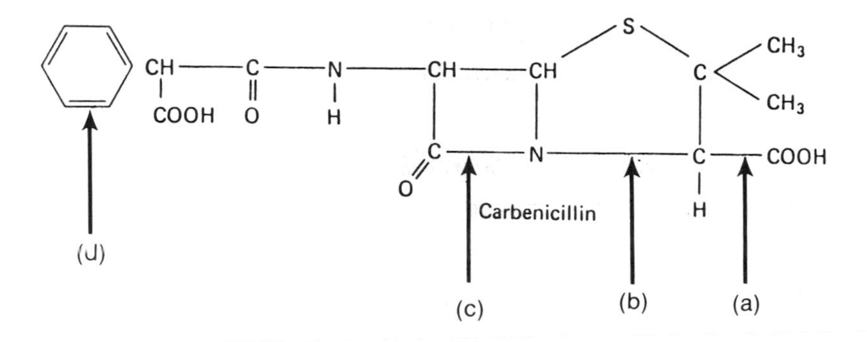

Figure 12-2

Essay Questions

Answer the following questions on a separate sheet of paper.

44. What is MIC?
45. Distinguish between bactericidal and bacteriostatic drugs.
46. By what mechanisms do microorganisms, such as bacteria, develop resistance to antimicrobial agents?

Answers

1. c	12. b	23. F	34. F
2. b	13. c	24. F	35. F
3. d	14. e	25. T	36. T
4. a	15. c	26. F	37. F
5. a	16. d	27. F	38. F
6. d	17. a	28. T	39. T
7. d	18. b	29. F	40. F
8. b	19. d	30. F	41. F
9. b	20. T	31. F	42. T
10. a	21. T	32. F	43. c
11. a	22. T	33. F	

44. MIC stands for minimal inhibitory concentration. It is the lowest quantity of a drug that will prevent the growth of a standardized suspension of the organism under study.

45. The use of several chemotherapeutic agents leads to irreversible injury of susceptible microorganisms and ultimately cell death. Effects of this type are referred to as *cidal.* A drug that causes the death of bacteria is termed *bactericidal.* Those chemotherapeutic agents that do not cause cell death but inhibit growth produce a *static* effect. Dilution or removal of such drugs enables microorganisms to resume growth and reproductive activities. A chemotherapeutic agent that inhibits bacterial growth is referred to as *bacteriostatic.*

46. The mechanisms by which organisms develop resistance to antimicrobial agents include (1) an enzymatic alteration of the drug, (2) a change in the selective permeability of the cell wall and membranes of organisms, (3) a change in the sensitivity of affected enzymes, and (4) an increased production of a competitive substrate.

VI. ENRICHMENT

Bacterial Resistance to Antibacterial Agents

The bacteria we carry outnumber our own body cells. Each human and other type of animal supports an ecosystem of competing strains and species of bacteria. Groups of hosts that exchange bacteria, such as patients in an intensive care unit or animals in a herd, form larger bacterial ecosystems, and eventually interconnect in a global bacterial ecosystem.

Many bacterial strains throughout these ecosystems are now able to withstand antibacterial agents that would earlier have killed them. Such resistant strains have come to carry genes encoding products that (1) inactivate those antibacterial agents, (2) keep them from reaching their target sites, or (3) provide alternatives for those sites that have been blocked. Some of these antibacterial resistance genes are located on the bacterial chromosome, but the majority are on extrachromosomal genetic elements called *plasmids* that may transfer themselves to other strains or bacterial species.

The transfer of a resistance gene on a plasmid to other strains and/or species spreads the gene to the niches in the ecosystems that the other strains are fit to occupy. It also favors encounters with, and eventually insertion of, the resistance gene into other plasmids. The other plasmids may then be able to transfer to additional niches. Spread of a resistance gene in this manner would also preserve it by linking it to other genes with a variety of survival values.

Each step in the sequence of spread would lead to a succeeding amplifying step and would be reinforced by the selection pressure provided through the use of antibacterial agents. Recognition of this cycle of antibacterial use has prompted attempts to employ antibacterial agents more selectively. The cycle has also stimulated the development of a succession of new antibacterial agents designed to be unaffected by the inactivating products of previously spread resistance genes. Unfortunately, genes that encode for the resistance to the newer antibacterial agents also have appeared after such agents have been used.

Antibacterial agents have undoubtedly saved and improved more lives than any other existing group or class of pharmaceuticals. Their use, however, has set in motion the biggest intervention in population genetics seen to date on this planet. The effects of that intervention are seen in the distributions of antibiotic-resistance genes throughout the world's bacterial populations.

PART IV
MICROBIOLOGY TRIVIA PURSUIT
(Chapters 11 and 12)

It is time again to examine your attention to detail, and your fact-gathering ability. There are *four* challenging parts to this section. Remember, a certain number of questions must be answered in each Part before you proceed up the MICROBIOLOGY TRIVIA PURSUIT trail. The answers and directions for continuing are given at the end of each Part. Try Part 1.

PART 1

Completion Questions

Provide the correct term or phrase for the following. Spelling counts.

1. ______________________ is the oldest chemical disinfectant and is still used as the standard to compare new disinfecting agents.
2. A ______________________ ______________________ is used to be sure that a sterilization procedure has been effective.

3, 4, and 5. Acquired antibiotic drug resistance may result from (3) ______________________ mutations or from extrachromosomal genetic exchanges involving (4) ______________________ and (5) ______________________.

Directions

Check your responses with the *Answers* section, and add up your score. Enter the number of correct answers in the space provided.

Total Correct Answers for Part 1: ______________.

If your score was 4 or higher, proceed to Part 2.

Answers
1. phenol; 2. biological monitor; 3. chromosomal; 4. plasmids; 5. transposons.

PART 2

Very good! Congratulations on the successful completion of Part 1. Now try your hand with Part 2.

Completion Questions

Provide the correct term or phrase for the following. Spelling counts.

6. ______________________ is different from sterilization in that not all microorganisms in the treated area are killed.

7. This partial killing is usually a satisfactory treatment because ______________________ are usually easier to kill than organisms which do not cause disease.

8. ______________________ is the enzyme that connects the peptides of the peptidoglycan layer.

9 and 10. Imidazoles are effective against what two types of microorganisms? (9) ______________________ and (10) ______________________

Directions

Check your responses with the *Answers* section, and add up your score. Enter the number of correct answers in the space provided.

Total Correct Answers for Part 2: ____________ .

If your score was 4 or higher, proceed to Part 3.

Answers

6. Sanitization (or disinfection); 7. pathogens; 8. transpeptidase; 9. protozoa; 10. fungi.

PART 3

Excellent! You are doing very well. Here is a challenging Part 3.

Completion Questions

Provide the correct term or phrase for the following. Spelling counts.

11. One chemical which releases oxygen and is therefore very effective against anaerobic bacteria is ______________________ .

12. A modern procedure for disinfecting milk is the process of ____________ ____________ in which the milk is heated for only a few seconds.

13. The ______________________ technique is important because it gives a good estimate of how effective a given antimicrobial chemical will be under actual application conditions.

14. Interferon inhibits viral replication by interfering with the ______________________ process.

15. In Figure IV-1, which letter represents the penicillin nucleus? ______________________

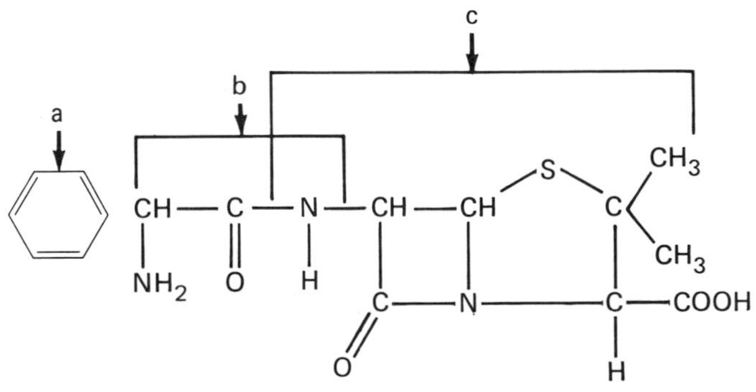

Figure IV-1

Directions

Check your responses with the *Answers* section, and add up your score. Enter the number of correct answers in the space provided.

Total Correct Answers for Part 3: ______________.

If your score was 5, proceed to Part 4.

Answers
11. peroxide; 12. flash pasteurzation; 13. usc-dilution; 14. protein translation; 15. c.

PART 4

You have come a long way. Now for the last and final challenge.

Completion Questions

Provide the correct term or phrase for the following. Spelling counts.

17, 18, and 19. Three recognizable effects of combining two antimicrobial drugs for the treatment of an infectious disease are (17) ______________________________, (18) ______________________________, and (19) ______________________________.

20. In Figure IV-2, what type of reaction is taking place between the two antibiotic disks? ______________

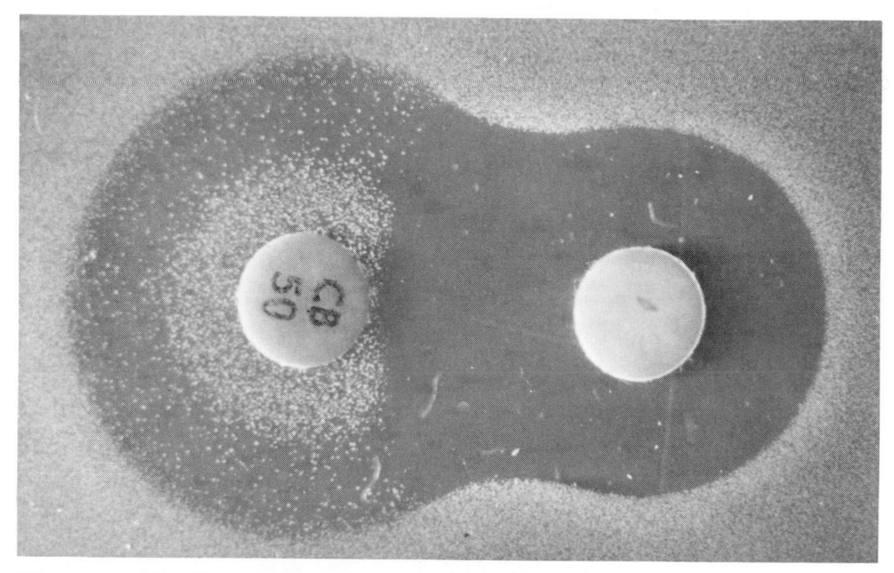

(Courtesy of Dr. E. Yourassowsky, *et al.*, Hospital Universitaire Brugman, Brussels, Belgium.)

Figure IV-2

Directions

Check your responses with the *Answers* section, and add up your score. Enter the number of correct answers in the space provided. Then add all your scores for Parts 1 through 4 together and find your score on the *Performance Score Scale.*

Total Correct Answers for Part 4: ______________.

Total Correct Answers for Parts 1 through 4: ______________.

Answers

16. ionizing radiation; 17. synergistic; 18. additive; 19. antagonistic; 20. synergistic.

PERFORMANCE SCORE SCALE

Number Correct	*Ranking*
12	You should have done better
13	Better
14	Good
16	Excellent
18	Outstanding! Keep it up

Note that each major section of this Study Guide has a *MICROBIOLOGY TRIVIA PURSUIT* section. If you did not do as well as you expected, there are many more opportunities ahead.

13

A Survey of Procaryotes

I. INTRODUCTION

The terms bacterium and procaryote are synonyms. Bacteria have been isolated from essentially every environment on the face of the earth and, as a group, possess the greatest diversity of morphological forms and metabolic diversity of any group of living organisms. It is therefore necessary to have systematic procedures to identify the various types of bacteria and to group them so that the most similar bacteria are placed together. This grouping of similar types is made on the basis of physical, chemical, and genetic characteristics. As more detailed analytical techniques are developed, it is becoming clear that those organisms which share large numbers of these characteristics are evolutionarily related to each other as well. This chapter will review the modern procedures for classifying organisms and then briefly summarize the major groups of procaryotes.

II. PREPARATION

Read Chapters 2, 3, 7, and 13. Knowing the following terms will be helpful in understanding the subject matter of this chapter. Refer to the glossary and appropriate chapters of your text if you are uncertain of any of them. A pronunciation guide for selected terms is provided as a learning aid.

1. 16s RNA
2. akinete
3. anneal
4. archaeobacteria
5. autotroph
6. axial filament
7. *Bergey's Manual of Systematic Bacteriology*
8. binary fission
9. budding
10. chemoautotroph
11. chemoheterotroph
12. chemolithotroph
13. chemotaxonomy
14. DNA hybridization
15. flagella
16. fruiting body
17. gliding
18. gram-negative
19. gram-positive
20. heterocyst
21. heterotroph
22. homology

23. hybrid
24. *Mycoplasma*
25. oxygenic photosynthesis
26. parasite
27. % G + C
28. photoheterotroph
29. procaryote
30. ribosomal
31. *Rickettsia*
32. RNA libraries
33. RNA sequencing
34. serology
35. sheath
36. spirochete
37. stalk
38. sterol
39. vector

III. PRETEST

Correct answers to all questions can be found at the end of this section. Write your responses in the appropriate space provided.

Completion Questions

Provide the correct term or phrase for the following. Spelling counts.

1. The ______________________________ is a measurement of the proportion of G-C base pairs in a DNA molecule.
2. DNA hybridization requires heating to (a) ______________________________ the DNA and cooling to (b) ______________________________ it.
3. DNA strands can also be hybridized with ______________________________.
4. The sequencing of ______________________________ has been useful in determining taxonomic relations.
5. The study of chemical characteristics of cells for the purpose of classification is called ____________________.
6. The four divisions of the Kingdom Procaryotae are (a) ____________________, (b) ____________________, (c) ______________________________, and (d) ______________________________.
7. The three types of bacteria which can use light to derive energy for cellular processes are the (a) ________, (b) ______________________________, and the (c) ______________________________.
8. Bacteria with no cell walls that may form characteristic "fried-egg" colonies are placed in the division ______________________________.
9. ______________________________ ______________________________ are found in spirochetes and are used for cellular motility.
10. The name of the bacterium which attacks other bacteria and penetrates their cell walls is ______________________________.
11. The type of bacteria which possesses chlorophyll a is called ______________________________.

12. The method of motility displayed by bacteria on moist surfaces is called ________________.

13. The ________________ are obligate intracellular parasites which cause serious diseases such as typhus.

14. The three groups of archaeobacteria are (a) ________________, (b) ________________, and (c) ________________.

Matching

Select the answer from the right-hand side that corresponds to the term or phrase on the left-hand side of the question sheet. An answer may be used more than once. In some cases, more than one answer may be required.

______ 15. Anoxygenic photosynthesis	a.	16s RNA
______ 16. Archaeobacteria	b.	actinomycetes
______ 17. Axial filaments	c.	annealing
______ 18. DNA-rRNA hybridization	d.	*Bdellovibrio*
______ 19. DNA hybridization	e.	*Caulobacter*
______ 20. Form fruiting bodies	f.	cyanobacteria
______ 21. Form stalks	g.	denaturing
______ 22. Lack cell walls	h.	green bacteria
______ 23. Mainly transferred by arthropods	i.	halophiles
______ 24. May reproduce by budding	j.	*Hyphomicrobium*
______ 25. Most require sterols	k.	methanogens
______ 26. Nucleotide libraries	l.	*Mycoplasma*
______ 27. Oxygenic photosynthesis	m.	myxobacteria
______ 28. Parasitic on other bacteria	n.	photoheterotroph
______ 29. Purple bacteria	o.	radioactive label
______ 30. Rocky Mountain spotted fever	p.	*Rickettsia*
______ 31. Typhus fever	q.	sheathed bacteria
	r.	spirochetes

Answers

1. percent homology
2a. denature
2b. anneal
3. ribosomal RNA
4. 16s RNA
5. chemotaxonomy
6a. Gracilicutes
6b. Firmicutes
6c. Tenericutes
6d. Mendosicutes
7a. purple bacteria
7b. green bacteria
7c. cyanobacteria
8. Tenericutes
9. axial filaments
10. *Bdellovibrio bacteriovorus*
11. cyanobacteria
12. gliding
13. *Rickettsia*
14a. halophiles
14b. methanogens
14c. thermoacidophiles
15. h, o
16. i, k
17. r
18. p
19. c, g
20. m
21. e
22. l
23. p
24. j
25. l
26. a
27. f
28. d
29. n
30. p
31. p

IV. CONCEPTS AND TERMINOLOGY

Classification and Taxonomy of Bacteria

As described in the textbook, the principles involved in modern methods of microbial taxonomy have advanced tremendously in recent years. Of particular note are those techniques which are based on molecular considerations, all of which ultimately reflect the sequence of bases in the DNA of the organism.

Percent G + C of DNA

It is only in the last decade that technical advances have allowed the rapid sequencing of bases in DNA. In the two decades before that, the most common approach was the much simpler measurement of the relative proportions of G-C base pairs and A-T base pairs in a DNA molecule. The actual parameter measured is called the % G + C, which indicates the percentage of G-C pairs in the total DNA molecule. Although % G + C does not indicate base *sequence,* it is very useful in comparing two DNA molecules (for example, from two different bacteria). If the DNA from two different bacteria has very different values of % G + C, then it is almost certain that they have quite different base sequences and are therefore not closely related. However, it should be noted that the reverse is not necessarily true. That is, two DNA molecules with similar or even identical % G + C values may have quite different *sequences* of bases. We may summarize % G + C as a measurement which is likely to indicate true differences between organisms, but which provides much less certain indication of similarities.

DNA Hybridization

This procedure is more directly involved with the base sequence of the DNA. The primary observation is that gentle heating causes the two strands in a DNA molecule to separate (denature), and that subsequent slow cooling will allow them to reassociate (anneal). If DNA molecules from two different sources are mixed together and cooled, there is a certain chance that the final annealed molecule is a *hybrid,* meaning that it contains one strand from the DNA of one organism, and one strand from the DNA of the second organism. The tightness of binding in such hybrids is reflected in the amount of heat necessary to separate them. The chance that such a hybrid will be formed depends on the similarity in base sequence of the two original DNA molecules for the following reasons. In any DNA molecule, the two strands are held together by hydrogen bonds between G and C bases and between A and T bases. The larger the number of G-C and A-T pairs which are formed, the tighter the binding between the two strands. Therefore, in order to have a tight association between two DNA strands, there must be a large number of bonds between the two strands. This requires the simultaneous pairing of many G bases on one strand with the C bases of the other, and similarly for A and T. In a large molecule such as DNA with hundreds of thousands of base pairs, such extensive pairing requires that the base *sequences* be very similar. The matching of sequences is termed homology and the degree to which the sequences are the same is called percent homology. Two strands from the same DNA molecule will have 100% homology, while two strands from the DNA molecules of completely unrelated organisms may in principle have 0% homology.

DNA-Ribosomal RNA Hybridization

All RNA molecules are formed by transcription from DNA; therefore, they may form hybrids with DNA strands of the appropriate sequence. Ribosomal RNA molecules are the largest RNA molecules within a cell and are not altered after they are formed. These two characteristics make them good candidates for hybridization studies. In addition, they represent a portion of DNA for which the function is known with some certainty. Therefore, their use in hybridization studies can provide powerful and fairly specific information concerning the relation of different organisms. The procedure is similar to the DNA hybridization method discussed above, except that the RNA molecules are radioactively labelled. Single strands of DNA are allowed to anneal with the labelled RNA and any residual unbound RNA is removed by enzymatic digestion. The remaining double-stranded molecules are hybrids of DNA and RNA. The exact amount of RNA bound in this procedure determines the percent homology by the level of radioactivity in the hybrid.

Sequencing of 16s Ribosomal RNA

The two hybridization procedures rely on the sequence of the DNA strands involved, but they do not give direct information as to what those sequences are. As mentioned in the previous section, ribosomal RNA molecules are very useful to study for taxonomic relations, since they are formed by transcription from specific, fairly limited portions of the organism's DNA. Furthermore, although direct sequencing of DNA is becoming technically less difficult every year, it is still easier and quicker to examine the much smaller RNA molecules. However, it is important to remember that the ultimate value of studying base sequences in RNA is that it gives us information about base sequences in at least portions of the DNA. The procedure is to cleave the ribosomal RNA molecules into fragments of 1 to 20 nucleotides (bases) in length. These small portions can be sequenced and used to form an RNA "library." If two organisms have a large number of these fragments in common, then it is likely that they have a large amount of base sequence homology. As the number of entries in these libraries increases, it is possible to see more details of the taxonomic and evolutionary relations between different organisms.

Chemotaxonomy

This approach to microbial classification relies on chemical analysis of different portions of bacterial cells. For example, differences in the amino acids present in peptidoglycan, in the lipids which compose cell membranes, or in the antigenic features of cell structures have all been used in chemotaxonomic determinations (see Chapter 19 for details concerning antigenic features). One advantage to all of these analyses is that they are much quicker and easier to perform than the molecular techniques of hybridization and sequencing. They are therefore used extensively for the *rapid* identification of bacterial types, for example, in clinical situations where reliable fast methods are extremely important. However, it should be remembered that all of these chemical characteristics are also ultimately related to the sequence of bases in a cell's DNA. The synthesis of the specific chemical structure is controlled by the enzymes which are made by the cell to express its genetic information (see Chapter 6 for review).

Kingdom Procaryotae

The major reference which summarizes characteristics used for the identification and classification of bacteria is *Bergey's Manual of Systematic Bacteriology*. This *Manual* has recently undergone a significant reorganization which has altered some of the principles used to group bacteria.

Since the presentation in Chapter 13 of the textbook already contains organized descriptions, the groups and characteristics of individual bacterial types will not be repeated here. However, it should be noted that the classification scheme presented in Chapter 13 is based on the traditional characteristics of cell shape, Gram stain, method of motility, nutritional needs, and relation to oxygen. The modern molecular approaches are gaining in use but have still not been used widely enough to serve as the basis for an entire classification system of bacteria.

V. CHAPTER SELF-TEST

Continue with this section only after you have read Chapters 2, 3, 7, and 13 and have reviewed all illustrations. A score of 80 percent or better is good. If your score is less than 65 percent, reread the chapter.

Correct answers to all questions can be found at the end of this section. Write your responses in the appropriate space provided.

Completion Questions

Provide the correct term or phrase for the following. Spelling counts.

1. The % G + C measurement is better at indicating that two organisms are (a) ______________________

   than it is at showing that they are (b) ______________________.

2. The __________ __________ is a measure of the similarity in base sequences.
3. Hybridization studies with DNA and ribosomal RNA use molecules which are labelled __________ .
4. 16s RNA sequencing involves the formation of (a) __________ of RNA fragments from 1 to (2) __________ nucleotides in length.
5. Cell features which are used in chemotaxonomy include the (a) __________ , the (b) __________ , and cellular (c) __________ .
6. Chemotaxonomy is useful in clinical identification largely because it is __________ .
7. Features used in chemotaxonomy are expressions of the __________ information of the bacteria.
8. The category __________ contains bacteria with no cell walls, many of which are pathogenic.
9. Axial filaments are used by bacteria in the category __________ for motility.
10. Bacteria which can use light for energy and organic compounds for cellular carbon are called __________ .
11. The bacteria with the ability described in question 10 are in the group of __________ __________ .
12. The bacteria which can perform oxygenic photosynthesis are called __________ .
13. Myxobacteria are gliding bacteria which may form complex structures called __________ __________ .
14. Arthropods have been shown to be important in the transmission of many diseases caused by __________ , such as Rocky Mountain spotted fever.
15. The archaeobacteria which require high concentrations of salt for growth are the __________ .

Matching

Select the answer from the right-hand side that corresponds to the term or phrase on the left-hand side of the question sheet. An answer may be used more than once. In some cases, more than one answer may be required.

______	16. Anoxygenic photosynthesis	a. % G + C
______	17. Cell wall structure	b. 16s RNA sequencing
______	18. Gliding bacteria	c. *Caulobacter*
______	19. Heterocysts	d. chemotaxonomy
______	20. Lipid composition	e. cyanobacteria
______	21. Nitrogen fixation	f. DNA hybridization
______	22. Percent homology	g. green bacteria
______	23. Reproduction by budding	h. *Hyphomicrobium*
______	24. Stalk-forming bacteria	i. myxobacteria

Essay Questions

Answer the following questions on a separate sheet of paper.

25. Briefly explain why the molecular techniques of hybridization and nucleic acid sequencing are gaining in popularity as procedures for classifying bacteria.

26. Which properties of cyanobacteria distinguish this group of microorganisms from eucaryotic algae?

Answers

1a. different
1b. similar
2. percent homology
3. radioactively
4a. libraries
4b. 20
5a. cell wall
5b. cell membrane
5c. antigens
6. rapid
7. genetic
8. *Mycoplasma*
9. spirochetes
10. photoheterotrophs
11. purple bacteria (or purple nonsulfur bacteria)
12. cyanobacteria
13. fruiting bodies
14. *Rickettsia*
15. halophiles
16. 6
17. d
18. i
19. e
20. d
21. e
22. a
23. h
24. c

25. These procedures are directly related to the base sequences of the DNA in the bacteria, while the traditional chemotaxonomy methods are much less directly related to this basic genetic information. It is clear that the best knowledge about an organism's evolutionary relations will come from considerations of its DNA base sequence.

26. The cellular properties of the cyanobacteria are clearly unlike those of any eucaryotic algal group. Cell wall composition, ribosome structure, features of protein synthesis, and certain nucleic acid properties in cyanobacteria all serve to distinguish these procaryotes from algae.

VI. ENRICHMENT

The Spiroplasmas

The spiroplasmas, first recognized as helical, cell-wall-less procaryotes in 1972 by R. E. Davis and his associates, are unique among cell-wall-less microorganisms in their capacity to maintain a helical shape and to exhibit several types of motility in thick fluid environments. The shape and motility of spiroplasmas have attracted considerable attention because of the existence of the helical and similarly sized *spirochetes.* The shape and motility of spirochetes appear to depend on a peptidoglycan wall and periplasmic fibrils (axial filaments). Since the spiroplasmas lack these components, their helicity (springlike appearance) and motility must have a different structural basis.

Spiroplasmas are now grouped into a specific genus. They can be readily classified as mycoplasmas because of their ability, under many conditions, to form fried egg (umbonate) colonies on solid media, their ability to pass through filters with 0.22 μm pores, their complete failure to revert to cell-walled forms, their absolute resistance to the antibiotic penicillin, and their absence of cell wall or murein layer precursors. Several properties separate spiroplasmas from other mycoplasmas, including the former's (1) gram-positivity, (2) helical structure and motility, and (3) moderate sensitivity to drugs such as amphotericin B (an antibiotic used in the treatment of certain fungus infections).

Most, if not all, spiroplasmas known today are associated in some way with arthropods. Some spiroplasmas are symbiotes that may regulate sex ratios in populations of the fly *Drosophila.* Others are maintained in cycles in the conductive tissues of plants (phloem) and the bodies of sap-sucking insects that transmit them. Some spiroplasmas occur in ticks, and still others are found on flower surfaces where they are probably deposited by insects. Most spiroplasmas are known or suspected to be pathogenic in animals other than humans or plants.

Examples of disease states caused by these microorganisms include citrus stubborn disease, corn stunt, honey bee spiroplasmosis, and suckling mouse cataract syndrome. Recent autopsy studies of brain specimens from patients who died of the neurological Creutzfeldt-Jacob disease suggest a possible association with spiroplasmas. It is quite apparent that members of *Spiroplasma* occur in a wide range of habitats and are pathogenic to plants, invertebrates, and vertebrates.

14

Microbial Ecology and Environmental Activities

I. INTRODUCTION

Microorganisms clearly play significant roles in most natural processes that affect our environment. There are many examples of the direct effect of microorganisms, such as nitrogen fixation in legume nodules and acid production in mine wastes. More subtle, indirect effects include the recycling of minerals from organic remains (the so-called "decomposer activity") and the production of vitamins within the intestines of mammals.

Since microorganisms occur in all environments, microbial ecology is not a very unified field. The subject is usually divided on the basis of either a habitat or a process. Examples of habitat-oriented microbial ecology studies are aquatic microbiology, soil microbiology, and ruminant microbiology. However, each broadly defined habitat of this type contains so many different microorganisms and activities that is is not possible to study all the microbially-related aspects at one time. Therefore, most investigations follow the pattern of this chapter, which is to consider only individual microbial groups and rather narrowly defined activities at any one time. The combination of the results of a series of these focused studies hopefully leads to understanding of the habitat-level processes. A common approach is to consider microbial action in the cycling of nutrients. Our biosphere continually receives energy in the form of sunlight. However, the chemical elements that store that energy and synthesize cellular components are present in limited amounts. It may be said that we have an "open" system with respect to energy, but a "closed" system with respect to biological matter. An unavoidable consequence of such a closed system is that a great deal of recycling must occur. That is, the atoms that compose the molecules of a cell are reused over and over by many organisms. This concept of recycling goes far beyond decomposer activities. The abundance of the different forms of nitrogen, sulfur, carbon, and oxygen is strongly controlled by microbial (largely bacterial) activities. These activities have been briefly mentioned in discussions of autotrophy, photosynthesis, and respiration. In this chapter, we examine the ecological consequences of these reactions and their interrelationships.

II. PREPARATION

Chapters 13 and 14 should be read before continuing. The following terms are important for you to know. Refer to the glossary and the appropriate chapters if you are uncertain of any of them. A pronunciation guide for selected terms is provided as a learning aid.

1. abiotic
2. acidophilic (ah-SID-oh-fill-ick)
3. alkaliphilic (al-kah-LEE-fill-ick)
4. ammonification
5. barophilic (bar-oh-FILL-ick)
6. biocentration

7. biogeochemistry
8. bioconversion
9. biodegradation
10. biosphere
11. biotic
12. carbon cycle
13. cometabolism
14. commensalism (koh-MEN-sal-izm)
15. community
16. competition
17. consumer
18. decomposer
19. denitrification
20. *Desulfovibrio* (de-sulf-OH-vib-ree-oh)
21. desulfurylation
22. ecology
23. ecosystem (eh-koh-SIS-tem)
24. ectosymbiosis (EK-toh-sim-bye-oh-sis)
25. endosymbiosis (EN-doe-sim-bye-oh-sis)
26. habitat
27. halophilic (HAL-oh-fill-ick)
28. hydrostatic pressure
29. methanogenesis
30. microbial infallibility
31. microhabitat
32. mineralization
33. mutualism (MEW-tu-al-izm)
34. niche
35. nitrification (nye-treh-feh-KAY-shun)
36. *Nitrobacter* (nye-troh-BACK-ter)
37. nitrogen cycle
38. nitrogen fixation
39. *Nitrosomonas* (NYE-troh-soh-mon-us)
40. oxygen cycle
41. parasitism
42. phosphorus cycle
43. population
44. predation
45. producer
46. psychrophilic (SYE-krow-fill-ick)
47. recalcitrant
48. recycling
49. *Rhizobium* (rye-ZOH-bee-um)
50. sulfate reduction
51. sulfur cycle
52. sulfur oxidation
53. symbiosis (sim-bye-OH-sis)
54. syntrophism (sin-TROFF-izm)
55. thermophilic (THER-moh-fill-ick)

III. PRETEST

Correct answers to all questions can be found at the end of this section. Write your responses in the appropriate space provided.

Completion Questions

Provide the correct term or phrase for the following. Spelling counts.

1. The portion of the earth's environment that sustains life is called the ______________________.
2. Living systems can be subdivided into habitats or (a) ______________________. These units can be divided into (b) ______________________, which are composed of (c) ______________________ of the same or similar species.

3. The important factors for a given organism in its environment are the physical or (a) ________________ and the biological or (b) ________________ .
4. The exact location and set of conditions where an organism grows are referred to as its (a) ________ . If two organisms both require a nutrient in the environment, one will usually be selected against (b) ___ .
5. Each organism usually performs one of three major roles in nature: (a) ________________ , (b) ________________ , or (c) ________________ .
6. Organisms that grow best at very cold temperatures are called (a) ________________ ; those that prefer high temperatures are (b) ________________ .
7. Some organisms require high concentrations of salt and are called ________________ .
8. Organisms that prefer low pH are (a) ________________ ; those that grow best at high pH are (b) ________________ .
9. Some organisms from the bottom of the ocean that require high hydrostatic pressures are called (a) ___ . Since the bottom of the ocean is very cold, these organisms are usually also (b) ________________ .
10. A special kind of relation between two organisms that have metabolic interactions but no close physical contact is called ________________ .
11. The general phenomenon of organisms living together is called ________________ .
12. If this association involves one organism living inside the tissue of the other, the relation is called (a) ___ . If the two organisms merely contact each other externally, the relation is called (b) ________________ .
13. If this relation benefits both organisms, it is (a) ________________ ; if neither is helped or harmed, it is (b) ________________ .
14. If one organism gains by damaging the other, it is called ________________ .
15. The transformations of matter necessary for continued life are called ________________ .
16. Since these reactions are biologically mediated and relate to the chemistry of the earth, their study is called ________________ .
17. The reactions can be divided into five cycles of individual elements: (a) ________________ , (b) ________________ , (c) ________________ , (d) ________________ , and (e) ________________ .
18. The process by which gaseous nitrogen is converted to ammonia is called ________________ .
19. An important symbiotic example of this process involves plants of the (a) ________________ group and bacteria of the genus (b) ________________ .
20. (a) ________________ oxidizes ammonia to nitrite and (b) ________________ oxidizes nitrite to nitrate.
21. The complete oxidation process is called ________________ .
22. The complete oxidation of organic materials to inorganic compounds is called ________________ .
23. The reduction of nitrate to nitrogen gas by bacteria is called ________________ .

24. Another type of anaerobic respiration is sulfate reduction, which is performed by ____________ bacteria.
25. The oxidation of sulfide and sulfur is done by bacteria of the genus ____________.
26. The theory that microorganisms can eventually degrade any organic compound is called ____________.
27. There is evidence that many pesticides are not degraded and may actually accumulate through a process called ____________.
28. However, some molecules can only be degraded if another, easily metabolized compound is also present. This is called ____________.
29. The biological production of a useful product, often from waste materials, is called (a) ____________. One example is the anaerobic reduction of carbon dioxide by (b) ____________.
30. Nutrients can be added with ____________ to increase their mixing with crude oil as it is being degraded.
31. Two examples of damaging microbiological conversions are the partial breakdown of (a) ____________ to more toxic forms and the methylation of (b) ____________, which is then concentrated in the food chain.

Answers

1. biosphere
2a. ecosystems
2b. communities
2c. populations
3a. abiotic
3b. biotic
4a. microhabitat
4b. competition
5a. producer
5b. consumer
5c. decomposer
6a. psychrophilic
6b. thermophilic
7. halophiles
8a. acidophiles
8b. alkaliphiles
9a. barophiles
9b. psychrophiles
10. syntrophism
11. symbiosis
12a. endosymbiotic
12b. ectosymbiotic
13a. mutualism
13b. commensalism
14. parasitism
15. recycling
16. biogeochemistry
17a. carbon
17b. oxygen
17c. nitrogen
17d. sulfur
17e. phosphorus
18. nitrogen fixation
19a. legume
19b. *Rhizobium*
20a. *Nitrosomonas*
20b. *Nitrobacter*
21. nitrification
22. mineralization
23. denitrification
24. *Desulfovibrio*
25. *Thiobacillus*
26. microbial infallibility
27. biocentration
28. cometabolism
29a. bioconversion
29b. methanogenesis
30. paraffin
31a. pesticides
31b. mercury

IV. CONCEPTS AND TERMINOLOGY

Microbial Ecology and Laboratory Microbiology

The study of the ecology of any group of organisms is more difficult than is the act of observing them in the controlled conditions of the laboratory. This difference is even more pronounced in microbial ecology because the laboratory culture of microorganisms is done so easily. The examination of natural microbial populations, however, has great technical difficulties associated with it. The source of most of these problems is the small size of the organisms, which makes it virtually impossible ever to identify precisely the relevant habitat. We are therefore in the unusual position of knowing many physiological features of the organisms quite precisely, but not knowing how many of these features are relevant in nature.

Levels of Organization

The primary divisions for ecological studies are based on physical distinctions. However, for studies of microbial ecology, the distinctions quickly change from physical to functional. These two aspects will now be considered in more detail.

Physical

The portion of the earth that can support life is called the *biosphere.* This loosely defined area includes portions of the land, water, and atmosphere. Within the biosphere are several ecosystems, each of which is usually characterized by a certain type of habitat, such as forest, lake, or ocean. The habitat types are usually identified on the basis of large-scale physical features, as is obvious from the examples just given. Of more concern to a given organism is its *microhabitat,* which refers to those physical and biological factors that react immediately with the organism. The microhabitat is not the same as the *niche,* which is not a physical description of the organism's environment, but is rather the sum of all factors affecting the organism. A niche therefore includes physical (e.g., temperature, water availability), chemical (e.g., nutrient availability), and biological (e.g., competition, predation) factors. It may be seen that not only is the concept of niche difficult to define, also the actual niche for an organism is constantly changing. The factors exerting control are frequently divided into *biotic* and *abiotic* for the convenience of the investigators, but this distinction is probably not very meaningful to organisms in nature.

This discussion has considered organisms as individuals. Although individual organisms are of course important, it is usually more meaningful to discuss *populations,* especially for microorganisms. A population is a collection of identical or very similar organisms living in the same habitat. Usually, a habitat will support more than one type of population. These assemblages of populations are called *communities.* The interactions of populations within communities are among the most significant biotic factors.

Functional

Each population has its own specific functions. However, it is useful to simplify the many possible activities into three groups: producer, consumer, and decomposer, of which the last two are very closely related. Each of these activities is observed in microbial populations.

The *producers* are those organisms that convert (fix) inorganic carbon (CO_2) into organic carbon, usually in the form of biomass. The largest group of producers on the earth is the higher plants. However, microbial producers are also significant in certain habitats. The single-celled algae, diatoms, and cyanobacteria all fix CO_2 and release O_2, as do plants. The purple and green bacteria use light to fix CO_2, but they do not evolve O_2. The lithotrophs described in Chapter 5 fix CO_2, but they do so by oxidizing inorganic sulfur or nitrogen compounds or H_2 in the absence of light. These latter two groups of bacteria are usually not as important quantitatively, but must be considered in a complete discussion of producers.

The *consumers,* as the name implies, are those organisms that ingest and metabolize the organic material synthesized by the producers. Common examples of consumers include carnivorous animals that eat meat, or animals that eat vegetation, such as grass. A less obvious but quantitatively important example is microorganisms (bacteria and fungi) that can metabolize compounds exuded from plant roots. These compounds are usually amino acids and simple sugars. Different members of the consumer community also have metabolic relations. For example, some organisms degrade cellulose to glucose. Others use the glucose and ferment it to lactic acid. Still others can oxidize the lactic acid and create acetic acid. Still others can perform the final oxidation of acetic acid to CO_2. This hypothetical sequence of different organisms is not at all unreasonable and probably reflects some actual ecosystem transformations (see the discussion of syntrophism that follows).

The *decomposer* organisms are often misunderstood because they are treated as a distinct group. It is probably best to consider the decomposers as a special subgroup of the consumers. The major difference is that the decomposers ingest dead plant and animal (and microbial) material and their excretion products. The chemical similarities between the substrate (food) for the decomposers and the consumers are great. The distinction between the two groups on the basis of living or dead food is rather arbitrary. For example, the bacteria and fungi already mentioned that eat root exudates can easily be called either consumers or decomposers.

An additional function of decomposers is to serve as recyclers. As will be developed later, this term has several meanings, but in its simplest sense, it means that elements that are contained in living organisms are returned (recycled) back to their inorganic form. For example, the carbon in organic compounds is converted back to CO_2; the nitrogen in amino acids to ammonia or nitrate; the sulfur in amino acids to hydrogen sulfide or sulfate; and so on.

The subdivision of ecological activities is clearly difficult. The specific plan chosen for organization is largely determined by which aspects of the system are most important to the investigator. It is important to realize that each plan, however valuable, is merely an analytical tool and should not be accepted as an absolute.

It is also important to note that the physical and chemical factors considered in Chapter 11 affect the growth and survival of microorganisms in nature, just as they do in the laboratory. It is not always possible to measure these effects accurately in nature, but they are important nonetheless.

Environmental Extremes

As a group, microorganisms have a broader tolerance for extreme environmental conditions than do any higher forms of life. For example, some bacteria are so tolerant of high levels of salt that they actually require solutions which are nearly saturated with sodium chloride, such as are found in the Great Salt Lake or the Dead Sea. Others are able to thrive in deep ocean waters where the hydrostatic pressure is several hundred times greater than we experience at the earth's surface. Similar examples are known for extremes of temperature and pH. It is important to remember that these organisms are individually quite limited in the environmental conditions which they can tolerate. That is, no single organism grows well at both pH 1 and ph 10 or at both 5°C and 85°C. Tolerance of these extreme conditions is impressive but is only made possible by a number of specific physiological adaptations by these organisms during the course of evolution. Each adaptation which increases diversity in one way lowers it in another.

Syntrophism Versus Symbiosis

Syntrophism means "nourished together" and implies no physical contact between the interacting organisms. The group metabolism of cellulose described earlier under the discussion of consumers is an example of syntrophism. Another is the bacteria and fungi that consume plant exudates.

Symbiosis means "living together." Symbiotic relations are those in which there is physical contact between the interacting organisms. This physical relationship is described as either endosymbiotic or ectosymbiotic. *Endosymbiotic* relations are those in which one member of the pair is actually inside the tissues of the other. *Ectosymbiotic* relations, which are far more common, are those in which the two members are close to each other, but do not actually penetrate each other's tissues.

The more detailed distinctions among types of symbiosis are made on functional grounds. There are four main types:

1. *Commensalism* is the name given to the relation in which one organism benefits from the association, while the other organism is neither harmed nor helped. The bacteria and fungi that live on plant roots and ingest the root exudates are probably commensal. The microorganisms gain obvious nutritional advantage, but they do not harm the plant.

2. *Mutualism* is less common. In this relation, both members of the association benefit. A clear example is the lichen, which is a mutualistic association of an alga or cyanobacterium and a fungus. The fungus obtains excreted organic compounds from the alga, while the alga obtains water and possibly inorganic nutrients from the fungus, which is more efficient at absorption.

3. *Parasitism* is widespread in all biological systems. In this relation, one member (the parasite) is benefited, while the other (called the host) is harmed. All viruses are parasites. Many human diseases are caused by parasitic bacteria, fungi, and viruses. Often, the host is killed by the parasite, but not always.

4. *Predation* is usually the easiest of the symbiotic relations to identify because it generally involves the engulfment of a smaller organism (the prey) by a larger one (the predator). There are numerous animal examples, of course, but there are also many microbial examples. Some of the earliest experimental work relating to predation was performed with protozoa that preyed on bacteria.

Dynamics of Nutrient Cycling

Ecological studies continually reveal the dynamic interrelationships of natural activities, of which nutrient cycling is one of the best examples. The individual atoms that compose the molecules of organisms are constantly converted from their organic form to basic constituents and back. Many of these conversions are controlled by microbial activities. One of the clearest examples is the relative abundance of nitrogen in the ammonia or nitrate form. This proportion is important to plants and other organisms with specific requirements and is controlled directly by microbial (mostly bacterial) activities. The role of microorganisms in an ecosystem's complex ecology should not be overlooked. The simple description of microorganisms as decomposers ignores many important features. Recent studies have shown the far-reaching effects of these activities.

Nutrient Cycles

The five nutrient cycles are quite interrelated, in particular, the oxygen cycle to the other four. Four of the cycles are redox cycles, which means that the various conversion steps are oxidations or reductions and therefore involve potentially large energy changes, either consumptions or releases. The phosphorus cycle is not a redox cycle and is therefore simpler and less directly related to the other four.

Oxygen Cycle

The oxygen cycle is the simplest of the redox cycles because it requires only two steps, which can be performed by many organisms. (Oxygen is mainly present as H_2O [water] or O_2. Oxygen atoms are found in various organic compounds, but they are not part of the major cyclic reactions.) Water is oxidized to O_2 by plants, algae, and blue-green bacteria during their photosynthesis. This biological reaction is the source of essentially all O_2 in the earth's atmosphere. The second and last of the oxygen cycle steps is the reduction of O_2 to H_2O. This reaction is performed by all aerobic respiring organisms, including the plants which did the original H_2O oxidation.

Carbon Cycle

The carbon cycle is very closely related to the oxygen cycle, although we may consider them separately for convenience of analysis. Carbon has several possible oxidation states, but we can simplify the discussion by considering only the two major forms: CO_2 and organic compounds. The carbon cycle really only has two steps: reduction (or fixation) of CO_2 into organic material and the oxidation of organic material to CO_2. These reactions can occur in several ways, but by far the predominant number are directly connected to the oxygen cycle. The photosynthetic organisms oxidize H_2O to O_2 as they reduce CO_2 to organic matter. The aerobic respirers oxidize organic compounds to CO_2 while they reduce O_2 to H_2O. There are exceptions to these general reactions, some of which we will consider below. But the close connection between carbon and oxygen transformations is of major importance.

Nitrogen Cycle

This element has three stable oxidation states in nature: N_2, NH_3 (ammonia), and NO_3^- (nitrate). The conversions of these forms are done almost exclusively by bacteria. Nitrogen is often a limiting nutrient for growth of plants and bacteria even though N_2 composes 79 percent of our atmosphere. The difficulty is that N_2 contains a triple covalent bond that only a small group of bacteria is capable of breaking. All these bacteria use the enzyme nitrogenase to reduce (fix) N_2 to NH_3. Some organisms, including the N_2-fixing bacteria, can use the NH_3 directly to make amino acids and the other nitrogen-containing organic compounds of the cell. Other organisms, including most plants, prefer NO_3^- as their nitrogen source. The oxidation of NH_3 to NO_3^- is called *nitrification* and is performed in two steps by bacteria of the genera *Nitrosomonas* and *Nitrobacter*. Other bacteria are capable of using NO_3^- as an alternate electron acceptor in anaerobic respiration (see Chapter 5). These bacteria reduce NO_3^- to N_2 and complete the nitrogen cycle. Since this reduction, called *denitrification*, removes nitrogen from an ecosystem, it is clearly an undesirable process in agriculture. Another important nitrogen cycle reaction, which is not a redox change of the nitrogen, is the conversion of organic nitrogen compounds to NH_3 in a process called *ammonification*.

The nitrogen cycle is related to the oxygen and carbon cycles as follows: Nitrification only occurs in the presence of O_2 and denitrification only occurs in the absence of O_2. Therefore, physical features of the environment that affect oxygen concentration and diffusion are important. The nitrifying bacteria are chemosynthetic autotrophs and therefore are also important in the carbon cycle. The denitrifying bacteria usually oxidize organic compounds and represent another connection between the carbon and nitrogen cycles.

Sulfur Cycle

The reactions of the sulfur cycle are similar to those of the nitrogen cycle, the major difference being that there is no major gaseous form of sulfur analogous to N_2. There are three important stable forms of sulfur: H_2S (sulfide), S° (elemental sulfur), and SO_4^{2-} (sulfate). Sulfide, a toxic compound to most organisms, is oxidized spontaneously to S° or microbiologically by *Thiobacillus* or photosynthetic bacteria to S° and SO_4^{2-}. Sulfate, the preferred form for most plants, may be reduced by *Desulfovibrio* in anaerobic respiration. The product of this reduction is sulfide. Most sulfur in organic compounds in an organism is at the oxidation level of sulfide. Therefore, decomposition releases sulfide in a process called *desulfurylation*.

The sulfur cycle is connected to the oxygen, carbon, and nitrogen cycles. Sulfate reduction only occurs under anaerobic conditions. Sulfide and sulfur oxidation can occur both aerobically and anaerobically. The sulfate reducers oxidize organic compounds while reducing sulfate. Some sulfate reducers have recently been shown to be nitrogen fixers as well. Sulfur oxidizers of the genus *Thiobacillus* are capable of denitrification and some species can fix nitrogen.

The conclusion from this brief discussion is that the redox cycles of these four elements are amazingly intertwined. This close relation provides a good example of how difficult it is to directly apply laboratory physiology results to field studies.

Phosphorus Cycle

The phosphorus cycle is much simpler than the others but involves only the formation and cleavage of one form of phosphorus, namely, phosphate (PO_4^{3-}). This compound is assimilated in its inorganic form and is incorporated by ester bonds into nucleic acids and some coenzymes. The decomposition of organisms converts the organic phosphate back to its inorganic form. This compound is often a limiting nutrient in the growth of plants and microorganisms, but its availability is controlled almost entirely by physical and chemical factors, not by biological cycling.

Most microbial reactions can be classified under one or more of these cycles. However, there are some reactions that cannot and others that have special significance to us. We therefore will consider several microbial reactions separately. These reactions will fall into the two general areas of pollution and reactions of economic importance.

Pollution

The emphasis here is on reactions that microorganisms perform with materials introduced into the environment by humans. Some of these reactions help to solve pollution problems; others make the pollution more serious.

Pesticides

Manufactured pesticides designed to control weeds and animal pests have only been in wide use since the end of World War II. These compounds have produced dramatic benefits for commercial agriculture, but they have also had some damaging effects. Some of these compounds and their derivatives are toxic to several forms of wildlife that were not the intended target. The role of microorganisms in this situation is significant. It was originally thought that normal soil bacteria would be able to completely degrade these organic pesticides after their agricultural purpose was fulfilled. One expression of this opinion is the term *microbial infallibility,* which basically means that all organic compounds can eventually be degraded by some microorganism or group of microorganisms. Microbial degradation of pesticides is not as simple as was once thought. Some are only attacked when the organism has another energy source available. This requirement leads to the concept of *cometabolism.* Some molecules are degraded very slowly or only partially; they are called *recalcitrant molecules.* In fact, some pesticides are partially degraded and thereby are converted into even more toxic compounds. The concept of infallibility should probably be modified to consider only naturally occurring compounds. The pesticides have simply been in the environment for too short a time to allow microorganisms to develop enzyme systems to degrade them.

Mercury

Another example of microbial activity that has unfortunate consequences is the methylation of mercury. Mercury is released into the environment by a number of industries. Several microorganisms are capable of adding a methyl group to create methyl mercury. This methylated compound is much more toxic than is the native metal, itself a powerful nerve poison.

Both the pesticide alteration and mercury methylation are made worse by the process of *biocentration.* This term refers to the fact that certain nondegradable materials accumulate in the tissues of organisms that ingest them. This accumulation occurs at each step of the food chain. The farther up the food chain an organism is, the more extensive the accumulation.

Petroleum

The last area of pollution we will consider is the microbial oxidation of petroleum. Significant oil spills occur around the world each year. Crude oil is a complex mixture of materials, but it will usually be degraded if there are sufficient nitrogen and phosphorus nutrients. Potential difficulties arise from microorganisms in this area as well. Several purified petroleum products, such as gasoline and jet fuel, are readily oxidized by microorganisms. Since these materials are of considerable value, great care must be taken in their storage to protect them from spoilage.

Economic Reactions

Microorganisms have been economically valuable to a number of industries for many years—food fermentations and antibiotic production are two examples. Recently, microorganisms have been used in ways that are different from traditional industrial microbiology.

Thiobacillus

Many ores (copper, uranium, nickel) occur as the metal sulfides. Members of the genus *Thiobacillus* can oxidize the sulfides and create sulfuric acid. The metals are quite soluble in this acidic, sulfide-depleted condition. This process of metal "mining" is gaining widespread use.

Methanogenesis

Our society needs combustible fuel. A likely source for some of the needed material is the microbial process of converting waste material to methane under anaerobic conditions. The details of the microbial process are still being examined and industrial applications refined. One area in which methanogenesis has long been important is sewage treatment (see Chapter 16).

Manganese

Manganese, an important industrial material, is present in only trace levels in most environments. One potential commercial source of manganese is the nodules or spheres of this element found in oceans and deep lakes. Apparently, bacterial activities are significant in the formation of these nodules. Although the nature of the microbial involvement with manganese is not well understood, this source of the element has potential industrial and economic importance.

V. CHAPTER SELF-TEST

Continue with this section only after you have read Chapters 5, 13, and 14. A score of 80 percent or better is good. If your score is less than 65 percent, reread the chapter.

Correct answers to all questions can be found at the end of this section. Write your responses in the appropriate space provided.

Matching

Select the answer from the right-hand side that corresponds to the term or phrase on the left-hand side of the question sheet. An answer may be used more than once. In some cases, more than one answer may be required.

_____ 1. A type of consumer activity

_____ 2. The sum of all factors affecting an organism

_____ 3. The assemblage of populations

_____ 4. Fundamental physical division of the biosphere

_____ 5. Types of symbiosis

_____ 6. Growth made of viruses

_____ 7. Usually occurs between two similar populations

_____ 8. The physical environment of an organism

_____ 9. Composed of similar or identical organisms

_____ 10. An organism that eats dead material

_____ 11. Product of nitrification

_____ 12. Helped to refute microbial infallibility

_____ 13. Anaerobic respiration

_____ 14. Element cycle not involving redox reactions

_____ 15. Possibly useful as fuel source

_____ 16. Usually directly related to O_2 evolution

_____ 17. Sulfate reduction

a. population
b. predation
c. parasitism
d. competition
e. habitat
f. community
g. decomposer
h. niche
i. nitrogen fixation
j. *Desulfovibrio*
k. biocentration
l. aerobic respiration
m. petroleum
n. ammonification
o. CO_2 fixation
p. pesticides
q. phosphorus
r. mercury
s. denitrification
t. methanogenesis
u. desulfurylation
v. nitrate

_____ 18. Activities of sulfur-oxidizing bacteria

_____ 19. Process that releases sulfide

_____ 20. Pollution problems made worse by microorganisms

_____ 21. Major link between oxygen and carbon cycles

_____ 22. Often done in symbiotic association with legumes

_____ 23. Can be spoiled by microorganisms

_____ 24. Toxic compounds that accumulate through biological action

_____ 25. Release of inorganic nitrogen during decomposition

_____ 26. Loss of nitrogen as gas; detrimental to agriculture

Completion Questions

Provide the correct term or phrase for the following. Spelling counts.

27. An organism is affected by both the (a) _____ and (b) _____ of its environment.
28. The physical surroundings of an organism compose its _____.
29. The _____ is not a physical location but rather reflects all features that affect an organism.
30. Organisms that fix CO_2 are called (a) _____ and are members of the (b) _____ group.
31. Consumers are the organic material created by _____.
32. Decomposers may be considered a subgroup of _____.
33. A special metabolic interaction between consumers is _____.
34. Recycling is a function of _____.
35. If one organism enters the tissues of another and both derive benefit from the association, then it is both a(an) (a) _____ and a(an) _____.
36. There are two symbiotic relations in which one member benefits and the other is harmed. They are distinguished from each other because a(an) (a) _____ always kills its (b) _____, whereas (c) _____ rarely kills its (d) _____.
37. O_2 is released during photosynthesis by the oxidation of _____.
38. These organisms usually couple this oxidation directly to the reduction of _____.

39. Organisms that do this same reduction while oxidizing inorganic compounds are in the (a) ____________ and (b) ____________ cycles.

40. The complete conversion of organic compounds to their organic constituents is called ____________ .

41. Sulfide is released into the environment from two microbiological processes: (a) ____________ and (b) ____________ .

42. The process in which a pollutant is degraded only in conjunction with some other compound is called ____________ .

43. One pollution problem usually helped by microorganisms is the degradation of ____________ .

44. One conversion of waste material to usable fuel done by bacteria is ____________ .

45. Mining operations for some metals make use of ____________ bacteria.

Essay Questions

Answer the following questions on a separate sheet of paper.

46. What is the difference between a microhabitat and a niche?

47. Which nitrogen cycle activities are carried out only by bacteria?

Answers

1. b or c
2. h
3. f
4. e
5. b and c
6. c
7. d
8. e
9. a
10. g
11. v
12. p
13. s
14. q
15. t
16. o
17. j
18. i, o, and s
19. u
20. p and r
21. l
22. i
23. m
24. k
25. n
26. s

27a. biotic
27b. abiotic
28. habitat (microhabitat)
29. niche
30a. autotroph
30b. producer
31. producers
32. consumers
33. syntrophism
34. decomposers
35a. mutualism
35b. endosymbiosis
36a. predator
36b. prey
36c. parasite
36d. host
37. H_2O
38. CO_2
39a. nitrogen
39b. sulfur
40. mineralization
41a. sulfate reduction
41b. desulfurylation
42. cometabolism
43. petroleum
44. methanogenesis
45. *Thiobacillus*

46. A microhabitat is a physical, very small area in which microorganisms live. A niche is not a location but, rather, is a sum of all biotic and nonbiotic factors that affect an organism.

47. Nitrification, denitrification, and nitrogen fixation are all carried out, although each process is done by different organisms.

VI. ENRICHMENT

Mutualism

Some of the more interesting examples of mutualism involve microorganisms and higher organisms. The textbook chapter described the legume-*Rhizobium* nitrogen-fixing mutualism and the rumen-microorganism mutualism. There are two other examples of considerable interest.

Mycorrhiza is the name given to the intimate association between fungi and the roots of certain plants. The advantage of the fungus is probably the obvious gain of root exudates. The plant appears to benefit much as does the alga in a lichen. The fungi are very efficient absorbers of water and nutrients and apparently are able to transfer the absorbed materials to the plant. In many cases, the plants will barely grow without this fungal assistance. Two specific examples of mycorrhizal relation are pine trees and orchids, which have quite specific fungal partners.

The termite is an insect that eats wood. Since wood is about 60 percent cellulose, it is surprising to learn that the termites have no cellulose enzymes. It is now clear that the termites survive through the action of the microbial community in their digestive tracts. These microorganisms degrade the cellulose and release products that are suitable for assimilation by the termites. In addition, some members of this intestinal community can fix nitrogen so that the termite, unlike almost all other animals, can grow on a pure cellulose diet.

PART V
MICROBIOLOGY TRIVIA PURSUIT
(Chapters 13 and 14)

Let's examine your attention to detail, and your fact-gathering ability. There are *four* challenging Parts to this section. Remember, a certain number of questions must be answered in each Part before you proceed up the MICROBIOLOGY TRIVIA PURSUIT trail. The answers and directions for continuing are given at the end of each Part. Try Part 1.

PART 1

Completion Questions

Provide the correct term or phrase for the following. Spelling counts.

1. An unusual method of motility, very useful in classifying those organisms which display it, is ______________________ .
2. Some filamentous cyanobacteria form specialized, oxygen-impermeable cells called __________ .
3. Molecules which resist microbial degradation are said to be ______________________ .
4. In ecological terms, a community is composed of ______________________ .
5. A specialized type of consumer organism, which eats dead material, is called a __________ .

Directions

Check your responses with the *Answers* section, and add up your score. Enter the number of correct answers in the space provided.

Total Correct Answers for Part 1: __________ .

If your score was 4 or higher, proceed to Part 2.

Answers
1. gliding; 2. heterocysts; 3. recalcitrant; 4. populations; 5. decomposer.

PART 2

Very good! Congratulations on the successful completion of Part 1. Now try your hand with Part 2.

Completion Questions

Provide the correct term or phrase for the following. Spelling counts.

6. The group ______________________ contains three very unusual types of bacteria which grow best under extreme conditions.

7. The sequence bases in nucleic acid molecules can be compared by heating and annealing in a process called ____________________.
8. A predator kills its prey, while a ____________________ keeps its host alive, occasionally for very long periods of time.
9. Classifying bacteria on the basis of their chemical and nutritional characteristics is called ______.
10. Into which Division would you place the organism that forms fried-egg appearing colonies such as the ones shown in Figure V-1? ____________________

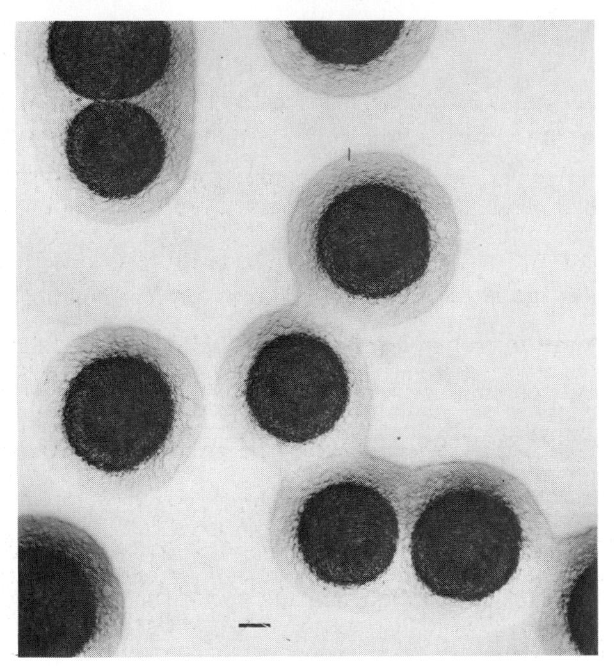

The bar marker = 1 mm.

Figure V-1

Directions

Check your responses with the *Answers* section, and add up your score. Enter the number of correct answers in the space provided.

Total Correct Answers for Part 2: __________.

If your score was 4 or higher, proceed to Part 3.

Answers

6. chemotherapy; 7. archaeobacteria; 8. hybridization; 9. parasite; 10. Tenericutes.

PART 3

Excellent! You are doing very well. Here is a challenging Part 3.

Completion Questions

Provide the correct term or phrase for the following. Spelling counts.

11 and 12. Bacteria which grow best at very cold temperatures are said to be (11) ____________________ , while those which require high salt concentrations are called (12) ____________________ .

13. The relatedness of two organisms can be estimated by comparing the ____________________ content of their DNA.

14. Determination of the sequence of bases in ____________________ RNA has given great detail concerning the evolution of bacterial groups.

15. Into which Division would you place the organism shown in the transmission electron micrograph (Figure V-2)? ____________________

(This bacterium is *Thermotoga maritima,* an anaerobic, extremely thermophilic bacterium with an optimum growth temperature of 80° C. It was isolated from the geothermally heated sea floors in Italy and the Azores.)

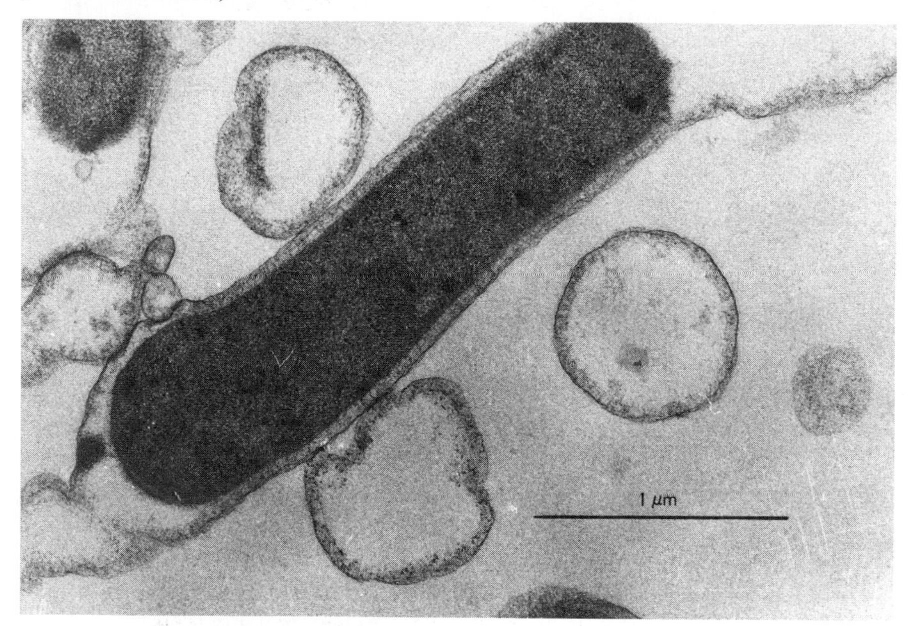

(From R. Huber, *et al., Arch. Microbiol.* **144:**324-333 [1986].)

Figure V-2

Directions

Check your responses with the *Answers* section, and add up your score. Enter the number of correct answers in the space provided.

Total Correct Answers for Part 3: ____________ .

If your score was 5, proceed to Part 4.

Answers

11. psychrophilic; 12. halophilic; 13. % G + C; 14. ribosomal; 15. Mendosicutes.

PART 4

You have come a long way. Now for the last and final challenge.

Completion Questions

Provide the correct term or phrase for the following. Spelling counts.

16. Two organisms which are nutritionally related are said to be in a ________________ association.

17 and 18. The production of nitrogen gas by bacteria is called (17) ________________, while the conversion of nitrogen gas to ammonia is called (18) ________________.

19 and 20. The (19) ________________ is the physical environment in which an organism lives, while the (20) ________________ is a description of the characteristics of that environment.

21. Spirochetes are able to move by the action of their ________________ ________________.

Directions

Check your responses with the *Answers* section, and add up your score. Enter the number of correct answers in the space provided. Then add all your scores for Parts 1 through 4 together and find your score on the *Performance Score Scale.*

Total Correct Answers for Part 4: ________________.

Total Correct Answers for Parts 1 through 4: ________________.

Answers

16. syntrophic (*not* symbiotic); 17. denitrification; 18. nitrogen fixation; 19. habitat; 20. niche; 21. axial filaments.

PERFORMANCE SCORE SCALE

Number Correct	*Ranking*
12	You should have done better
13	Better
14	Good
16	Excellent
18	Outstanding! Keep it up

Note that each major section of this Study Guide has a *MICROBIOLOGY TRIVIA PURSUIT* section. If you did not do as well as you expected, there are many more opportunities ahead.

15

Industrial Microbiology and Biotechnology Applications

I. INTRODUCTION

Industrial microbiology has traditionally involved the use of microorganisms for economic advantage, either through the encouragement of a desired activity (such as the synthesis of some product) or the avoidance of an undesired activity (such as spoilage of valuable material). Industrial microbiology therefore has combined the talents of scientists and engineers to achieve desired goals.

Modern industrial microbiology has developed along two very different lines. On the one hand, there have been significant improvements in the control of microbial processes used for many years, such as the production of alcoholic beverages or antibiotics. On the other hand, there are developments of an entirely different type. These are new uses of microorganisms in the realm of genetic engineering, or the action of putting specific desired genes into an organism to enhance the development of products such as insulin or synthetic vaccines. Another important aspect of genetic engineering is the use of these new techniques in the clinical laboratory, allowing improvement in the speed and accuracy of identification of a microorganism isolated from an infected patient.

II. PREPARATION

Read Chapters 2, 6, and 15 before continuing. The following terms are important for you to know. Refer to the glossary and the textbook if you are uncertain of any of them. A pronunciation guide for selected terms is provided as a learning aid.

1. amylase (AM-ee-lace)
2. antibiotics
3. electrophoresis
4. fermentation
5. hybridization
6. immobilized cells
7. lipase (LYE-pace)
8. microbial insecticide
9. Northern blot
10. pectinase (PECK-tin-ace)
11. protease (PRO-tee-ace)
12. Southern blot
13. submerged culture
14. Western blot

III. PRETEST

Correct answers to all questions can be found at the end of this section. Write your responses in the appropriate space provided.

Completion Questions

Provide the correct term or phrase for the following. Spelling counts.

1. Anaerobic processes in industrial microbiology are ____________________.
2. Aerobic processes are performed by ____________________.
3. The production of alcoholic beverages is done by ____________________.
4. The genus of yeast used in alcohol production is ____________________.
5. (a) ____________________ are antimicrobial compounds created by (b) ____________________ and (c) ____________________.
6. Examples of food products made by aerobic microbiological processes include (a) ____________________, some (b) ____________________, and some (c) ____________________.
7. Two alcohols of industrial importance are (a) ____________________ and (b) ____________________, which are produced by bacteria of the genus (c) ____________________ in a(an) (d) ____________________ process.
8. Specific actions can be brought about industrially by the use of microbial (a) ____________________. The major classes of action are (b) ____________________, (c) ____________________, (d) ____________________, and (e) ____________________.
9. A major modification in using these molecules is the creation of ____________________, allowing easier separation of the product.
10. An important agricultural application of industrial microbiology is (a) ____________________. These agents are (b) ____________________ to agricultural pests and are (c) ____________________ in their action.
11. ____________________ ____________________ are single-stranded segments of DNA synthesized by gene machines.
12. The importance of these segments is their specificity, since the investigator has determined the desired ____________________.
13. Several blotting procedures exist in which samples undergo (a) ____________________ in an electric field and then are (b) ____________________ with the material mentioned in question 11.

Answers

1. fermentations
2. submerged cultures
3. fermentation
4. *Saccharomyces*

5a. antibiotics
5b. fungi
5c. bacteria
6a. vinegar
6b. amino acids
6c. vitamins
7a. butanol
7b. isopropanol
7c. *Clostridium*
7d. fermentation
8a. enzymes
8b. protease
8c. lipase
8d. amylase
8e. pectinase

9. immobilized enzymes

10a. microbial insecticides
10b. pathogenic
10c. specific

11. DNA probes
12. sequence

13a. electrophoresis
13b. hybridized

IV. CONCEPTS AND TERMINOLOGY

Fermentation

The textbook points out that the term *fermentation* is often mistakenly applied to aerobic microbiological processes. Note that discussion in this chapter is divided into fermentations and submerged cultures, a specific separation of anaerobic and aerobic processes.

The oldest and most common uses of microbial fermentations involve food processes, particularly the manufacture of alcoholic beverages. The details of the manufacture were presented in the textbook. These processes may all be summarized as the fermentation of sugars to ethanol by yeasts, usually of the genus *Saccharomyces*. The major advances in this production in the last 100 years have been the use of pure yeast cultures, which makes the process more reproducible and reliable, and improved preservation methods pioneered by Pasteur. Other examples of the production of foodstuffs through microbial activity will be discussed in Chapter 16.

Several industrial processes also depend on microbial fermentations. The production of alcohols such as butanol and isopropanol uses the fermentative powers of *Clostridium butylicum* and *Clostridium acetobutylicum*. These organic compounds can also be created by chemical processes using petroleum as a starting material. Since petroleum is becoming increasingly rare and expensive, the microbial processes using relatively cheap and renewable substrates are gaining in economic significance.

One microbial fermentation of considerable importance—the production of silage—is not performed by defined pure cultures. Silage is a naturally fermented vegetable product made from grass or hay. The material is incubated anaerobically at an appropriate moisture level in an enclosed chamber, usually a silo. The desired reactions are the conversion of sugars in the vegetable material to lactic acid and other reduced organic compounds by the action of lactic acid bacteria. The silage fermentation is usually conducted under relatively uncontrolled conditions and does not involve inoculation with pure bacterial cultures. It is remarkable that the lactic acid bacteria are usually only a very minor component of the starting plant material ($< 100/g$) but are the predominant members of the final silage product ($> 10^7/g$). This natural selection of a specific physiological form under anaerobic conditions is the key feature to silage production. There are two agricultural benefits to this fermentation. First, the process results in a pH of only 4 to 5 in the silage, which greatly restricts the growth of other organisms and acts as a significant preservation. Second, the biochemical changes during the fermentation increase the nutritional value of the material for the livestock that eat it. One limitation to silage use, however, is that many animals refuse to ingest the acidic material. Certain additives are sometimes used to increase its palatability.

Submerged Cultures

Several food products are created by aerobic microbial cultures—for example, vinegar, some amino acids, and some vitamins. However, the largest effort in the industrial use of aerobic submerged cultures is in pharmaceuticals. We will also consider some less extensive applications.

Antibiotics

The largest class of pharmaceutical items created microbiologically is antibiotics, specific antimicrobial compounds created by other microorganisms. Some antibiotics are currently manufactured by organic chemical procedures, but the vast majority are still synthesized microbiologically.

The technology involved in antibiotic production has evolved logically and efficiently. The starting factor is that certain bacteria and fungi produce compounds that are lethal to other microorganisms. An early step in the industrial process was the selection of mutants that produce more antibiotics than the original isolates. There has been considerable development in the technology of building, sterilizing, and precisely controlling culture vessels of the enormous size (typically 50,000 gallons) used in antibiotic production. After culture growth has ceased, the soluble antibiotics must be separated from the culture. To achieve this separation on such a large scale, usually by continuous centrifugation, was also a major technological development. A more recent development in the antibiotic industry has been the investigation of methods to make the antibiotics more resistant to inactivation. Since antibiotics are natural products, it is not surprising that many microorganisms have developed enzymatic mechanisms to destroy or modify the antibiotics. Several antibiotics currently used are chemical modifications of the microbially produced compounds. These modified compounds ideally retain their antimicrobial activity while being more difficult to inactivate.

It can therefore be seen that not only is the antibiotic industry significant in itself, but it also caused the development of major technological advances in culture and purification techniques which, in turn, have found wider application.

Enzymes

One of the newer classes of microbial products of industrial significance is enzymes. Several classes of enzymes—primarily proteases, lipases, and polysaccharidases such as amylase and pectinase—are prepared in large quantity for commercial application. The major advantage of enzyme production is that it allows a given catalytic activity to be applied over a broad range of areas. Huge cultures and sophisticated technology are not needed to achieve a given chemical conversion. Of course, it must be remembered that the application of these enzymes is limited to rather simple chemical processes, typically single steps; complicated cellular processes such as antibiotic production cannot be recreated by isolated enzymes. The textbook reviewed the uses of specific enzymes in tanning, baking, and linen production. An additional application is in laundering. In the early 1970s, presoaks and detergents containing enzymes were widely used commercially and domestically. However, their use has diminished greatly for two reasons. First, enzymes are proteins and therefore have considerable antigenic properties, which caused significant allergic responses. Second, these products require high levels of phosphate to be active. The phosphate serves as a buffer and chelates metal ions such as calcium, which could inhibit the enzymes. The use of phosphates in laundry products has been banned in most of the United States because they contributed significantly to blue-green blooms and eutrophication problems (see Chapter 16). Therefore, laundry products containing enzymes are used very little today.

Microbial Insecticides

The use of microorganisms and microbial products to control agricultural pests is a different type of industrial microbiology. Unlike fermentations and antibiotic productions, this process is not done under controlled laboratory conditions. The appropriate organisms are grown in large numbers and applied in field situations that are impossible to completely control. Therefore, the effects of microbial insecticides are not as dramatic in the field as they are in the laboratory. The use of these materials is still of economic benefit to agriculture, and research continues to determine ways of increasing their effectiveness.

Microbial insecticides are pathogenic to agricultural pests. The application of these materials merely speeds up the pathogenic process to diminish the numbers of the pests. The textbook described the mechanism of action of several microbial insecticides and also discussed the features of an ideal insecticide. There are two major advantages to a microbial insecticide. First, the action is usually quite specific against the pest, so there are no undesirable effects on the crops or on beneficial (or commensal) animals. Second, the microbial insecticides do not persist for long periods of time, as do many human-made compounds (see Chapter 15). At first it may seem that this lack of persistence is a drawback, for it would require repeated application. This argument is valid, but it is also true that the relatively rapid disappearance of these agents means that we have much better control over them.

Genetic Engineering

As discussed in the text, there are a number of new aspects of industrial microbiology involving genetic engineering. The key point of these procedures is that they make use of "constructed" organisms, that is, an organism whose genetic composition has been chosen by humans for a specific purpose. The simplest form of this construction is the insertion of genes from one organism into another. The recipient is almost always a bacterium, since bacteria, being simpler, are easier to culture under defined conditions. The inserted genes may come from virtually any desired source and, in fact, other bacteria, fungi, plants, animals, and humans have been used. The recipient bacteria can then be grown on a large scale in a submerged culture and will produce quantities of the desired product. Examples of such products are animal growth hormones, human insulin, and the antiviral compound interferon.

An important new application of genetic engineering has been the use of the new techniques to enhance the identification of specific microbial pathogens in clinical situations. There are several variations, but all involve the use of DNA probes. These probes are synthesized by specialized automated techniques. The key point is that the sequence of bases in each probe is determined in advance by the experimenter. It is almost always possible to identify some specific DNA sequence in the cell or virus of interest so that the DNA probe will identify that target sequence specifically in the appropriate organisms.

All of the DNA probe techniques have two important steps in common. First, samples are subjected to electrophoresis, or movement in an electric field. This step is necessary to separate the often complicated mixture of cellular proteins or nucleic acids. Second, the separated components are transferred to nitrocellulose filter paper where they are allowed to interact with the DNA probe. The interactions will be specific due to the base sequence, as discussed in Chapter 13. The probe is usually labelled radioactively so that any hybrids which are formed can be easily identified. The first such technique used DNA fragments and was called Southern blotting, after Dr. E. Southern, the scientist who devised the procedure. Subsequent modifications have been made by using DNA-binding proteins or RNA fragments, procedures called Western and Northern blotting, respectively. Note that these latter names are merely making use of the compass theme brought about accidentally by virtue of of Dr. Southern's name. The "Western" and "Northern" titles have no inherent meaning of their own.

V. CHAPTER SELF-TEST

Continue with this section only after you have read Chapters 2, 6, and 15. A score of 80 percent or better is good. If your score is less than 65 percent, reread the chapter.

Correct answers to all questions can be found at the end of this section. Write your responses in the appropriate space provided.

Matching

Select the answer from the right-hand side that corresponds to the term or phrase on the left-hand side of the question sheet. An answer may be used more than once. In some cases, more than one answer may be required.

Topic: Applications of Industrial Microbiology

______	1. Butanol	a. microbial insecticide
______	2. Submerged culture	b. tanning
______	3. Animal feed	c. vinegar
______	4. Antibiotic producers	d. silage

_____ 5. Water pollution problem
_____ 6. Pathogen
_____ 7. Fairly quickly degraded
_____ 8. Alcohol
_____ 9. *Acetobacter*
_____ 10. Enzyme use advance
_____ 11. Specific action
_____ 12. Lipase
_____ 13. Lactic acid bacteria
_____ 14. Amylase
_____ 15. Isopropanol
_____ 16. DNA probe
_____ 17. Genetic engineering
_____ 18. Nitrocellulose filters

e. fungi
f. immobilization
g. *Clostridium*
h. *Aspergillus*
i. *Saccharomyces*
j. fermentation
k. laundering enzyme
l. bacteria
m. aerobic process
n. blotting
o. electrophoresis
p. vaccine production

Completion Questions

Provide the correct term or phrase for the following. Spelling counts.

19. Industrial microbiology processes are either anaerobic or aerobic and are performed in (a) __________ or (b) __________, respectively.
20. Most industrial microbiology processes use __________ cultures to ensure reproducible results.
21. One exception is the use of natural populations in (a) __________ formation, in which the (b) __________ are selected for.
22. They create a(an) (a) __________ pH and both (b) __________ and increase the (c) __________ value of the vegetable material.
23. __________ production has encouraged a great technological advance in industrial microbiology.
24. The production of these specific compounds has been increased by __________ selection.
25. Specific microbial activities are carried out by isolated (a) __________, especially when they are (b) __________.
26. __________ are important agricultural aids.
27. They are (a) __________ to pests and are rather quickly (b) __________.
28. __________ is the creation of bacteria with specific desired characteristics.

29. Bacteria constructed in this way can be used to produce pharmaceutical compounds such as (a) ________ , (b) ________________, and (c) ________________.
30. These human-designed bacteria may now be ________________, since they represent activities they do not normally carry out.
31. There are three blotting techniques used with DNA probes. (a) ________________ blotting uses RNA fragments, (b) ________________ blotting involves DNA fragments, while (c) ________________ blotting gives information concerning DNA-binding proteins.
32. The use of these techniques is important in clinical situations where identification of a pathogenic microorganism must be made (a) ________________ and (b) ________________.

Essay Questions

Answer the following questions on a separate sheet of paper.

33. Describe a major advantage to preparing fermented foods.
34. Describe the important advantages to using DNA probes in identifying clinical isolates.

Answers

1. g
2. m
3. d
4. e and l
5. k
6. a
7. a
8. i
9. c
10. f
11. a
12. b
13. d
14. k
15. g
16. n
17. p
18. n

19a. fermentation
19b. submerged culture
20. pure
21a. silage
21b. lactic acid bacteria
22a. low (acid)
22b. preserve
22c. nutritional
23. antibiotics
24. mutant
25a. enzymes
25b. immobilized
26. microbial insecticides
27a. pathogenic
27b. degraded
28. genetic engineering
29a. antibiotics
29b. hormones
29c. vaccines
30. patented
31a. Northern
31b. Southern
31c. Western
32a. rapidly
32b. reliably

33. These foods are usually acidic and therefore are easier to preserve. In many cases, the taste of the material is enhanced as well.
34. These probes are very specific so that the identification made by using them is quite likely to be correct, therefore indicating the proper treatment. Also, the determination can be made quickly, thereby improving the chances for proper treatment.

VI. ENRICHMENT

Microbiology and Politics

It is rare that a direct connection can be made between a specific discovery and a significant political event. Perhaps the clearest example of such a connection involves the production of acetone by *Clostridium acetobutylicum*. This fermentation ability was discovered by Jewish scientist Chaim Weizmann, working in England in 1910. In World War I, the production of large amounts of acetone was necessary to continue the operation of British naval guns, which Weizmann's discovery supplied. Weizmann was an active Zionist and a close friend of David Balfour, the British foreign secretary who, in November 1917, sponsored the Balfour declaration, which stated British support for the establishment of an independent Jewish homeland in Palestine. Chaim Weizmann was the first president of the nation of Israel, founded in 1948.

16

Food, Dairy, and Water Microbiology

I. INTRODUCTION

It may come as a surprise to learn that the food we eat every day is far from sterile. Even carefully baked meats contain a number of easily detectable bacteria. The microorganisms we continually ingest do not, of course, usually cause disease. Many food products are tested for the presence of disease agents and are processed appropriately to make them safe to eat. The most common form of such processing is heat treatment through cooking. Moreover, several foods are rendered safe by human-controlled chemical treatments or microbial fermentation reactions. Through fermentation, specific microbial enzymes create the characteristic flavors, aromas, and features of the well-known foods, such as cheese, yogurt, pickles, sauerkraut, fermented sausage, and various related products.

It should be clear that microorganisms are present in every habitat on the earth. Some of their activities are beneficial to humans, while others are damaging. Our discussion of food has emphasized these dual roles of microorganisms. There is an additional area in which microbial activities are intimately involved with daily human events. That area is water. As with food, microorganisms in water bring about changes which affect humans and we alter the environments which affect the microorganisms. Furthermore, some of these interactions are potentially great threats to humans, such as the spread of infectious disease, while others are indispensable to our protection, such as the microbial cleaning of waste water.

II. PREPARATION

Read Chapter 16 before continuing. The following terms are important for you to know. Refer to the glossary and the textbook if you are uncertain of any of them. A pronunciation guide for selected terms is provided as a learning aid.

1. activated sludge
2. aeration
3. bacteriophage (back-TEER-ee-oh-fayj)
4. blooms
5. BOD
6. Camembert (KAM-em-bair)
7. chlorination
8. coagulation
9. coliform
10. compost
11. curd
12. direct acidification
13. filtration
14. hydrophobic (high-droh-FOE-bick)

15. indicator organism
16. lactobacillus (lack-toh-bah-SILL-us)
17. membrane filter
18. ozone
19. preservative
20. *Propionibacterium* (pro-pee-on-ee-back-TEER-ee-um)
21. ripening
22. Roquefort (ROWK-fourt)
23. *Saccharomyces cerevisiae* (sack-ah-row-MY-seas sair-ee-VIS-ee-ee)
24. sauerkraut
25. sedimentation
26. single cell protein (SCP)
27. starter culture
28. thermal exhaustion
29. trickling filter
30. water activity
31. whey (way)
32. yogurt (YOH-gurrt)

III. PRETEST

Correct answers to all questions can be found at the end of this section. Write your responses in the appropriate space provided.

Completion Questions

Provide the correct term or phrase for the following. Spelling counts.

1. Microorganisms can promote four benefits in food products: (a) ________________, (b) ________________, (c) ________________, and (d) ________________.
2. The major microbial activity in foods is (a) ________________, specifically that of (b) ________________.
3. Some of these foods are prepared with natural populations of microorganisms, but most use carefully defined ________________.
4. The use of microorganisms directly as food is ________________.
5. This material is important because it represents the conversion of ________________.
6. However, many animals will not eat this material alone, and it must be used mostly as a feed ________.
7. Food spoilage is controlled by the basic processes of (a) ________________, (b) ________________, (c) ________________, and (d) ________________.
8. (a) ________________ inhibits bacterial growth without killing, whereas (b) ________________ reduces the number of organisms on a food.
9. Specific defects associated with the characteristics of starter organisms used in cultured dairy foods include (a) ________________, (b) ________________, (c) ________________, and (d) ________________.
10. (a) ________________ and (b) ________________ are used as the starter culture for yogurt.

11. The three texture types of cheeses are (a) ______, (b) ______, and (c) ______.
12. The action of the appropriate bacterial culture on starter material causes the formation of a firm (a) ______ and a watery fluid portion (b) ______.
13. Salt is added during cheese production to (a) ______, (b) ______, and (c) ______.
14. ______ is the method of choice for determining bacterial counts in water and beverages.
15. The symbol ______ is used to represent the water content of foods.
16. Domestic sewage treatment is designed to lower (a) ______ and reduce the numbers of (b) ______.
17. (a) ______ is a measurement of the organic content of the system in terms of the amount of (b) ______ needed for its oxidation.
18. Additional problems in the treatment of industrial waste are caused by (a) ______ agents in the waste, such as (b) ______.
19. Anaerobic treatment helps kill (a) ______ and produces (b) ______ for energy use.
20. Another procedure for the microbial decomposition of wastes is (a) ______, which is slow and uses much (b) ______.
21. Pathogens are screened for by attempting to grow ______.

Answers

1a. taste
1b. preservation
1c. vitamin production
1d. digestibility
2a. fermentation
2b. lactic acid
3. starter cultures
4. SCP (single cell protein)
5. waste products
6. additive
7a. drying
7b. heat
7c. oxygen removal
7d. chemical additives
8a. drying
8b. heat
9a. insufficient acid development
9b. abnormal flavors
9c. gas accumulation
9d. bitterness
10a. *Lactobacillus bulgaricus*
10b. *Streptococcus thermophilus*
11a. soft
11b. semisoft
11c. hard
12a. curd
12b. whey
13a. reduce moisture
13b. prevent growth of unwanted bacteria
13c. contribute to cheese flavor
14. membrane filtration
15. a_w
16a. BOD
16b. pathogens
17a. BOD
17b. O_2
18a. toxic
18b. heavy metals
19a. viruses
19b. methane
20a. composting
20b. land
21. indicator organisms

IV. CONCEPTS AND TERMINOLOGY

Microorganisms and Food

This chapter will discuss the major aspects of the relations between microorganisms and food: (1) the roles of microorganisms in the creation and preparation of some foods, (2) the use of microorganisms themselves as food, and (3) the spoilage of food by microorganisms. Attention is also given to methods used to detect microorganisms in food and alternate methods of dairy food production that eliminate the use of microbial cultures.

Microbial Food Products

Microorganisms can have four major effects when used in food manufacture:

1. They affect taste. The bacterial production of lactic acid in sauerkraut fermentation is a clear example.
2. They aid in preservation. The production of yogurt involves a lowering of the pH to near 4, which greatly restricts the growth of other microorganisms.
3. A vitamin increase often occurs. Yogurt has increased levels of B vitamins as a consequence of the bacterial fermentation.
4. In some cases, the digestibility of a food is enhanced by microbial activity because the microorganisms possess some enzymatic abilities that humans do not. The degradation of lactose in milk and of cellulose in some vegetable products are examples.

The predominant microbial activity in these foods is *fermentation.* Lactic acid fermentation is particularly important. Most of the starting materials (milk, vegetables) contain enough sugar or polysaccharide so that the lactic fermentation can readily occur.

The fermentations are sometimes performed by organisms naturally associated with the foods. More often, however, pure *starter cultures* are used. The advantages of the pure cultures were discussed in Chapter 15. The most important advantage is that it enables an inoculated product to develop quickly, consistently, and reproducibly.

Microorganisms are involved in more than just the acid fermentation step of these food productions. An instructive example is cheese manufacture. The first step is the formation of curd from the milk. Curd formation is a biological process involving bacterial growth and enzymes from cattle. The second step is the acid fermentation. The final step in the making of many cheeses is an additional microbial fermentation. The characteristic flavors and textures of cheeses and related foods are due to the types of microorganisms used in fermentation and subsequent processes. A number of species belonging to the bacterial genera of *Lactobacillus, Propionibacterium,* and *Streptococcus* are used in cultured dairy foods for acid production, flavor development, and flavor production. Other dairy foods produced by the enzymatic action of microbes include butter, buttermilk, sour cream, cottage cheese, and yogurt.

Changes in normal fermentation activity occur. These are called defects and include insufficient acid development, abnormal flavors, accumulation, and bitterness. Defects may be caused by antibiotics, chemicals, mutations of culture organisms, and the destructive action of bacteriophages.

Alternate approaches to the acid fermentation step involving microorganisms are receiving increasing attention. One such method is direct acidification. With the use of food-acceptable acids, artificial flavors, and stabilizers, not only is the need for acid production by lactic acid starter cultures eliminated, but reliable imitation fermented dairy products can be made easily.

The advantages to the direct acidification method include significant reduction in equipment and personnel, fewer controls, elimination of defects, and the consistent production of a reliable product.

The food products described in the textbook and briefly considered here have been created and consumed by humans for thousands of years. The understanding of the microbiological nature of these processes in the last century has led to improved food products and their output on a very large scale. The advantage is not just the use of pure cultures. The basic growth factors such as pH, O_2, water availability, and temperature are now understood and carefully controlled for optimum product development.

Single Cell Protein

Single cell protein (SCP) is the use of microorganisms as food (protein) rather than indirectly in food preparation, as in the preceding examples. The advantages are quite impressive. It is possible to convert material with no human nutrition value (oil, corn wastes, feedlot wastes) into a high-protein food with accompanying increases in vitamin content. In our crowded world, SCP has the additional advantage of using considerably less space than would be necessary for the equivalent agricultural production.

As promising as it is, SCP is not used directly as human food because of an unpleasant taste and aroma. In fact, it has not even been possible to use it alone as animal feed. Through genetic engineering, additional modifications may possibly remove these difficulties, but they will certainly increase the cost of the material. The predominant use of SCP is as an additive to animal feed, allowing conservation of high-protein food for other uses.

Various techniques are used to determine the numbers and kinds of microorganisms and their products in foods. These include direct microscopic counts, standard plate counts, staining procedures, immunologic tests for enzymes and other microbial products, isolation, and biochemical identification. (Most of these techniques are described in Chapters 3, 4, and 11.)

Food Spoilage

The most widespread study of microorganisms in food involves the prevention of food spoilage. In the introduction to this chapter, it was said that food products are never sterile. Many foods, such as vegetables, acquire significant numbers of microorganisms from the environment where they are grown. Other foods, such as meat and dairy products, do not contain indigenous microorganisms but do become inoculated with microorganisms from the air and nonsterile equipment during handling and processing. This exposure to microorganisms does not in itself cause food spoilage. For foods to spoil, they must not only be inoculated, they must also be kept under conditions that favor growth. The most important of these factors is temperature. Normal refrigeration slows growth tremendously and greatly retards spoilage. In addition to having an unpleasant taste, odor, and appearance, spoiled food may also be a potential cause of disease. For example, most cases of botulism come from improperly canned vegetables.

Commercial and home food-processing methods are designed to reduce microbial numbers without altering the food too greatly. Several empirical methods have been used for many years, but all preservation processes are applications of basic principles relating to growth control. These methods can be divided into four groups: drying, heat, oxygen removal, and chemical additions.

Drying

The oldest and simplest methods are those that limit the availability of water to the organisms. Simple drying in the sun or at slightly elevated oven temperatures is used for many fish, meats, and fruits. The addition of large amounts of salt to fish and meat absorbs water from the food and therefore inhibits the microorganisms that remain. Fruits are often canned in concentrated sugar solutions. Sugar acts much as salt does on meat products and removes water from the food. The newest method of water removal is freezing. The low temperatures are important, but the removal of water by freezing is more significant in long-term food preservation.

Heat

Unlike drying, heat does not just inhibit microbial growth; it actually kills microorganisms. Pasteurization was discussed in Chapter 18. The normal heat treatment involves much higher temperatures and longer exposure times than pasteurization does. This more extensive treatment, of course, alters the food more.

Oxygen Removal

Oxygen removal (anaerobiosis) is only partly effective in protecting food because anaerobes still grow. The usual method of creating anaerobic conditions is to completely seal the container, which also prevents atmospheric contamination by new microorganisms. An additional reason for caution is that the most serious food disease, botulism, is caused by the obligate anaerobic bacterium, *Clostridium botulinum.*

Chemical Additives

Chemical additives are the newest method and have proven successful in certain areas. They usually inhibit growth without killing the organisms, although some do disinfect to a degree. The gassing of dried fruits with sulfur dioxide (SO_2) increases their storage time considerably. One must be very careful, of course, to be sure that the chemical additives are not harmful in themselves. Other examples of antimicrobial chemical agents used as chemical preservatives include acetic acid (bakery goods, beverages, cheeses), benzoic acid (soft drinks, preserves), citric acids (soft drinks, preserves), lactic acid (soft drinks, cottage cheese), sodium nitrate (meats and fish), sodium nitrite (meats and fish), and sorbic acid (bakery goods, syrups).

A final consideration in limiting food contamination and spoilage is cleanliness. Clean preparatory facilities and cooking utensils limit places where contaminating microorganisms can accumulate. Without these reservoirs of organisms, much contamination is avoided.

Drinking Water Treatment

We usually take the safety of our supplies of drinking water for granted, with the details of treatment only being of interest when something goes wrong. However, it is important to understand that we are so fortunate because of carefully developed procedures for treating water before it is distributed to our homes and businesses. These procedures are both chemical and microbiological and are important for removing toxic (poisonous) chemicals, as well as disease-causing microorganisms.

Most domestic water comes from rivers, lakes, and ground water. In general, these systems may be described as nutrient poor, having very low concentrations of organic material (food) and inorganic minerals necessary for microbial growth. Therefore, microorganisms usually grow quite slowly in these natural waters. Pathogenic microorganisms are especially disadvantaged in these environments which are very different from the warm nutrient-rich habitats in the human body, which they prefer. Nonetheless, as the human population increases, there are increasing additions to waterways, largely from domestic and industrial waste water. In addition, in some areas the runoff from agricultural operations can also be significant.

As the textbook describes, the treatment of drinking water varies greatly, depending on the source of the water supply and the likelihood of different types of contamination. Most of the steps are physical and/or chemical processes. However, one step, aeration, is designed to stimulate microbial activity with the intent of removing organic materials through biological oxidation. After this biological step, there are more physical actions to remove the microorganisms and residual organic material. The final step is a chemical treatment to accomplish three objectives simultaneously: (1) removal of chemicals which may give the water undesirable tastes and odors; (2) removal of iron and manganese compounds which may damage pipes; and (3) disinfection to ensure that no pathogenic microorganisms remain in the water. The traditional method for this last step has been the addition of chlorine, a powerful oxidizing agent. However, it has recently been demonstrated that chlorine has a significant chance of forming potential cancer-causing compounds by reacting with the organic molecules in the water. Therefore, some treatment facilities have begun using the oxygen compound ozone (O_3), which is also a strong oxidizing agent but will not form the cancer-causing compounds.

Waste Treatment

Most waste treatment is organic decomposition ultimately carried out by microorganisms. We will present here the use of microorganisms in controlled decomposition processes, by far the widest application of which is in sewage treatment.

There are two major problems to be solved in domestic waste water treatment: organic decomposition and pathogen removal. The organic load to be treated is referred to as *biochemical oxygen demand* (BOD). This term refers to the manner in which the load is determined. Measured samples are placed in sealed bottles and in five days, the amount of O_2 consumed by microbial oxidation is measured. The importance of the BOD concept is that it is a fair approximation of the amount of organic matter present in a sample. When much organic material exists, much O_2 is consumed and the BOD is high. Similarly, low organic contents consume little O_2, hence, a low BOD. Therefore, BOD is a convenient, easily measured parameter indicating the organic content of the waste water.

Both organic decomposition and pathogen removal are achieved by aerobic treatment, usually either the activated sludge or trickling filter processes described in the textbook. These processes bring large microbial communities in contact with highly oxygenated organic waste to create maximum oxidation, which can remove up to 95 percent of the BOD under ideal conditions. The pathogens are outcompeted by other microorganisms in the community so that the final product of activated sludge treatment contains a large number of bacteria but relatively few pathogens. Unfortunately, viruses are not removed as efficiently.

Industrial wastes also require treatment. These materials are often very high in BOD, especially those from food industries such as potato processing plants. Although this waste has few, if any, human pathogens, it often contains toxic components such as heavy metals. Most industries with large wastes have their own treatment plants, necessary because the wastes from each industry (chemical, food), being quite different from domestic sewage, cannot be treated efficiently by municipal sewage plants. The industrial treatment plants quickly develop a microbial community that is specifically adapted to the compounds in that waste. For example, phenol is a potent antibacterial agent (see Chapter 11). However, the treatment plants at chemical industries usually employ activated sludge microorganisms that have been selected for rapid degradation of this compound.

Anaerobic sewage treatment is also important. The anaerobic processes occur more slowly and difficult molecules are more likely to degrade. More important, many viruses are inactivated by passage through an aerobic digester. The exact causes are not clear, but the relatively long retention time and usual high temperature (40°-50°C) are probably responsible. The final value from anaerobic sewage treatment is the production of methane (CH_4), a natural gas that can be used to create electricity or run furnaces. Although only small amounts of methane are produced today, it is quite likely that this process can be improved to contribute significant amounts of energy to our society.

Composting is another method of waste treatment that is increasing in use. This process is usually used for solid waste (trash) degradation, but it can also be used for the sludge resulting from sewage treatment. The process simply involves piling the material to be degraded in long, carefully shaped rows to degrade. The microbiological processes that occur are both aerobic and anaerobic. The key feature of the composting is the long time involved, usually months. Composting has two disadvantages: it consumes substantial energy, and it requires large amounts of land. The energy use comes from the need to constantly mix and rearrange the piles, usually with power equipment such as tractors and bulldozers. Since the process is slow and new waste is constantly created, the amount of land required must constantly increase.

Indicator Organisms

It is important to know if a river or lake or soil has been contaminated with pathogens from sewage. It is not possible to directly examine an area to determine contamination; it is necessary to perform microbiological analysis. Most of the large number of pathogens are somewhat difficult to grow in culture. Therefore, a general procedure is necessary. This procedure involves a sample of the material in question to look for an *indicator organism*. There are three major requirements for an organism to be useful as an indicator: (1) it must always be associated with pathogens; (2) it must be easy to culture; and (3) it must live longer than the pathogen. This last requirement means that sometimes a sample will be judged contaminated because of indicator growth when all pathogens are dead. This mistake is obviously in the direction of safety and is a prudent measure. The group of bacteria called coliforms fits these characteristics very well. The textbook details the procedures for detecting coliforms.

V. CHAPTER SELF-TEST

Continue with this section only after you have read Chapter 16. A score of 80 percent or better is good. If your score is less than 65 percent, reread the chapter.

Correct answers to all questions can be found at the end of this section. Write your responses in the appropriate space provided.

Matching

Select the answer from the right-hand side that corresponds to the term or phrase on the left-hand side of the question sheet. An answer may be used more than once. In some cases, more than one answer may be required.

_____	1. Use of syrups	a.	anaerobiosis
_____	2. Vitamins	b.	lactic acid
_____	3. Growth inhibition	c.	heat
_____	4. Made from water	d.	SCP
_____	5. Pasteurization	e.	drying
_____	6. Starter cultures	f.	botulism
_____	7. Freezing	g.	oxygen
_____	8. Animal feed supplement	h.	yogurt
_____	9. Sauerkraut	i.	trickling filter
_____	10. Effect of salt	j.	methane
_____	11. Inactivated during anaerobic digestion	k.	composting
		l.	heavy metals
_____	12. Measure of organic content	m.	BOD
_____	13. Industrial waste problem	n.	chlorination
_____	14. Produced during anaerobic digestion	o.	coagulation
_____	15. Possible solid waste disposal	p.	ozone
_____	16. Lowered during activated sludge process	q.	aeration
		r.	viruses
_____	17. Aerobic digestion		
_____	18. Disinfects		
_____	19. Removes particles		
_____	20. Stimulates microbial oxidation		

Completion Questions

Provide the correct term or phrase for the following. Spelling counts.

21. Microorganisms sometimes improve foods by making them more _______________ through special enzymatic activities.

22. Many microbial food fermentations create (a) ________________ from the (b) ________________ already present in the food.
23. Commercial and domestic food methods use ________________ for controlled, reproducible products.
24. The direct use of microorganisms as food is ________________.
25. This material improves nutrition because it is made from ________________.
26. The main use for this product is as a(an) (a) ________________ for (b) ________________.
27. Three commonly used bacterial genera for cultured dairy products are (a) ________________, (b) ________________, and (c) ________________.
28. Changes in normal fermentation activity associated with starter organisms are referred to as (a) ________________ and include (b) ________________, (c) ________________, (d) ________________, and (e) ________________.
29. Three steps in cheese making are (a) ________________, (b) ________________, and (c) ________________.
30. The process that eliminates the need for acid production by lactic acid starters is (a) ________________, (b) ________________, (c) ________________, (d) ________________, and (e) ________________.
31. Food preservation usually involves ________________ microorganisms.
32. Seven techniques used to determine the numbers and kinds of microorganisms and their products in foods are (a) ________________, (b) ________________, (c) ________________, (d) ________________, (e) ________________, (f) ________________, and (g) ________________.
33. Once a food is prepared, an important way to prevent spoilage is by ________________.
34. The oldest and simplest preservation methods involve ________________.
35. Two antimicrobial chemicals used as the preservatives in meats are (a) ________________ and (b) ________________.
36. Three non-biological steps in drinking water treatment are (a) ________________, (b) ________________, and (c) ________________.
37. The addition of ________________ has traditionally been the final step in water treatment.
38. This addition brings about ________________ by killing pathogenic microorganisms.
39. It may also cause the formation of (a) ________________ and is therefore being replaced in some areas by (b) ________________.
40. The largest waste that is treated microbiologically is ________________.
41. The (a) ________________ of the material is lowered and the number of (b) ________________ is greatly reduced.

42. In industrial wastes treatment, there is the additional problem of ______________________________ components.
43. The possible release of pathogens is monitored by searching for ______________________________.
44. Some solid wastes are treated microbiologically in ______________________________.

Essay Questions

Answer the following questions on a separate sheet of paper.

45. Can lowering the water content of certain foods serve as an effective way to control food spoilage? Why or why not?
46. Waste treatment causes the growth of a large number of microorganisms. Explain why this is not bad.

Answers

1. e
2. h
3. e
4. d
5. c
6. h
7. e
8. d
9. b
10. e
11. r
12. m
13. l
14. j
15. k
16. m
17. i
18. n and p
19. o
20. q
21. digestible

22a. lactic acid
22b. polysaccharide
23. starter cultures
24. SCP
25. waste materials
26a. supplement
26b. animal feed
27a. *Lactobacillus*
27b. *Propionibacterium*
27c. *Streptococcus*
28a. defects
28b. insufficient acid development
28c. abnormal flavors
28d. accumulation of gas
28e. bitterness
29a. curd production
29b. moisture removal
29c. ripening

30a. direct acidification
30b. reduction in equipment and personnel
30c. fewer controls
30d. elimination of microbial-associated defects
30e. consistent production of reliable products
31. inhibiting
32a. direct microscopic counts
32b. standard plate counts
32c. staining
32d. isolation
32e. biochemical identification
32f. immunological tests for enzymes
32g. membrane filtration
33. refrigeration
34. drying
35a. sodium nitrate
35b. sodium nitrite
36a. coagulation
36b. filtration
36c. chlorination
37. chlorine
38. disinfection
39a. carcinogens
39b. ozone
40. sewage
41a. BOD
41b. pathogens
42. toxic
43. indicator organisms
44. composting

45. Lowering the water content of certain foods can be an effective control of some food spoilage microorganisms. The reasons for this is that microorganisms and enzymes need water to be active. Limiting the availability of water could interfere with microorganisms and their activities. It should be noted, however, that organisms such as molds, yeasts, and salt-tolerant bacteria can grow in foods with low water content.

46. As the organisms grow, they consume large amounts of organic matter, which is the main objective of the treatment. Also, the organisms that grow usually overcome the pathogen through competitive means and thereby contribute to disinfection as well as organic oxidation.

VI. ENRICHMENT

The origin of fermented dairy products dates back to the dawn of civilization. The ancient Sanskrit scriptures of India describe the food dadhi, a fermented milk product similar to modern yogurt. Further evidence for the existence of soured milk as a food in early times is corroborated by the Bible. The historical, geographical, ecological, and dietary patterns in various regions of the world are reflected in the diversity, variety, and types of fermented milks in vogue today (see Table 16-1). These products are generally produced by the intense activity of lactic acid bacteria added as pure cultures. In addition, certain yeasts may be a part of the fermenting microflora producing low levels of alcohol in the product. Furthermore, the use of milk of various animals adds another dimension to the variety of flavor, body, and texture of fermented milk foods.

TABLE 16-1 Examples of Fermented Dairy Products Consumed in Various Regions of the World

Product Name	*Major Country/Region*	*Source of Milk Used*
Cream cheese	United States, various European countries	Cow
Cottage cheese	United States	Cow
Tvorog	U.S.S.R.	Cow
Sour cream or cultured cream, Smetana	United States, U.S.S.R., Central Europe	Cow
Cultured half-and-half	United States	Cow
Cultured buttermilk	United States	Cow
Ymer	Denmark	Cow
Taettmelk	Norway	Cow
Filmjolk	Sweden	Cow
"Long" milk	Scandinavia	Cow
Pitkapiima	Finland	Cow
Viili	Finland	Cow
Lactofil	Sweden	Cow
Acidophilus milk	United States, U.S.S.R.	Cow
Yákult	Japan	Cow
Yogurt (yoghurt, yoghaurt, yoghourt, yahourth, yaaurt, yourt, jugart, yaert, yaoert)	United States, Europe, Asia	Cow, goat, sheep
Dough or abdoogh	Afghanistan, Iran	Cow, buffalo
Eyran	Turkey	Cow
Leben raib	Egypt	Cow, buffalo
Dahi	Indian subcontinent	Cow, buffalo
Mazurn	Armenia	Cow
Kisselo maleko	Balkans	Cow
Gioddu	Sardinia	Cow
Kefir	U.S.S.R.	Cow, goat, sheep
Koumiss	U.S.S.R.	Mare
Kurunga	Western Soviet Asia	Cow
Chal	Turkmenistan	Camel
Quarg	Germany	Cow

PART VI
MICROBIOLOGY TRIVIA PURSUIT
(Chapters 15 and 16)

It is time again to examine your attention to detail, and your fact-gathering ability. There are *four* challenging Parts to this section. Remember, a certain number of questions must be answered in each Part before you proceed up the MICROBIOLOGY TRIVIA PURSUIT trail. The answers and directions for continuing are given at the end of each Part. Try Part 1.

PART 1

Completion Questions

Provide the correct term or phrase for the following. Spelling counts.

1. The term ______________________________ is often used to describe submerged cultures, even if the process is aerobic.
2. Some modern techniques of industrial microbiology do not use liquid cultures of bacteria, but rather use ______________________________ .
3. When a food product is inoculated with a known microbial type, this inoculum is called a ____ .
4. Many of the effects of bacteria in food fermentations can be duplicated by the process of _____ .
5. The organism which is responsible for the commercial production of beer is the yeast ________ .

Directions

Check your responses with the *Answers* section, and add up your score. Enter the number of correct answers in the space provided.

Total Correct Answers for Part 1: ________________ .

If your score was 4 or higher, proceed to Part 2.

Answers
1. fermentation; 2. immobilized cells; 3. starter culture; 4. direct acidification; 5. *Saccharomyces cerevisiae.*

PART 2

Good show! Congratulations on the successful completion of Part 1. Now try your hand with Part 2.

Completion Questions

Provide the correct term or phrase for the following. Spelling counts.

6. The separation of nucleic acid fragments in an electric field is called ________________________ .
7. *Bacillus thuringiensis* is a bacterium which has been used widely as a ___________ ___________ .

8. ______________________________ are commercial microbial products of great medical importance.
9. The process which uses DNA fragments with DNA probes is called the __________ __________ .
10. This procedure has been used for the rapid clinical ______________________________ of bacterial isolates.

Directions

Check your responses with the *Answers* section, and add up your score. Enter the number of correct answers in the space provided.

Total Correct Answers for Part 2: ______________ .

If your score was 4 or higher, proceed to Part 3.

> ***Answers***
>
> 6. electrophoresis; 7. bacterial pesticide; 8. antibiotics; 9. Southern blot; 10. identification.

PART 3

Great performance! You are doing very well. Here is a challenging Part 3.

Completion Questions

Provide the correct term or phrase for the following. Spelling counts.

11. The use of microorganisms directly as food for humans or animals is called ______________ .
12. When drinking water is treated, the microbial oxidation of organic matter is stimulated by _____ .

13 and 14. The final step in water treatment is disinfection through the addition of some oxidizing chemical. The traditional choice has been (13) ____________________________, but this compound has the potential danger of forming (14) ____________________________ compounds.

15. Most large cities treat their waste water through the highly oxygenated system known as _____ .

Directions

Check your responses with the *Answers* section, and add up your score. Enter the number of correct answers in the space provided.

Total Correct Answers for Part 3: ______________ .

If your score was 5, proceed to Part 4.

> ***Answers***
>
> 11. single cell protein; 12. aeration; 13. chlorine;
> 14. carcinogenic (cancer-causing); 15. activated sludge.

PART 4

You have come a long way. Now for the last and final challenge.

Completion Questions

Provide the correct term or phrase for the following. Spelling counts.

16. Waste water treatment also has a significant anaerobic component which produces methane and inactivates ________________ that might survive the aerobic stages of treatment.

17 and 18. The two oldest methods for food preservation are the use of (17) ________________ and (18) ________________.

19 and 20. Hops are added to beer in order to enhance its (19) ____________ and (20) ____________.

Directions

Check your responses with the *Answers* section, and add up your score. Enter the number of correct answers in the space provided. Then add all your scores for Parts 1 through 4 together and find your score on the *Performance Score Scale.*

Total Correct Answers for Part 4: ____________.

Total Correct Answers for Parts 1 through 4: ____________.

Answers

16. viruses; 17. heat; 18. drying; 19. color; 20. flavor.

PERFORMANCE SCORE SCALE

Number Correct	*Ranking*
12	You should have done better
13	Better
14	Good
16	Excellent
18	Outstanding! Keep it up

Note that each major section of this Study Guide has a *MICROBIOLOGY TRIVIA PURSUIT* section. If you did not do as well as you expected, there are many more opportunities ahead.

PART VII PRINCIPLES OF IMMUNOLOGY

17

Introduction to Immune Responses

I. INTRODUCTION

The immune system is concerned with the body's responses to foreign substances and, to a lesser degree, with responses to substances from the body that it does not recognize as a normal component. Much of the human body's resistance to disease is closely linked to factors—cellular, chemical, mechanical, and microbial—that prevent the penetration and establishment of a disease agent in body tissues. This chapter considers several of these factors. The features of immunological responses, the properties of immunogens (foreign matter), and the properties of specific protein substances—immunoglobulins—that are produced by the body in response to antigens (immunogens) also are presented.

II. PREPARATION

Chapters 17 and 18 should be read before continuing. The following terms are important for you to know. Refer to the glossary and the text if you are uncertain of any of them. A pronunciation guide for selected terms is provided as a learning aid.

1. abscess (AB-sess)
2. agranulocyte (a-GRAN-u-loh-site)
3. AIDS
4. albumin (al-BUE-min)
5. anamnestic (an-am-NES-tik) response
6. antigenic determinant
7. autoantigen
8. B lymphocyte
9. cancer
10. carbuncle (KAR-bung-kel)
11. carrier
12. cell-mediated response
13. C region
14. chemotaxis (key-moh-TAK-sis)
15. chemotherapy (key-moh-THER-ah-pee)
16. complement (KOM-plee-ment)
17. cytotoxic T cell
18. differential count
19. DNA recombinant technology
20. edema (ee-DEE-mah)
21. epithelial (ep-ee-THEE-lee-all)
22. erythrocyte
23. fever
24. fibrinogen
25. globulin
26. granulocyte (GRAN-u-loh-site)

27. hapten (HAP-ten)
28. heavy chain
29. hemagglutination
30. host
31. humoral (HUE-mor-al) response
32. idiotype (id-ee-OH-type)
33. immunity (ee-MEW-neh-tee)
34. immunoelectrophoresis (im-u-no-ee-LEK-troh-fo-ree-sis)
35. immunoglobulin
36. inflammation
37. interferon
38. Küppfer (KOOP-ferr) cell
39. lactoferrin (lak-TOH-ferr-in)
40. leukocyte (LOO-koh-site)
41. light chain
42. lymphatic system
43. lymphocyte (LIM-foe-site)
44. macrophage
45. major histocompatibility complex (MHC)
46. memory cell
47. microbiota
48. monocyte
49. mononuclear phagocyte system
50. normal flora
51. nucleoprotein
52. passive immunity
53. placental transmission
54. opportunist
55. opsonization (op-soh-neh-ZAY-shun)
56. pathogen
57. phagocyte
58. phagocytosis (fag-oh-sye-TOE-sis)
59. phagosome (FAG-oh-some)
60. plasma (PLAZ-mah)
61. plasma cell
62. polymorphonuclear leukocyte
63. polypeptide
64. polysaccharide
65. properdin (PRO-per-din)
66. pyrogen (PIE-row-jen)
67. secretary immunoglobulin
68. sensitized
69. serum
70. thymus (THIGH-mus) gland
71. titer (TIE-ter)
72. T lymphocyte
73. toxoid (TOK-soid)
74. transferrin (trans-FERR-in)
75. V region

III. PRETEST

Correct answers to all questions can be found at the end of this section. Write your responses in the appropriate space provided.

Completion Questions

Provide the correct term or phrase for the following. Spelling counts.

1. The various components of the ______________________________ system provide protection against pathogens and foreign matter that enter the body.

2. Three factors that can affect species resistance are (a) ______________, (b) ______________, and (c) ______________.
3. The respiratory system is protected against pathogens and foreign particles by several mechanisms, which include (a) ______________, (b) ______________, (c) ______________, and (d) ______________.
4. Two protective mechanisms for the mucous membranes of the female genitourinary tract are (a) ______________ and (b) ______________.
5. Tears contain the bactericidal substance ______________.
6. Organisms that take advantage of a host's weakened defenses to cause disease are called ______________.
7. Normal flora, which includes organisms ranging from commensals to pathogens, are commonly called ______________.
8. Nine anatomical locations that as a rule contain few or no microbes are (a) ______________, (b) ______________, (c) ______________, (d) ______________, (e) ______________, (f) ______________, (g) ______________, (h) ______________, and (i) ______________.
9. Three mechanisms used by the immune system to cope with disease agents are (a) ______________, (b) ______________, and (c) ______________.
10. The liquid portion of blood is called ______________.
11. The special type of calibrated tube used to determine portions of cells and plasma in a blood sample is called ______________.
12. The fluid appearing after blood clots within a tube is called ______________.
13. Three factors known to provoke (stimulate) immunoglobulin production are (a) ______________, (b) ______________, and (c) ______________.
14. Red blood cells are formed in the ______________.
15. The two recognized groups of white blood cells are (a) ______________ and (b) ______________.
16. ______________ are present in greater number in individuals with parasitic infections or allergies.
17. An increase in white blood cells is called ______________.
18. The procedure used to determine the proportions of the various white blood cells in a blood sample is called a(an) ______________.
19. Two specific functions of macrophages are (a) ______________ and (b) ______________.
20. Ingestion of bacteria by a phagocyte in the absence of detectable antibody is called ______________.
21. Specific migration of phagocytic cells toward bacteria or particulate matter due to chemical stimuli is an example of ______________.

22. The characteristic cardinal signs of inflammation are (a) ______, (b) ______, (c) ______, (d) ______, and (e) ______.
23. A localized region or sac containing pus, microorganisms, and the various factors needed for blood coagulation is called a(an) ______.
24. The disintegration products of microbes, called ______, are thought to be a cause of fever during infections.
25. The stimulation of antibody production is termed ______.
26. Substances that cause antibody production and combine with such antibodies are called ______.
27. The large molecules to which drugs or small reactive molecules can attach that subsequently become immunogenic are called ______.
28. The specific groups of atoms on the surface of an antigen that both stimulate and react with antibodies are called ______.
29. The number of determinant antigenic sites on an antigen's surface is known as its ______.
30. The sudden secondary rise in antibody titer upon the second injection of an immunogen some time after the first immunogen exposure is called a(an) ______.
31. The major types of cells associated with immunoglobulin formation are the (a) ______ and (b) ______.
32. The three basic events of the immune response are (a) ______, (b) ______, and (c) ______.
33. The production of immunoglobulins by an individual in response to an immunogenic stimulus represents a(an) ______ state of immunity.
34. The transfer of antibodies from an immunized donor to a nonimmune recipient is an example of ______ immunity.

True or False

Indicate all true statements with a "T" and all false statements with an "F".

______ 35. The mouth and lower intestine are major habitats for indigenous microorganisms.

______ 36. Development of an individual's indigenous flora begins with the normal birth process.

______ 37. Different microorganisms predominate in specific anatomical areas.

______ 38. Normal flora helps to protect the body against pathogens.

______ 39. Indigenous organisms help to maintain mechanisms for antibody production.

______ 40. Lactobacilli stabilize the flora of the respiratory tract.

______ 41. Lymphocytes and basophils are examples of granulocytes.

______ 42. The causative agents of tuberculosis, typhoid fever, and influenza cause a leukocytosis in infected individuals.

______ 43. Complement is a lipid macromolecule found in the sera of mammals.

______ 44. Complement can cause the lysis of certain bacterial cells.

______ 45. Lysozyme is found in tears and exhibits bactericidal action against gram-positive cells.

______ 46. Microorganisms are not effective immunogens.

______ 47. The majority of lipids are immunogenic.

______ 48. Immunogens can be made synthetically.

______ 49. Haptens are immunogenic.

______ 50. Antibodies produced against normal body tissues are called autoantigens.

______ 51. Forssmann antibodies are present in the sera of persons with infectious mononucleosis.

______ 52. Immunoelectrophoresis can be used to separate and identify body fluids.

______ 53. The dosage of immunogens used for immunization does not affect the extent of antibody response.

______ 54. The reaction to the first injection of an immunogen is called an anamnestic response.

______ 55. Antibody response is better in animals lacking a thymus gland.

______ 56. The immune system of the newborn begins to mature about one year after birth.

______ 57. B lymphocytes control cell-mediated immunity.

______ 58. In the selective hypothesis for immunoglobulin formation, immunogens serve directly as patterns for the formation of the active sites of immunoglobulins.

Answers

1. immune
2a. diet
2b. stress
2c. temperature
3a. mucus
3b. cilia lining portions of respiratory tract
3c. coughing reflex
3d. sneezing reflex
4a. thick secretion
4b. local acidity
5. lysozyme
6. opportunists
7. microbiota
8a. blood
8b. larynx
8c. trachea
8d. nasal sinuses
8e. upper urinary tract
8f. posterior genital tract
8g. stomach
8h. bronchi
8i. esophagus
9a. removal of disease agents
9b. neutralization of pathogens and their toxic products
9c. destruction of foreign cells
10. plasma
11. hematocrit
12. serum
13a. pathogens
13b. vaccine preparations
13c. foreign proteins such as pollens
14. bone marrow
15a. agranulocytes
15b. granulocytes
16. eosinophils
17. leukocytosis
18. differential count
19a. phagocytosis
19b. preparing substances for the formation of antibodies
20. nonimmune or surface phagocytosis
21. chemotaxis
22a. heat
22b. pain
22c. redness
22d. swelling
22e. loss of function
23. abscess
24. pyrogens
25. immunogenicity
26. immunogens
27. carriers
28. antigenic determinants
29. valence
30. anamnestic response
31a. B lymphocytes
31b. plasma cells
32a. recognition
32b. activation
32c. differentiation
33. active
34. passive
35. T
36. T
37. T
38. T
39. T
40. F
41. F
42. F
43. F
44. T
45. T
46. F
47. F
48. T
49. F
50. F
51. T
52. F
53. F
54. F
55. F
56. F
57. F
58. F

IV. CONCEPTS AND TERMINOLOGY

Humans and other vertebrates are protected in varying degrees from disease-causing microorganisms and cancer cells by a surveillance mechanism referred to as the *immune system.* The components of the system provide protection by imposing barriers to invasion by microorganisms and other disease agents and by selectively eliminating foreign invaders that do find their way into the body.

Among the body's defense mechanisms are those that provide *specific immunity* against particular microorganisms and their products. Other defense mechanisms are used against any and all disease-causing agents. This form of immunity is called *nonspecific resistance.* Some contributing factors in nonspecific resistance are species or racial factors, mechanical and chemical barriers, phagocytosis, inflammation, and various antimicrobial chemical products of the body.

The Immune System

The immune system has several means of coping with disease agents, including (1) removing them from the body, (2) neutralizing infectious organisms and biologically active molecules, and (3) destroying foreign cells. Several immune system and normal blood components are described in Table 17-1.

Nonspecific Resistance

Species Resistance

Species resistance is determined by physiologic and anatomic properties of a particular animal species and is inheritable. Demonstrable antibodies (protective protein molecules) are not associated with this state of resistance; rather, it depends on the interplay of many factors, not all of which are known.

Mechanical and Chemical Barriers: The Body's First Line of Defense

Several systems of the body act as barriers to potential disease-causing agents. Their effectiveness depends on the physiologic or pathologic state of the host. Alcoholism, poor nutrition, aging, fatigue, and prolonged exposure to extreme temperatures or immunosuppressive therapy are among the conditions that contribute to establishment of a disease process. Table 17-2 lists several of these barriers and their properties.

Indigenous Microbiota or Normal Flora

The microorganisms indigenous to humans are often referred to simply as *normal flora. Flora* refers to all microscopic life; *normal* in this context is a statistical term and should not be construed to mean nonpathogenic, for many organisms found on and in the body can pose problems.

Normal flora are commonly referred to as *microbiota,* ranging from beneficial commensals to pathogens. These organisms are obligately parasitic for humans or other animals but are not necessarily pathogenic. They are found at least as often in the absence of as in the presence of disease. The indigenous microorganism may flourish in the general region of tissue damage and act as *opportunists,* taking advantage of a host's weakened defenses to cause disease.

As a rule, few or no microorganisms are found in the following anatomical locations: larynx, trachea, nasal sinuses, bronchi, esophagus, stomach, upper intestinal tract, upper urinary tract (including the posterior urethra), and posterior genital tract (passage above cervix included). The major habitats for indigenous microorganisms include the skin and contiguous mucous membranes, conjunctivae, portions of the upper respiratory tract, the mouth, lower intestine, and most external and internal parts of the reproductive system.

Because the infant is bathed during gestation (pregnancy) in a sterile amniotic fluid, development of the indigenous flora begins with the normal birth process. As the baby passes through the birth canal, it picks up organisms, many of which may remain with it throughout its lifetime. Additional microorganisms are acquired by contact with the air, hospital personnel, and the mother. These organisms may be transient (temporary) or may become permanent members of the flora.

TABLE 17-1 Components of the Immune System and Normal Blood

Components	*Composition and Brief Description*	*Activities or Functions*
Blood	Consists of cellular elements in a fluid substance called plasma. Cells and cell-like components are often referred to as the formed elements. These include *erythrocytes* (red cells), *leukocytes* (white cells), and *platelets. Chylomicrons,* which are visible minute fat globules, are also suspended in the plasma portion of blood. The cellular elements comprise approximately 45% of the blood, and plasma constitutes the remaining portion.	Transports food and hormones, removes cellular waste products, assists in the regulation of body temperature, and aids in the removal and, in certain situations, the destruction of foreign substances and invading microorganisms. Can play an important role in the transmission, production, diagnosis, cure, and prevention of many conditions caused by microorganisms.
Serum	Yellow fluid formed during blood clotting. All elements of the clotting mechanism are involved in serum formation. The serum contains several types of proteins, including albumin and globulins. Globulins occur in three major groups: alpha, beta, and gamma.	Albumin is important to the maintenance of adequate cellular nutrition and a normal osmotic pressure. Alpha and beta globulins are involved with the transport of other proteins, whereas the gamma globulins are of greatest importance to immunity. Most immunoglobulins or antibodies are found in the gamma globulin fraction. Immunoglobulins provide humoral immunity and are produced by the body in response to (1) infectious agents; (2) vaccine preparations of killed, weakened, or attenuated organisms or their products; or (3) other foreign protein substances, such as pollens.
Plasma	Formed in blood treated with an anticoagulant. Cellular elements settle to bottom of container holding blood, leaving clear fluid (plasma) above. Plasma is a complex mixture of substances, including carbohydrates, lipids, proteins, gases, inorganic salts, hormones, and water. The pH of plasma normally is slightly alkaline, approximately 7.4.	Generally the same as serum.
Erythrocytes (red cells)	Formed in the bone marrow and measure about 7.5 to 7.7 μm in diameter and 1.9 to 2 μm in thickness. In mammals, they are not nucleated when mature and appear as biconcave disks. The red cells are composed of a membrane which is in close association with the iron-containing protein compound hemoglobin.	Red cells are specialized for the transport of oxygen, gas, and possess the major, minor, and Rh blood factors.

TABLE 17-1 Components of the Immune System and Normal Blood ***(continued)***

Components	*Composition and Brief Description*	*Activities or Functions*
Leukocytes	Classified into two groups, granulocytes and agranulocytes. Granulocytes contain distinct cytoplasmic granules and have irregular and multilobed nuclei. Like erythrocytes, they are formed in the bone marrow. Staining reveals three types of granulocytes: *eosinophils,* which contain granules that react with acid dyes and become red; *basophils,* which contain granules that react with basic dyes and become blue; and *neutrophils* or polymorphonuclear leukocytes (PMNLS), which contain granules that react with neutral dyes or acid and basic dye mixtures and become orange colored. The agranulocytes do not possess granules, and their nuclei are rounded rather than lobed. The two general cell types of this group are the *lymphocyte* and the *monocyte.* Lymphocytes are principally produced in the appendix, lymph nodes, spleen, thymus, and other lymphoid tissues. Monocytes are larger than lymphocytes, and their nuclei are generally kidney shaped. They are formed in the bone marrow.	Eosinophils combat certain parasitic worm infections. They are present in greatest number in individuals with parasitic infections or allergies. Basophils are phagocytic and thus are capable of engulfing and digesting invading foreign particles. They are also known to play a role in immunity to virus infection and in rejection of grafts. These cells also contain chemical substances that have powerful effects on the body's blood vessels and pulmonary system. Neutrophils are also phagocytic. The number of neutrophils in the blood increases rapidly in the early stages of bacterial infection. Certain lymphocytes and monocytes are involved with specific antibody formation. When in tissues, monocytes also participate in phagocytosis.
Lymphatic system	Components of this system include lymphatic vessels, lymph fluid, lymph nodes, and lymphocytes. The lymphatic vessels exist throughout most of the body. They pass through lymph nodes, which are nodular accumulations of lymphocytes and macrophages. Macrophages are derived from circulating blood monocytes.	The lymphatic system is of particular value in the transport of fatty acids, proteins, and white blood cells; it is also important during times of infection or other tissue injury. It removes material that accumulates at the site of tissue damage, such as foreign cells and their products, white cells, and tissue debris. Pathogens and their products are carried by the lymphatics to the lymph nodes, where the immune response by the host's macrophages and lymphocytes are initiated. Macrophages are capable of several immunologically related functions, including clearing and degrading foreign substances within the body (phagocytosis) and preparing foreign substances in the antibody formation process.
Mononuclear (reticuloendothelial) system	Consists of a variety of cells that ingest and digest foreign and host substances.	Phagocytic activity in regions such as capillaries, spleen, bone marrow, and lymph nodes.

TABLE 17-2 Physical and Chemical Barriers of the Body

Barrier	*General Properties*
Intact skin	Serves as an excellent mechanical barrier that most microorganisms cannot penetrate. In addition, certain bactericidal secretions are formed by skin glands. Injuries to the skin, such as abrasions, lacerations, or burns, permit microorganisms to pass through this first line of defense.
Mucous tissues and respiratory tract	The mucous membranes of the respiratory system are covered with a thick, slimy secretion known as mucus, which traps dust, foreign particles, and various microorganisms. Parts of the respiratory passages are also lined with cilia, whose rhythmic motion moves particles trapped by mucus upward toward the back of the throat, where they are swallowed. Several immune mechanisms and substances function at mucosal surfaces to prevent microbial invasion or damage. Predominant among immune factors are specific antibodies, the secretory immunoglobulins (sIgA) found in fluids that bathe mucous membranes. Functions of sIgA include inhibition of bacterial attachment, neutralization of toxins and viruses, and prevention of the uptake of foreign substances by epithelial cells.
Coughing and sneezing	These activities help to eliminate foreign particles.
Genitourinary system	Secretions of the female genitourinary tract trap certain invading organisms. In addition, the acidity of the vagina discourages some infectious agents. The outward flow of urine and its acidity contribute to the defense of the urinary tract; however, various pathogens, including those causing gonorrhea and syphilis, are able to invade the body through this portal.
Eyes	Protection is afforded by the mechanical motion of the eyelids, the eyelashes, and eyebrows and by the washing effects of tears, which contain lysozyme, a bactericide.
Gastrointestinal system	The composition and acidity of gastric juice provides considerable protection to the stomach. Mucus, certain enzymes, bile, and phagocytosis are important factors contributing to the body's defense in the small intestine. The large intestine usually harbors many microorganisms (indigenous flora) that are important in maintaining a "normal" balance.

Since each anatomical area varies in pH, oxygen content, nutrients, moisture, and bactericidal factors, different organisms will predominate. While the microbiota persist in their respective locations, saprophytic organisms and many parasitic microorganisms are destroyed or excreted. Conditions in these locations can change as a result of maturation of the individual, alteration in dietary habits, or chemotherapy. Thus, microorganisms may be temporary or permanent, depending upon the conditions that exist in the body.

Probably the main benefit derived by humans from their microbial inhabitants is protection from disease. As a rule, the normal flora occupy their own niches and inhibit foreign organisms invading from other portions of the body or from the external environment. Such inhibition is brought about by competition for food, by the production of antibiotics or other inhibitory substances, or by changes in environmental conditions, such as oxygen content or pH.

Another significant role played by indigenous organisms is helping to maintain mechanisms for antibody production. Amphibionts act as a constant source of antigens or stimulation to the antibody-producing systems of the body and thereby permit a more rapid immunological response when it is needed.

The role of the microbiota in nutrition is the subject of considerable research. Some of these organisms synthesize a variety of vitamins in excess of their own needs, thus supplying additional nutrients for the host. These vitamins include biotin, pyridoxin, pantothenic acid, and vitamins K and B_{12}.

Antimicrobial Substances

Chemical substances capable of *in vitro* antimicrobial activity have been isolated from various animal fluids and tissues. The full extent of their *in vivo* effectiveness is not known. The properties of several of these antimicrobial substances are listed in Table 17-3.

TABLE 17-3 Representative Antibacterial Substances in Animal Tissues and Fluids

Substance	*Common Source(s)*	*General Chemical Composition*	*Types of Microorganisms Affected*	*Other Properties*
Complement	Sera of most warm-blooded animals	Believed to be a protein-carbohydrate-lipoprotein complex	Gram-negatives	Complement concentrations do not rise in response to immunization. Reacts with a wide variety of antigen-antibody to produce: the destruction of erythrocytes as well as other tissue cells; the initiation of inflammatory changes; the lysis of certain bacterial cells; and enhancement of phagocytosis involving some opsonized particles. Heat labile (inactivated by heating to 56° C for 30 minutes).
Interferons[a]	Virus-infected cells	Protein	Various viruses and certain protozoa	Stable at low pH and temperatures of 50° C (at times higher). Sources of interferon include virus-infected cells and mitogen-stimulated T-type lymphocytes. Interferes with the synthesis of viral proteins.
Lactoferrin	Various body secretions, including tears, breast milk, bile, etc.	Glycoprotein	Bacteria and fungi	Iron-containing substance.
Lysozyme	Include leuko-cytes, saliva, perspiration, tears, egg whites	Protein	Mainly gram-positives	Thermostable; activity increased by low antibody concentrations.
Properdin[b]	Serum	Protein	Gram-negatives and certain viruses	Consists of at least four protein components; produces hemolysis; provides protection against total body irradiation; aids phago-cytosis.
Spermidine, spermine	Prostate and pancreas	Basic polyamines	Gram-positives	Tissues containing spermine-activating enzyme (spermine oxidase) more resistant to tubercle bacilli.
Transferrin	Serum and tissue and organ spaces	Glycoprotein	Bacteria and fungi	Iron-containing substance.

[a] Refer to Chapters 10 and 12 for additional details.

[b] Refer to Chapter 20 for additional details.

Other Lines of Defense in Nonspecific Resistance

Phagocytosis

Phagocytosis is the ingestion and subsequent digestion of particles by single cells. This process is carried out by circulating granulocytes and fixed macrophages such as Küppfer cells. Foreign matter is ingested by extensions of phagocytic cells. Cells become susceptible to phagocytosis through a process called *opsonization,* in which a specific antibody combines with microbial cells in the presence of a complex group of proteins known as complement. Opsonizing antibodies combine with the surface antigens of a cell (antigen-antibody complex) and prepare it for ingestion by the phagocyte. Ingestion of bacteria can occur in the absence of detectable antibody by means of *nonimmune* or *surface phagocytosis,* wherein the pathogen is trapped against tissue surfaces.

If phagocytosis of the invading organism is complete, the disease state is either averted or cured. However, in the event the pathogens manage to escape ingestion and intracellular destruction, they can reproduce and cause serious infections.

Destruction of microorganisms by phagocytosis occurs in three stages: (1) contact between the phagocyte and the particle to be ingested, (2) ingestion, and (3) intracellular killing and destruction (digestion). The contact stage can be either random or specific. In the more specific response, phagocytic cells migrate toward bacteria or particulate matter, drawn by chemotaxis, a reaction to chemical stimuli frequently produced by microbes.

The ingestion stage of phagocytosis is similar to food intake by amoeba. Bacteria are ingested through an invagination of a leukocyte's cytoplasmic membrane. In general, bacteria or particles that have been opsonized (coated) with antibody molecules are more readily ingested. In the ingestion process, a phagocytic vacuole forms, engulfing the bacterial cell.

Both lysosomal components and metabolic products contribute to the bactericidal activity in the intracellular destruction stage. As a result of the burst of metabolic activity following ingestion, lactic acid (which lowers the pH in the vacuole) and the strong oxidizing agent hydrogen peroxide are produced. These, together with histones, lysozyme, phagocytin, and various digestive enzymes released from lysosome granules, degrade the dead bacterium.

Several types of phagocytic dysfunction are known. Many of them are genetic disease states.

Inflammation

Inflammation is the body's second line of defense against infection. Inflammation can be produced by infectious disease agents and by irritants such as chemicals, heat, and mechanical injury. The *cardinal* (characteristic) *signs* of inflammation are heat, pain, redness, swelling, and loss of function. *Pus formation* may also result from inflammation. After the phagocytes have destroyed the microbial cells and engulfed the tissue debris, they become degranulated and die. In the involved area, a central mass or fluid—pus—is formed by remains of damaged tissue cells, dead phagocytes, and microbial casualties.

Fever

Fever, an elevation of body temperature above normal, is a symptom of many disease states. Temperatures of 38.5° to 39.0°C speed the destruction of disease agents by increasing immunoglobulin production and phagocytic activity. Fever is thought to be caused by the action of toxic substances, known as pyrogens, entering the bloodstream. Pyrogens are products of the disintegration of microorganisms or other cells in an injured area of the body.

Specific Immune Responses

Immunity is a specific form of resistance that depends on various factors, including the presence of protective substances known as immunoglobulins, or antibodies, produced in response to antigens or immunogens (the foreign substances that provoke antibody formation). The immune system has several means of defending against such foreign substances. These include the *humoral response,* the formation of antibodies or *immunoglobulins,* which function in the specific recognition of antigens, and the *cell-mediated response,* regulated by a group of differing antigen-reactive cells, the *T lymphocytes.*

Antigens and Immunogens

Antigens are macromolecules that can react with antibodies but do not necessarily stimulate their production. *Immunogens* are substances that provoke the formation of antibodies and can combine with them. The stimulation of antibody production is called *immunogenicity.* Both natural and synthetic substances can stimulate antibody production. Macromolecules of such substances can be: (1) free proteins; (2) combinations of proteins and other substances, including nucleoproteins, lipoproteins, and glycoproteins; and (3) polysaccharides. Entire microorganisms and most of their parts are immunogenic. Several drugs and small reactive drugs can become immunogenic by being chemically bonded (coupled or joined) to larger molecules called *carriers.*

Antigenic determinant sites are specific groups of atoms on the surface of an immunogen that both stimulate production of antibodies and react with them. The number of such sites on the surface of a molecule is known as its *valence.*

Immunogens provoke antibody production because the antibody-forming tissues of an animal recognize them as foreign matter. The greater the incompatibility between an immunogen and a recipient's tissues, the greater the immune response. Other factors that affect the immune response include the species of animal receiving the immunogen, the degree to which the animal's immune mechanisms are functioning, the route of inoculation, and the use of substances (adjuvants) that, when mixed with the immunogen, prolong and increase the intensity of the antibody-provoking stimulus.

Certain lipids and polysaccharides of animals and bacterial cells do not cause antibody production unless linked (coupled) to large proteins called carriers. Such substances, called *haptens,* are nothing more than single antigenic determinants.

The antibody-forming (*immunopoietic*) tissues of an animal recognize its individual cells as belonging to the body or "self" and not as foreign matter or "not self." In some cases, *autoantibodies* are produced against normal body components, or *autoantigens,* in a process known as *autoimmunization* or *autoallergy.* Table 17-4 lists representative autoimmune diseases and the immunogenic substances believed to be responsible for each condition.

TABLE 17-4 Some Autoimmune Disease States with the Incriminated Immunogenic Substances

Autoimmune Disease	*Immunogenic Substance*
Acquired hemolytic anemia	Red blood cells
Allergic encephalomyelitis	Myelin from the central nervous system
Aspermatogenesis	Spermatozoa
Idiopathic thrombocytopenic purpura[a]	Blood platelets
Rheumatoid arthritis	Immunoglobulins[b] (IgG)
Systemic lupus erythematosis (LE)	Deoxyribonucleic acid
Thyroiditis (Hashimoto's Disease)	Thyroglobulin

[a] This disease state is characterized by bleeding in various tissues and the presence of a rash and purpura (little areas of hemorrhaging) in the skin. It is also called purpura hemorrhagica.

[b] Other causes are believed to be operative in this disease condition.

The red blood cells of individuals within the same species contain different antigens, called *isoantigens.* In addition, *isoantibodies*—antibodies capable of reacting with these antigens—also differ among individuals. The clumping or agglutination reaction that results from mixing antigenically different red blood cells such as blood types A, B, or AB from one individual with the serum of an individual having a different blood type is called *isohemagglutination.* Blood typing procedures are based on this phenomenon.

Antigens that stimulate production of antibodies effective against material unrelated to the original antigens are called *heterophile antigens.* Antibodies to these antigens will cross-react with the cells of various animal species and microorganisms. The best known example of the heterophile antigens is the *Forssmann antigen.* It has been found in other animals, including birds, cats, dogs, mice, and tortoises, and has been associated with certain bacterial species (e.g., *Bacillus anthracis, Streptococcus pneumoniae, Salmonella* spp., and *Shigella dysenteriae*).

Antibodies

Antibodies are members of the protein group known as *globulins.* The various proteins in blood serum can be distinguished on the basis of specific physicochemical and immunogenic properties, which include:

1. Chemical composition
2. Chromatographic features
3. Electrical charge and migration in an electrical field

4. Molecular weight
5. Relative solubilities in alcohol, electrolytes, and water
6. Sedimentation coefficients

Immunoelectrophoresis incorporates a gel or gel-like material and electric field (electrophoresis) to separate and identify antibodies and antigens in a mixture. The technique can be applied to a variety of body fluids, including amniotic fluid, cerebrospinal fluid, human plasma and serum, respiratory secretions, saliva, animal and plant tissue antigens, and microbial antigens.

In an electrophoretic analysis, antibodies are found in the slowest-moving fraction of serum (the gamma globulin), which has been found to contain five different classes of immunoglobulins (Ig) or antibodies. The immunoglobulin classes, together with some general properties, are listed in Table 17-5. Several subclasses of immunoglobulins have been found: IgG1 to IgG4, IgA1, IgA2, IgM1, and IgM2.

TABLE 17-5 The Immunoglobulins and Selected Properties

	Immunoglobulins					
Properties	***IgA***	***sIgA***[a]	***IgG***	***IgM***	***IgE***	***IgD***
H chain class	α	α	γ	μ	ϵ	δ
L chain class	κ and λ	κ and λ	κ and λ	κ and λ	κ and λ	κ and λ
J chain	Absent	Present	Absent	Absent	Absent	Absent
Secretory component	Absent	Present	Absent	Absent	Absent	Absent
Molecular weight (approximate, in Daltons)	160,000	400,000	150,000	900,000	190,000	180,000
Sedimentation coefficient[b]	7S	10S	6-7	19	8	7-8
Percentage of total serum antibody	15	—	75	10	0.004	0.2
Participation in classic complement fixation	None	None	c	d	None	None
Placental transfer	None	None	Yes	None	None	None
Antibacterial destruction (lysis)	c	c	c	d	Questionable	Questionable
Viral neutralization	d	d	c	c	Questionable	Questionable

[a] Secretory IgA.

[b] Refers to the speed at which a particle settles in an ultracentrifuge system and is influenced by both particle size and molecular weight.

[c] Low level of activity.

[d] High level of activity.

All normal immunoglobulins have the same basic molecular arrangement, a unit of four polypeptide chains. Two of these are identical H (heavy) chains and two are identical L (light) chains, represented by the formula $(L_2H_2)_n$. The sequence of amino acids in the polypeptide chains can be quite diverse. The chains are covalently linked by disulfide bonds (S-S). H chains are held together by one or more bonds, whereas each L chain is bound to an H chain by means of the disulfide bond. Immunoglobulins composed of more than one basic (monomeric) unit normally contain a J chain, which is involved with the binding together of certain components.

Each chain of a basic immunoglobulin molecule (text Figure 17-15) may be divided into the two portions called the variable or *V region* and the constant or *C region.* The V region, which is the immunoglobulin portion exhibiting the greater variation in structure, binds antigens. The constant region performs the elimination/destruction function.

Immunoglobulins can contain two different types of light chains, kappa (κ) and lambda (λ). A given antibody molecule contains either two κ or two λ chains (or multiples of two) but never one of each. Each class of immunoglobulin has a different H chain structure. The designations for each class are as follows:

α (alpha) for IgA
δ (delta) for IgD
ϵ (epsilon) for IgE
γ (gamma) for IgG
μ (mu) for IgM

Detailed Immunoglobulin Structure

Three different immunoglobulin subregions or domains based on chemical and physical treatments are recognized. Their properties are summarized in Table 17-6.

TABLE 17-6 Immunoglobulin Domains

Domain	*Description*
F_{ab}	Contains antigen-binding sites; composed of the chemically variable portions of both H and L chains
F_c	Participates in various biological activities or reactions, including complement fixation and skin sensitization; contains chemically constant (c) portion of the H chain
F_d	A portion of the heavy chain in the F_{ab} domain

Immunoglobulin Classification

Variations in immunoglobulin structures exist and may be conveniently divided into three general categories: *isotypes, allotypes,* and *idiotypes.* Isotypes are antigenic determinants (characteristic means of identification or markers) shared by all immunoglobulins within each of the five classes. Allotypes are markers that reflect genetically determined antigenic differences among immunoglobulins. Idiotypes are antigenic determinants directed against specific antigens.

Genetic Aspects of Ig Expression

Humans, as well as other mammals, have an elaborate system that not only uses gene segments on various chromosomes, but also takes advantage of a flexible mechanism for genetic exchanges within or between chromosomes to reach the maximum limits of antibody diversity.

Antibody Production

The extent of the antibody response to an immunogen is known to be affected by various factors, including (1) the nature of the immunizing material, (2) the dosage received, (3) the number and frequency of exposures, (4) the particular animal species, and (5) the individual involved.

The reaction exhibited by an individual to the first injection of an immunogen is called the *primary response.* After a period ranging from a few hours to several days, the antibody titer (level) reaches a peak or plateau when the rates of antibody production and antibody breakdown are approximately the same. Depending on the immunogenic stimulus and animal species involved, such antibody levels may remain for several months or longer and then slowly begin to decline, as antibody breakdown exceeds production.

The sudden secondary rise in antibody titer caused by another exposure to an immunogen is called the *anamnestic response.* Such reactions are produced by reimmunization with vaccines.

The Development of the Immunologic System (Immunocompetence)

It appears that the maturation of the human immune system begins *in utero* sometime during the second to third month of pregnancy, and involves the differentiation of cells that will carry out both specific and nonspecific immunologic activities. These cells appear to arise from a population of stem cells or hemocytoblasts located within the blood-forming tissues of the developing embryo (see Figure 17-1). Depending on the type of environment the differentiated cells enter, they will develop into either the hematopoietic or the lymphopoietic tissue. The former will result in production of blood components, such as erythrocytes, granulocytes, monocytes, and platelets. The latter can lead to still further differentiation.

Descriptions of different components of the immunologic system are summarized in Table 17-7.

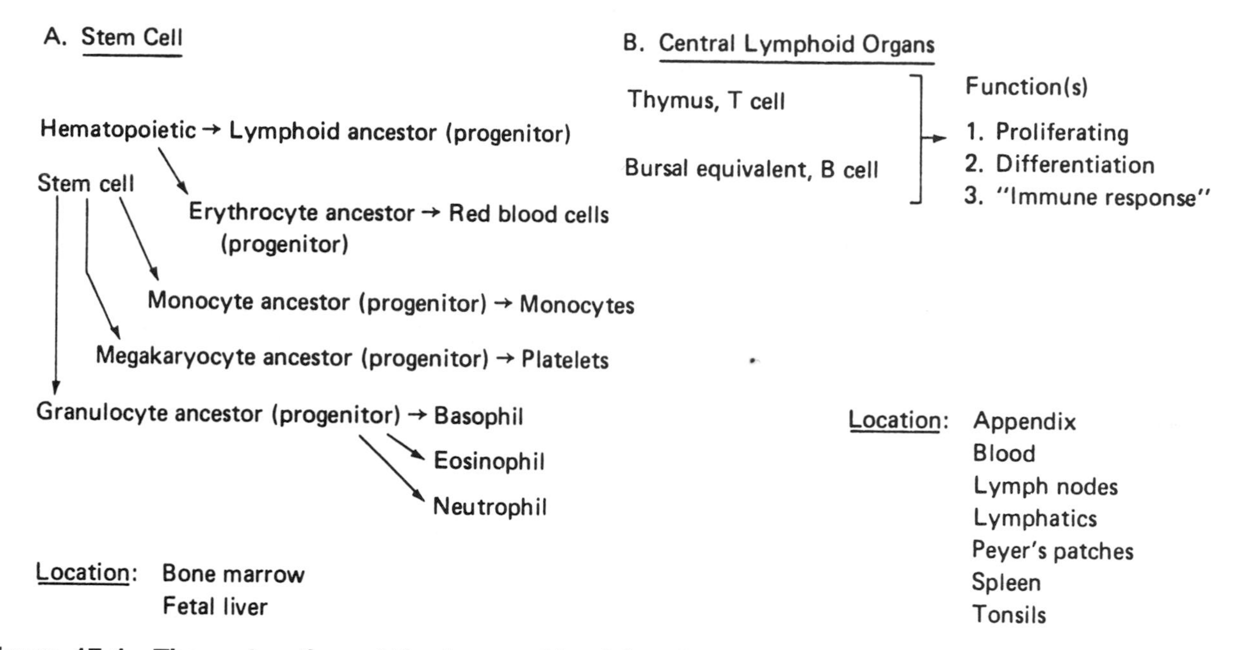

Figure 17-1 The maturation of the human blood-forming tissues and cells and the human immune system.

Major Histocompatibility Complex (MHC)

An important characteristic of developing T cells (thymocytes) is the ability to recognize the antigens (protein products) of closely linked genes in the region of DNA called the *major histocompatibility complex (MHC)*. The MHC gene products (antigens) are of three different kinds, *Class I MHC (host's self) protein,* a molecule present on the surface of all body-nucleated cells; *Class II MHC proteins,* molecules found primarily on the surface of specialized cells known as antigen-presenting cells; and *Class III MHC proteins,* certain components of complement (C2, C4, and factor B). Refer to Chapters 17 and 20 in the text for additional details.

Cells in the Immune Process

Both B and T cells occupy different areas within the same lymphoid tissues and are intimately associated with another form of white blood cell, the *macrophage.* Macrophages, which are also lymphoid tissues, process and present antigens to lymphocytes to initiate many immune responses. In addition, these cells synthesize and secrete chemical substances that regulate lymphocyte function.

The participation of macrophages, B and T cells, and plasma cells in antibody production involves three basic events of the immune response. In proper sequence, they are: (1) *recognition,* the binding of immunogen to specific receptor sites; (2) *activation,* the stimulation of a resting cell into an active cell; and (3) *differentiation,* the production of the specific plasma cells or memory cells that partake in an immune response. The mechanism involved in this process supports the clonal selection hypothesis for antibody production.

TABLE 17-7 Components of the Immunologic System

Component	*Description*
Thymus gland	During embryonic development, this gland is found between lungs and behind the breastbone. After birth, the thymus gland atrophies. Responsible for the multiplication of lymphocytes (T cells) containing theta antigen on their surfaces; also for the seeding of the lymph nodes, spleen, and related structures with these cells. The thymus synthesizes several substances that act as hormones and probably play a major role in the T cell regulation and differentiation. Such substances include thymosin, thymopoietin, and facteur thymique serique. Removal of the thymus from a newborn results in lowered antibody production.
B lymphocytes	Originate in the bone marrow and are associated with *humoral immunity*. B cells undergo differentiation into specific immunoglobulin-secreting cells known as plasma cells. B cells are distinguished from other lymphoid cells by the presence of immunoglobulins on their membrane surfaces.
T lymphocytes	T cells form under the influence of the thymus gland and are responsible for *cell-mediated immunity*. T cells are involved in essentially all immune reactions, either as effector (directly attacking) cells or as regulators of both humoral and cellular responses. Five functional T cell subsets (subpopulations) are recognized and named according to their associated activities. Their designations and functions are as follows: (1) T helper (T_H) cells, which work with B cells in immunoglobulin production; (2) natural killer (NK) cells; (3) cytotoxic T lymphocytes, which disrupt target cells by direct cell-to-cell contact; (4) T delayed-type hypersensitivity (T_{DTH}) cells, which bring about (mediate) inflammation and activate macrophages in delayed-type hypersensitivity reactions; and (5) T suppressor cells, which act to regulate or suppress the activities of the other types of T cells and B cells. The suppressive activities of T cells can be seen with newborn infants.

Existing hypotheses for antibody synthesis fall into two general categories: *template hypotheses* and *selective hypotheses*. These are summarized in the Chapter's Microbiology Highlights, *Mechanism of Antibody Formation*.

Conditions That Lower Host Resistance to Disease Agents

Various factors can lower host resistance to infectious diseases, a representative list of which is found in Table 17-8.

TABLE 17-8 Conditions That Lower Host Resistance to Infectious Disease Agents

Condition	*Selected Effect(s)*
Acute radiation injury	Alteration of the cellular defenses of the host
Age	Decreased efficiency of antibody synthesis and cell-mediated immunity at extremes of age; decreased levels of certain complement components during first 3 months of gestation
Agranulocytosis	Reduction or absence of phagocytosis by neutrophils
Alcoholism	Nutritional deficiencies; possible depression of the inflammatory response to bacterial infection
Altered lysosomes	Extensive or limited inability of macrophages and neutrophils to destroy ingested microorganisms
Atmospheric pollutants	Depressed immunological function of polymorphonuclear leukocytes
Circulatory disturbances	Localized destruction of tissues; congestion; accumulation of fluid in tissue
Complement deficiencies or defects	Limited or extensive inability to inactivate and destroy certain infectious disease agents

TABLE 17-8 Conditions That Lower Host Resistance to Infectious Disease Agents *(continued)*

Condition	*Selected Effect(s)*
Excessive or indiscriminate use of antibiotics	Elimination of natural flora that provide protection; overgrowth of resistant microbial forms; interferes with digestive process and vitamin utilization
Immunological deficiency	Interference with immunoglobulin production and/or cell-mediated immunity
Immunosuppression	Impairment of cell-mediated immunity mechanisms
Mechanical obstruction of body drainage systems (urinary, tear, and respiratory mechanisms or systems)	Interference with the mobilization and functioning of phagocytic cells
Nutritional deficiencies	Interference with and/or changes in several immune mechanisms, including antibody production; phagocytic activity; and integrity of mucous membranes and skin
Traumatic injury	Direct access to body tissues for opportunists and pathogens; possible interference with immunity mechanisms; possible obstruction of body drainage systems

V. CHAPTER SELF-TEST

Continue with this section only after you have read the appropriate chapters and have completed section IV. A score of 80 percent or better is good. If your score is less than 65 percent, reread the chapter.

Correct answers to all questions can be found at the end of this section. Write your responses in the appropriate space provided.

Completion Questions

Provide the correct term or phrase for the following. Spelling counts.

1. The greater the incompatibility between an immunogen and a recipient's tissues, the ________________ the immune response.
2. The clumping of red blood cells by isoantibodies is called ________________ .
3. Blood typing procedures are based on ________________ reactions.
4. Chemically, immunoglobulins are ________________ .
5. Known immunoglobulins are grouped in ________________ general classes.
6. The largest immunoglobulin is ________________ .
7. The immunoglobulin capable of passing through the placenta is ________________ .
8. Immunogens that stimulate the production of antibodies effective against material unrelated to the original antigens are called ________________ antigens.
9. Antibodies are members of the protein group known as ________________ .
10. Two of the four polypeptide chains in a normal immunoglobulin are identical (a) ________________ chains, and two are identicial (b) ________________ chains.

11. Antigen-binding sites are located on the immunoglobulin subregion known as the ______________ domain.
12. The sudden secondary rise in immunoglobulin titer after a second exposure to an immunogen is called a(an) ______________ response.
13. Immunoglobulin production is specifically associated with ______________ cells.

Matching

Select the answer from the right-hand side that corresponds to the term or phrase on the left-hand side of the question sheet. An answer may be used more than once. In some cases, more than one answer may be required.

Topic: Antibacterial Substances in the Body

______ 14. Provides protection against total body irradiation
______ 15. Imparts resistance to tissues against tubercle bacilli
______ 16. Involved with red blood cell destruction, and other tissue cells
______ 17. Bactericidal against *Bacillus anthracis*
______ 18. Bactericidal agent found in tears
______ 19. Antiviral protein

a. complement
b. properdin
c. spermine
d. interferon
e. phagocytin
f. lysozyme
g. none of these

Topic: Properties of Immunoglobulins

______ 20. The largest immunoglobulin
______ 21. Easily transferred through the placenta
______ 22. Major antibody associated with allergic reactions
______ 23. Initiates complement reactions
______ 24. Usually the first antibody formed after immunization
______ 25. Associated with the respiratory tract

a. IgG
b. IgM
c. IgA
d. IgE
e. none of these

True or False

Indicate all true statements with a "T" and all false statements with an "F".

______ 26. Nucleoproteins can be antigenic.
______ 27. IgD are generally found in colostrum.
______ 28. In animals from which the thymus is removed, foreign tissue grafts are readily destroyed.
______ 29. The immune system of a newborn human begins *in utero*.

_______ 30. The tuberculin skin response is an example of a cell-mediated immune reaction.

_______ 31. In the direct template hypothesis, immunogens act directly as patterns for the formation of the active sites of immunoglobulins.

_______ 32. All normal immunoglobulins have the same basic molecular arrangement.

_______ 33. Each class of immunoglobulin has the same type of heavy chain structure.

_______ 34. Variations in immunoglobulin structure are divided into the two general categories, allotype and isotype.

_______ 35. T lymphocytes are associated with cell-mediated immunity.

_______ 36. Natural killer B cells, macrophages, plasma cells, and memory cells may act together in antibody production.

_______ 37. The acidity of the vaginal environment encourages the growth of several infectious agents.

_______ 38. Amphibionts frequently take advantage of a host's weakened defenses to cause disease.

_______ 39. The amniotic fluid in which the fetus is bathed contains its own indigenous flora.

_______ 40. Amphibionts act as a constant source of immunogens to the antibody-producing systems of the body.

_______ 41. Amphibionts cause vitamin deficiencies in their hosts.

_______ 42. The fluid formed after blood clots in a tube is called *plasma.*

_______ 43. Pollens will provoke immunoglobulin production.

_______ 44. A neutrophil is an agranulocyte.

_______ 45. The lymphatic system is involved with the removal of materials that accumulate at sites of tissue injury.

_______ 46. Phagocytosis is the ingestion and subsquent digestion of particles by circulating granulocytes and fixed macrophages.

_______ 47. Organisms such as *Mycobacterium tuberculosis* and *Neisseria gonorrhoeae* tend to escape intracellular digestion by phagocytes.

_______ 48. Granules within phagocytes contain phagocytin as well as degradative enzymes.

_______ 49. Toxin-producing staphylococci are not killed by phagocytosis.

_______ 50. Pus formation may be associated with inflammation.

_______ 51. The symptoms of inflammation are thought to be caused by substances released from injured cells.

_______ 52. Carbuncles are isolated abscesses.

_______ 53. Body temperatures of 41° to 43°C favor immunoglobulin formation.

_______ 54. Complement levels rise in response to immunization.

_______ 55. Atmospheric pollutants can depress the immunological functioning of polymorphonuclear leukocytes.

_______ 56. Indiscriminate use of antibiotics generally does not affect the normal flora.

Essay Questions

Answer the following questions on a separate sheet of paper.

57. Why is the skin considered to be an excellent mechanical barrier to microorganisms?

58. What is inflammation?

59. Identify and describe the general properties of the most plentiful class of immunoglobulins found in the normal adult human.

60. What is the anamnestic response?

61. What are the functional T cell subsets or subpopulations?

Answers

1. greater
2. hemagglutination
3. hemagglutination
4. protein
5. five
6. IgM
7. IgG
8. heterophile
9. globulins
10a. heavy (H) chains
10b. light (L) chains
11. F_{ab}
12. anamnestic
13. plasma
14. b
15. c
16. a
17. g
18. f
19. d
20. b
21. a
22. d
23. a
24. b
25. c
26. T
27. F
28. F
29. T
30. T
31. T
32. T
33. F
34. F
35. T
36. F
37. F
38. F
39. F
40. T
41. F
42. F
43. T
44. F
45. T
46. T
47. T
48. T
49. T
50. T
51. T
52. F
53. F
54. F
55. T
56. F

57. Unbroken skin serves as an excellent mechanical barrier that most microorganisms cannot penetrate. In addition, certain secretions formed by skin-associated glands provide protection against various bacteria and fungi. For example, secretions from the sebaceous glands contain both saturated and unsaturated fatty acids that are both bactericidal and fungicidal.

58. Inflammation is the body's second line of defense against infection. Inflammation can be produced by infectious disease agents and by irritants such as chemicals, heat, and mechanical injury. The characteristic or *cardinal signs* of inflammation are heat, pain, redness, swelling, and loss of normal function.

59. The most plentiful class of immunoglobulin found in the normal adult human is IgG. It is the only immunoglobulin class that can cross the human placenta from the maternal circulation to that of the fetus. These maternal IgG molecules are responsible for the protection of the newborn during the first months of life.

    Other activities of the IgG class include neutralizing bacterial toxins, combining with (fixing) serum complement (a complex protein in serum), and attaching to macrophages, thereby arming these cells to function effectively in phagocytosis.

60. Some immunogens produce a sudden secondary rise in antibody titer when injected again some time after the first exposure. This effect is frequently called a *specific anamnestic response* from the Greek term *anamnesis,* meaning "recall." The antibody titers associated with a specific anamnestic response generally are higher than those produced by the primary reaction, occur with little or no lag period, and remain for long periods.

    Anamnestic reactions are produced by reimmunization with vaccines. The effectiveness of "booster shots" can be explained on this basis.

61. Five functional T cell subsets or subpopulations are recognized and named according to their associated activities. Their designations and functions are as follows: (1) T helper (T_H) cells, which work with B cells in immunoglobulin production; (2) natural killer (NK) cells; (3) cytotoxic T lymphocytes, which disrupt target cells by direct cell-to-cell contact; (4) T delayed-type hypersensitivity (T_{DTH}) cells, which bring about (mediate) inflammation and activate macrophages in delayed-type hypersensitivity reactions; and (5) T suppressor cells, which act to regulate or suppress the activities of the other types of T cells and B cells. The suppressive activities of T cells can be seen with newborn infants.

VI. ENRICHMENT

Immunosuppression

Immunoglobulin production can be suppressed in any of several ways—by the incorporation of chemical drugs, by the application of physical agents such as x-rays, or by the control of the antibody response by biological means. Biological suppression can result from inhibition of the immune response to one antigen by the introduction of a second antigen (antigen competition); immunoglobulin deficiency diseases in which there are pronounced losses in immunoglobulin levels as the result of deficits in B or T cells or a decrease in B and T cells functions; or a state of specific nonreactivity to a normally effective antigenic challenge created by a prior exposure to the antigen concerned. The latter condition is referred to as immunologic unresponsiveness, or *immunologic paralysis*.

Suppression of the immune response can be of great importance in situations such as in the survival of foreign tissue grafts (kidney and heart transplants) and in the control of autoimmune phenomena. Although application of immunosuppressive agents such as x-rays or certain drugs can serve a useful function, prolonged exposure to these agents can increase an individual's susceptibility to a variety of microbial pathogens. Moreover, serious infections with opportunistic microorganisms also occur among patients receiving intensive treatment with x-rays or chemical immunosuppressants.

18

States of Immunity and the Control of Infectious Diseases

I. INTRODUCTION

With the introduction of modern chemotherapy, immunization, and sanitation, it was assumed that many dreaded infectious diseases would be readily controlled, if not eliminated. This aim has been accomplished to a significant degree with various diseases, including polio, diphtheria, rubella (German measles), and smallpox, with smallpox being the first infectious disease to be eradicated through a vaccination program. This chapter examines the different states of resistance, the types of preparations used for immunization, and includes new developments such as synthetic vaccines.

II. PREPARATION

Chapters 17, 18, and 22 should be read before continuing. The following terms are important for you to know. Refer to the glossary and the appropriate chapters of the text if you are uncertain of any of them. A pronunciation guide for selected terms is provided as a learning aid.

1. acquired immunity
2. active immunity
3. adjuvant (AD-joo-vant)
4. AIDS
5. animal protection test
6. antitoxin
7. attenuation (ah-ten-u-AY-shun)
8. BCG
9. combined vaccines
10. complement
11. flocculation (flock-u-LAY-shun) test
12. genetic engineering
13. host
14. hypersensitivity
15. immune response
16. immunization
17. immunodeficiency
18. immunogen (IM-u-no-jen)
19. immunogenicity
20. immunoglobulin
21. innate
22. intravenous (in-trah-VEE-nus)

23. *in vivo* (in VEE-voh)
24. neutralizing antibody
25. opsonin (op-SOH-nin)
26. passive immunization
27. PPD
28. serum
29. skin challenging test
30. specific immune globulin
31. synthetic vaccine
32. toxin (TOK-sin)
33. toxoid (TOK-soid)
34. vaccine (VAK-seen)

III. PRETEST

Correct answers to all questions can be found at the end of this section. Write your responses in the appropriate space provided.

Matching

Select the answer from the right-hand side that corresponds to the term or phrase on the left-hand side of the question sheet. An answer may be used more than once. In some cases, more than one answer may be required.

______ 1. Tears
______ 2. Phagocytosis
______ 3. An injection of a toxoid
______ 4. Recovery from influenza
______ 5. Colostrum
______ 6. Injection of an inactivated virus preparation
______ 7. Placenta transmission
______ 8. An injection of DPT vaccine

a. artificially acquired active immunity
b. naturally acquired passive immunity
c. naturally acquired active immunity
d. native immunity
e. none of these

Topic: Immunization Materials

______ 9. Diphtheria
______ 10. Poliomyelitis
______ 11. Influenza
______ 12. Tetanus
______ 13. Rocky Mountain spotted fever
______ 14. Plague
______ 15. Yellow fever
______ 16. Tuberculosis
______ 17. Hepatitis B

a. attenuated pathogens
b. toxin
c. toxoid
d. heat-killed pathogens
e. formalin-killed pathogens
f. genetically engineered product
g. none of these

True or False

Indicate all true statements with a "T" and all false statements with an "F".

_______ 18. An attenuated vaccine contains living organisms.

_______ 19. The use of heat to kill organisms for vaccine reduces the immunizing potency of the preparation.

_______ 20. Rickettsial vaccines are generally heat-killed preparations of specific pathogens.

_______ 21. Heat- or formalin-inactivated toxins are called toxoids.

_______ 22. Almost all effective viral vaccine preparations used today contain attenuated pathogens.

_______ 23. Heat-killed vaccines produce a higher level of immunity than those obtained with attenuated preparations.

_______ 24. Combined vaccine preparations are less expensive to produce than are single-vaccine preparations.

_______ 25. BCG vaccination programs are of value for populations with a high frequency of influenza.

_______ 26. Passive immunization is permanent in its effect.

_______ 27. Antitoxins actually are immunoglobulins against specific toxins or toxoids.

_______ 28. Antitoxins are used for passive immunization.

_______ 29. Immune serum globulin preparations primarily consist of IgA.

_______ 30. Specific immune globulins, which are used more frequently, include those for mumps and the Rh_0 blood factor.

_______ 31. Measles was the first major disease to be eradicated through immunization.

_______ 32. The dosage of immunogens used for immunization does not affect the extent of antibody response.

_______ 33. The reaction to the first injection of an immunogen is called an anamnestic response.

_______ 34. Antibody response is better in animals with a thymus gland.

_______ 35. The immune system of the newborn begins to mature about one year after birth.

Answers

1. d
2. d
3. a
4. c
5. b
6. a
7. b
8. a
9. c
10. *a* and *d*
11. e
12. e
13. e
14. e
15. a
16. a
17. *e* and *f*
18. T
19. T
20. F
21. T
22. T
23. F
24. T
25. F
26. F
27. T
28. T
29. F
30. T
31. F
32. F
33. F
34. T
35. F

IV. CONCEPTS AND TERMINOLOGY

Immune Responses

Active immunization not only may result in the production of immunoglobulins directed against a disease agent or its toxic product, but it may also trigger cellular responses controlled by lymphocytes and macrophages. Both protective and nonprotective immunoglobulins are forms. The protective ones include *antitoxins,* which inactivate soluble toxic bacterial products; *opsonins,* which assist in the phagocytosis and intracellular digestion of bacteria; *lysins,* which interact with complement components to damage bacterial membranes; and *neutralizing antibodies,* which interfere with and prevent the replication of certain viruses.

Short-term protection or changing the outcome of certain infections may be accomplished through *passive immunization,* the administration of preformed immunoglobulins or immunoreactive cells. Although passive immunization is temporary, it is especially useful in the treatment of botulism, diphtheria, rubella, infectious hepatitis, and tetanus.

States of Immunity

The resistance to disease varies considerably among individuals because it is greatly affected by many innate or acquired factors (see Figure 18-1). In general, two major categories are recognized: *innate or native immunities* and *acquired immunities.* Innate immunity include species, race, and individual resistance to infection. Acquired immunity may be either *natural* or *artificial,* depending on the processes involved in producing the immunity. Immunization by the injection of a bacterial vaccine is an artificially produced contact with the organism, in contrast to an immunity produced by a natural infection.

Both categories are subdivided further into active and passive types. In the *active state,* an individual makes antibody in response to immunogenic stimulus; in the *passive state,* the antibody is acquired through transfer from an immunized individual (an outside source). The acquired states of immunity are described in more detail in Figure 18-1.

Adoptive Immunity

Adoptive immunity is a form of artificially acquired passive immune state; it is produced by transferring, from an immunized donor to a nonimmune recipient, cells capable of synthesizing immunoglobulins or of directly reacting with a specific antigen.

Immunization Preparations

The biological preparations used for immunization or diagnosis can be divided into three basic categories: (1) prophylactic agents for active immunization, which include bacterial and viral vaccines, toxins, and toxoids; (2) prophylactic preparations for passive immunization, mainly globulins (gamma globulins); and (3) diagnostic reagents designed to demonstrate hyperimmune states. Preparations used for immunization should be safe, free from unpleasant side effects, and simple to adminster.

Vaccines are preparations of either disease-causing microorganisms or certain of their component parts or products, such as toxins, that have been rendered unable to produce disease but are still antigenic. They are used to produce an artificially acquired active immunity. Vaccine-conferred immunity should provide as much or more resistance than would follow an actual infection.

Bacterial Vaccines

Bacterial vaccines are generally suspensions of killed bacteria in an isotonic (physiologic) salt solution. (One exception is BCG—Bacillus of Calmette and Guérin—vaccine, an attenuated, or weakened, preparation for immunization against tuberculosis). Administering such vaccines is the most common means for inducing active immunity. In the preparation of these materials, the organisms are cultured, harvested after a suitable incubation period, and then killed with heat or chemical agents such as acetone, formalin, merthiolate, phenol, and tricresol.

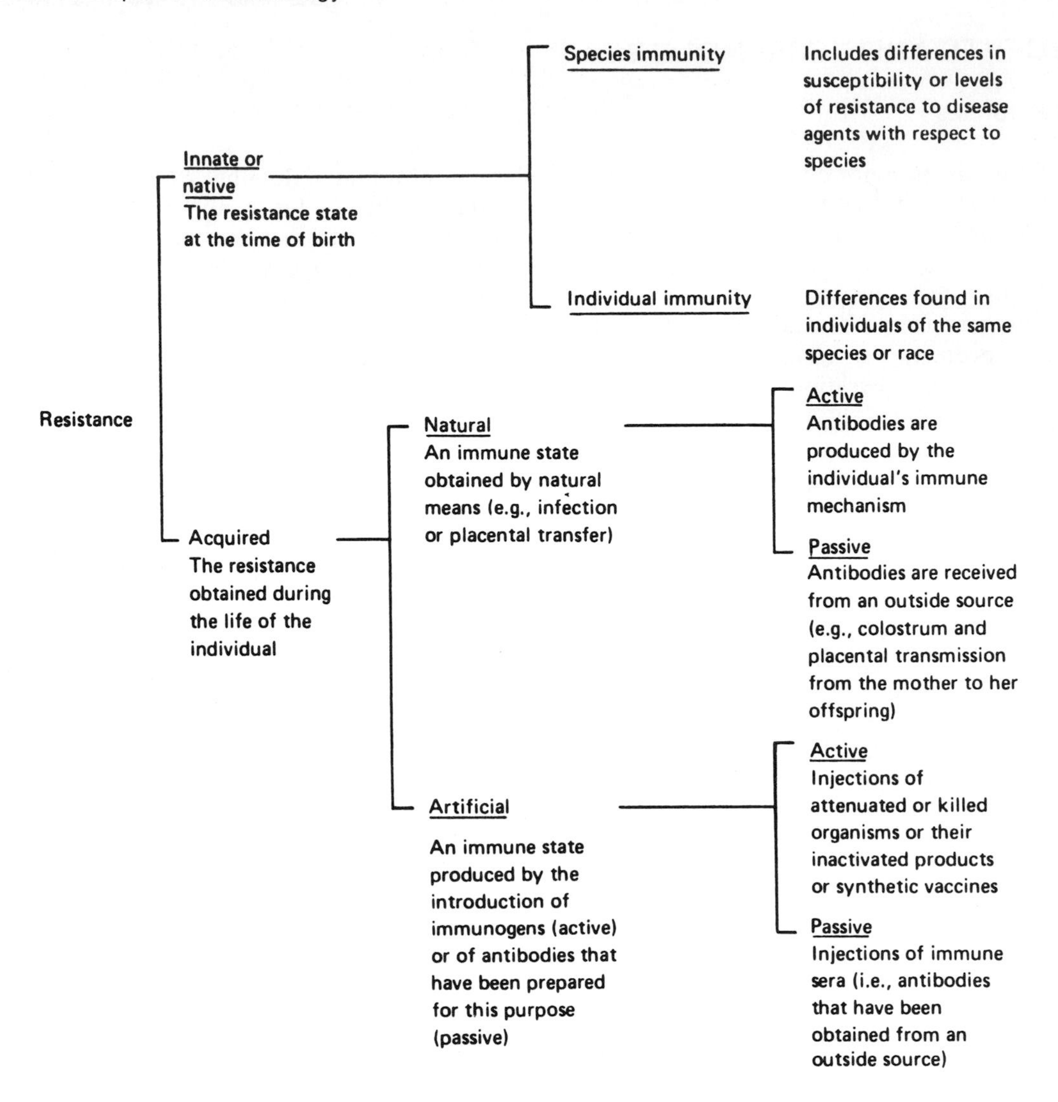

Figure 18-1 The different states of immunity.

Several preparations consisting of different parts of microorganisms are being tested for their immunizing properties. Examples include the pili of *Neisseria gonorrhoeae* and the ribosomes of *Escherichia coli* and *Vibrio cholerae.*

The toxins of several bacteria can be converted into nontoxic antigenic preparations called *toxoids*. Heat or formalin is used in the production of toxoids. In most commercial preparations, 0.2 to 0.4 percent formalin is added to a bacterial toxin such as diphtheria or tetanus. The resulting mixture is incubated until detoxification is complete. Inert protein is then removed. The resulting preparation is called natural fluid or plain toxoid. Another type of toxoid is prepared through the incorporation of certain aluminum compounds, such as alum, aluminum hydroxide, or aluminum phosphate. The resulting toxoid preparation is aluminum hydroxide or aluminum phosphate adsorbed toxoid.

Viral Vaccines

Several viral vaccines are composed of viruses whose virulence has been greatly reduced or eliminated. Such organisms are referred to as being *attenuated,* or weakened.

The virulence of viruses and other microorganisms can be reduced by several methods, including (1) cultivation at temperatures above normal for the organism; (2) desiccation; (3) the use of the unnatural host, for example, tissue culture or mice, for propagation of the microorganisms; and (4) continued and prolonged serial passages through laboratory animals.

Some viral vaccines consist of killed (inactivated) viruses. These include Salk polio vaccine, IPV (inactivated poliomyelitis vaccine), and influenza vaccine. Formalin is usually the chemical killing agent. Currently available vaccines and their means of preparation are summarized in Table 18-1.

TABLE 18-1 Preparations Currently Used for Active Immunizations (Vaccines)

Bacterial Disease	***Etiologic Agent(s)***	***Means of Vaccine Preparation***
Cholera	*Vibrio cholerae*	Heat-killed suspension of *V. cholerae* cells.
Diphtheria	*Corynebacterium diphtheriae*	Alum-precipitated toxoid preparation of *C. diphtheriae* toxin. This antigenic material is combined with tetanus toxoid and killed *B. pertussis* in the DPT vaccine.
Epidemic typhus	*Rickettsia prowazekii*	Formalin-killed suspension of *R. prowazekii* cultivated in chick embryo yolk sacs.
Meningitis	*Neisseria meningitidis*	Purified capsular polysaccharides.
Plague	*Yersinia pestis*	Formalin-killed and alum-coated suspension of *Y. pestis.*
Pneumococcal pneumonia	*Streptococcus pneumoniae*	Purified capsular polysaccharides.
Rocky Mountain spotted fever	*Rickettsia rickettsii*	Formalin-killed suspension of *R. rickettsii* cultivated in chick embryo yolk sacs or obtained from infected ticks.
Tetanus	*Clostridium tetani*	Alum-precipitated toxoid preparation of *C. tetani* toxin. See Diphtheria.
Tuberculosis	*Mycobacterium tuberculosis*	Bacillus of Calmette and Guérin (BCG). Prepared from a strain of *M. tuberculosis* var. *bovis* attenuated by continuous subculture on glycerol broth bile-potato media.
Tularemia	*Francisella tularensis*	Live attenuated strain of *F. tularensis.*
Typhoid and paratyphoid fevers	*Salmonella typhi, S. paratyphi, S. schottmuelleri*	Heat-killed, phenol-preserved, or acetone-dried preparations. Vaccines commonly contain *S. typhi* alone or in combination with *S. paratyphi* and *S. schottmuelleri* (TAB).
Whooping cough	*Bordetella pertussis*	Alum-precipitated or aluminum hydroxide- or aluminum phosphate-adsorbed, killed preparations of phase I *B. pertussis.* See Diphtheria.

Viral Disease	***Means of Vaccine Preparation***
Hepatitis B	Formalin-treated, purified viral (subunits) antigen. Genetically engineered yeast-derived and mammalian-derived preparations.
Influenza	Formalin-killed preparation of prevalent viral strains. The viruses are usually grown in embryonated chick eggs.
Measles (rubeola)[a,b]	Two attenuated vaccines are in use. One preparation employs the Edmonston strain in combination with measles-immune gamma globulin. The other vaccine utilizes the Schwartz strain of measles virus. A new chicken embryo preparation is being evaluated.
Mumps[b]	Formalin-killed preparation of virus obtained from chick embryo cultivation. An attenuated vaccine is currently in use. It is prepared by cultivation of the virus in embryonated chicken eggs and then in chick embryo cell cultures.
Poliomyelitis	Two vaccines are in use. The Salk vaccine is a formalin or ultraviolet irradiation-inactivated preparation of the virus. The Sabin, or oral, vaccine is an attenuated preparation. Both vaccines contain types 1, 2, and 3 polio viruses (trivalent).

TABLE 18-1 Preparations Currently Used for Active Immunizations (Vaccines) *(continued)*

Viral Disease	*Means of Vaccine Preparation*
Rabies (hydrophobia)	Several vaccines are available. These include the human diploid cell vaccine (HDCV), and duck embryo vaccine (DEV). The human diploid cell vaccine is prepared from virus grown in tissue culture and inactivated by either tri(*n*)-butyl phosphate or beta-propiolactone. The duck-embryo vaccine (DEV) is a beta-propiolactone-inactivated preparation of fixed virus obtained from duck embryo cultivation. Production of DEV has been discontinued but stocks may still be available.
Rubella (German measles)[b]	Three attenuated vaccines are in use: (1) $HPV_{77}D_5$, obtained from tissue cultures of duck embryo cells; (2) $HPV_{77}DK_{12}$, prepared from dog kidney cells; and (3) Cendehill, obtained from rabbit kidney cells.
Yellow fever	Attenuated strain of the virus (17D) cultivated in chick embryos.

[a] Inactivated preparations do not provide adequate protection.
[b] Combined vaccines. Measles, mumps, and rubella (MMR) virus vaccine is commonly used.

Live Versus Inactivated Preparations

The acceptability of a vaccine depends upon the general need for the preparation and the degree of protection it can provide. The general advantages and disadvantages or problems with live versus inactivated preparations are indicated in Table 18-2.

TABLE 18-2 Advantages and Disadvantages of Vaccines

Type of Vaccine	*Advantage(s)*	*Possible Disadvantage(s) and/or Potential Hazard(s)*
Live	Produces a high level of immunity usually in a single dose	(1) Infectious viruses will remain due to insufficient weakening; (2) the viruses will spread from the vaccinated individual to susceptible contacts; (3) contaminating viruses are present; (4) a genetic change will take place, resulting in a highly virulent viral strain; and (5) the vaccine will be inactivated by heat in facilities that lack adequate refrigeration.
Inactivated	Generally free of the hazards listed for live preparations	Requires large amounts of immunizing material to produce level of immunity obtainable with live preparations.

Vaccines Currently in Use: Combined Vaccines

The appearance of new vaccines makes the preparation of combined forms of immunizing vaccines desirable. Combined forms are simpler to administer, less expensive to produce, and more efficient than single-vaccine preparations. Examples of combined preparations include DPT (diphtheria-pertussis-tetanus); MMR (measles-mumps-rubella viruses vaccine); TOPV (trivalent oral poliomyelitis vaccine, which contains all three immunologically distinct strains of attenuated polio viruses); and TAB (typhoid and paratyphoid A and B).

Passive Immunization

Passive immunization is effective, acts immediately, and is temporary in its effect. The preparations used for passive immunization are of two basic types: antitoxins and antimicrobial sera (see Table 18-3). Many of these preparations are produced in laboratory or related animals.

TABLE 18-3 Preparations Used for Passive Immunization

Disease or Condition	*Product*	*Source of Product*
Black Widow spider bite[a]	Antivenin widow spider	Horse
Botulism	Antitoxin, types A, B, E	Horse
Diphtheria	Antitoxin	Horse
Hepatitis A	Immune globulin	Human
Hepatitis B	Immune globulin, and Hepatitis B immune globulin (HBIG)	Human
Measles	Immune globulin	Human
Pertussis	Pertussis immune globulin	Human
Rabies	Rabies immune globulin	Human
	Antirabies serum	Horse
Rh isoimmunization[a]	Rho (D) immune globulin	Human
Snake bite[a]	Antivenin: coral snake	Horse
	Antivenin: rattlesnake, copperhead, water moccasin	Horse
Tetanus	Tetanus immune globulin	Human
	Antitoxins	Horse or cow
Vaccinia (cowpox)[a]	Vaccinia immune globulin	Human
Varicella (chicken pox)[a]	Varicella-zoster immune globulin	Human

[a] Passive immunization also is applied to noninfectious disease conditions.

Antitoxins

Antitoxins are antibodies capable of neutralizing the toxin that stimulated their production. These protein substances constitute an immune serum. An antitoxin is specific in its action against its toxic counterpart, but it does not exert any effect on the microorganism that produced the toxin.

The preparation of antitoxins, such as those of diphtheria and tetanus, involves administration of toxin solutions to a horse or other suitable animal until sufficient antitoxin titers are produced. The animal is then bled and its serum or plasma processed to concentrate the antitoxin and to eliminate a large proportion of native horse serum protein.

Before antitoxin preparations are released, they are sterilized, usually by filtration, treated with a chemical preservative to maintain the sterility of the preparation, and standardized for dosage.

Immune Globulin (Gamma Globulin)

Immune globulin (IG), also referred to as gamma globulin, is derived from human blood, plasma, or serum and contains most of the antibodies found in whole blood. The concentrations of specific antibodies vary among different preparations. Immune serum globulin contains primarily IgG (γG) immunoglobulins.

The advantages of IG preparations include (1) the absence of serum hepatitis virus, (2) the presence of a large amount of antibodies in a small volume, and (3) stability during long-term storage. The value of human serum globulin has been demonstrated unequivocally in cases of measles and certain congenital immune-deficiency disease states.

Specific Immune Globulin (SIG)

Specific immune globulins (SIGs) have a higher concentration of specific antibody to the agent in question than immune globulin preparations. The specific immune globulins, which are used more frequently than gamma globulins, include those for measles, mumps, whooping cough, tetanus, shingles, and Rh_0 blood factor. These globulin preparations are made from sera obtained from individuals recently recovered from an active infection or from individuals who are hyperimmunized to a given material.

Vaccines Under Development

DNA technology is being applied to the synthesis of safe, highly specific, and inexpensive vaccines.

Administering Vaccines

Aseptic precautions must be taken during any type of inoculation procedure. All nondisposable equipment must be properly cleaned and sterilized, and the inoculator's hands must be washed and dried. The containers for the immunizing preparations must be disinfected. The cap or any other part of the container to be used should be wiped with 70 percent alcohol or other appropriate disinfectant.

For antibody production to occur, the immunogenic material must be introduced beneath the epithelial tissues. Among the possible injection routes for humans are the intradermal (intracutaneous), intramuscular, and subcutaneous injection routes. Oral administration of preparations such as the Sabin (live) poliomyelitis preparation is another functional method.

Complications Associated with Vaccinations

Undesirable, though apparently inevitable, side reactions to immunization procedures have been reported ever since Jenner introduced vaccination in 1796. Before any vaccine is administered, the patient's history of previous vaccination reactions should be obtained. This information is especially important where immunizations involve vaccines produced in eggs. Types of unfavorable reactions that can occur include anaphylaxis, inflammation of the central nervous system, fetal injury, high fever, infection, and toxic reactions.

Immunization for International Travel

Several infectious diseases are considered quarantinable under the World Health Organization's International Sanitary Regulations. These regulations require that individuals traveling to, through, or from certain specified regions must be immunized against specific diseases. Verification of vaccination is required by some countries at the time of entry and in certain situations may be needed for reentry.

V. CHAPTER SELF-TEST

Continue with this section only after you have read Chapters 17, 18, and 22. A score of 80 percent or better is good. If your score is less than 65 percent, reread the chapter.

Correct answers to all questions can be found at the end of this section. Write your responses in the appropriate space provided.

Completion Questions

Provide the correct term or phrase for the following. Spelling counts.

1. Immunizing preparations containing organisms freshly isolated from a patient are called ______________ .
2. Two specific properties of an attenuated vaccine are (a) ________________ and (b) ________________ .

3. Two bacterial diseases for which toxoids exist for immunization are (a) ______________________________ and (b) ______________________________.
4. Immunizing preparations used against virus diseases contain either (a) ______________________________ or (b) ______________________________ viruses.
5. The letters BCG stand for the ______________________________.
6. The production of immunoglobulins by an individual in response to an immunogenic stimulus represents a(an) ______________________________ state of immunity.
7. The transfer of antibodies from an immunized donor to a nonimmune recipient is an example of ________ immunity.

Matching

Select the answer from the right-hand side that corresponds to the term or phrase on the left-hand side of the question sheet. An answer may be used more than once. In some cases, more than one answer may be required.

Topic: Immunizations

______ 8. Rabies
______ 9. Tuberculosis
______ 10. Influenza
______ 11. Plague
______ 12. Meningococcal meningitis
______ 13. Hepatitis B
______ 14. Epidemic typhus
______ 15. German measles (rubella)
______ 16. Diphtheria
______ 17. Cholera

a. attenuated organisms
b. heat-killed organisms
c. formalin-killed organisms
d. toxoid
e. genetically engineered product
f. none of these

Topic: Immune States

______ 18. An injection of tetanus toxoid
______ 19. Intact skin
______ 20. An injection of immunoglobulins
______ 21. Recovery from syphilis
______ 22. Recovery from typhoid fever
______ 23. Injection of killed influenza virus vaccine

a. native immunity
b. artificially acquired passive immunity
c. artificially acquired active immunity
d. naturally acquired passive immunity
e. none of these

True or False

Indicate all true statements with a “T” and all false statements with an “F”.

_______ 24. Vaccines prepared with heat need not be tested for safety before they are released for human use.

_______ 25. DPT is the designation for a vaccine that includes diphtheria, plague, and tetanus toxoids.

_______ 26. Antitoxins are basically immunoglobulins.

_______ 27. Standardization of antitoxin preparations for human use is not necessary as long as the preparation will neutralize the toxin for which it is intended.

_______ 28. Immunization is the only means available to control certain communicable diseases.

_______ 29. Most bacterial vaccines are prepared by killing the microorganisms with heat or chemical agents.

_______ 30. The use of heat in the preparation of a vaccine interferes with its immunizing potency.

_______ 31. Inactivated bacterial suspensions are referred to as toxoids.

_______ 32. Immunizations necessary for foreign travel are generally determined by the countries or regions to be visited.

_______ 33. Specific immune globulin preparations contain a higher specific antibody concentration to the agent in question than that found in immune serum globulin preparations.

_______ 34. Multiple injections of killed vaccines are needed to produce levels of immunity comparable to those obtained with attenuated preparations.

Essay Questions

Answer the following questions on a separate sheet of paper.

35. Distinguish between active and passive forms of immunization.

36. What are the categories of biological preparations used for immunization or diagnosis?

37. Distinguish between a toxin and a toxoid.

Answers

1. autogenous vaccines
2a. antigenicity or immunogenicity
2b. low toxicity
3a. diphtheria
3b. tetanus
4a. killed
4b. attenuated
5. Bacillus of Calmette and Guérin
6. active
7. passive
8. f
9. a
10. c
11. c
12. f
13. e
14. c
15. a
16. d
17. b
18. c
19. a
20. b
21. e
22. e
23. c
24. F
25. F
26. T
27. F
28. T
29. T
30. T
31. F
32. T
33. T
34. T

35. There are two basic forms of immunization—*passive* and *active*. Passive immunization is the process by which whole serum or a fraction of serum containing immunoglobulin is extracted from a human or animal known to be immune or hyperimmune to a given disease and injected into an exposed susceptible host. Active immunity is the process by which an individual’s own immune system is stimulated to produce

immunoglobulins. This is achieved through inoculation with the offending organism or a product thereof that has been treated in such a way as to induce clinical immunity without producing the signs and symptoms of a full-scale disease.

36. The biological preparations used for immunization or diagnosis can be divided into three basic categories: (a) prophylactic agents for active immunization, which include bacterial and viral vaccines, toxins, and toxoids; (b) prophylactic preparations for passive immunization, mainly globulins (gamma globulins); and (c) diagnostic reagents designed to demonstrate hyperimmune states or to detect susceptibility to disease agents. The last category includes purified protein derivative (PPD) and diluted diphtheria toxin.

37. The toxin produced by a particular bacterium such as *Clostridium botulinum* causes a poisoning or intoxication of the victim. The toxin of this bacterial species as well as others can be converted into a nontoxic but still immunogenic preparation called a *toxoid.* Heat and formalin are used in the production of toxoids.

VI. ENRICHMENT

DPT Vaccine Trials

The combined vaccine against diphtheria, whooping cough (pertussis), and tetanus (DPT) protects children against these diseases. Unfortunately, the pertussis portion of the preparation has created its own victims. A small portion of vaccinated children have been reported to have developed permanent neurological impairment.

The current combined vaccine incorporates whole, killed cells of *Bordetella pertussis,* the causative bacterial agent. Side effects in immunized individuals range from a few days of pain, swelling, and fever to, in rare instances, lasting brain damage. The need for a safer vaccine has been quite obvious for several years. With such a preparation, the danger of side effects would become practically nonexistent, and would reduce or eliminate the estimated 51 cases per year of brain damage in U. S. children receiving the current vaccine.

A new vaccine developed in Japan looks exceptionally promising. The preparation uses two partially purified proteins from *B. pertussis.* Field testing of the vaccine in both Japan and Sweden produced results indicating a greater level of safety. In two small studies conducted in the United States in 1986, the side effects observed with the new vaccine were significantly fewer than with the older preparation. Both the new and older versions of the vaccine boosted the formation of pertussis antibodies to approximately comparable levels. Additional trials are continuing.

19

Diagnostic Immunologic and Related Reactions

I. INTRODUCTION

Immunologic and related laboratory tests to evaluate the immune responses of individuals to microorganisms, worms, or substances that cause allergies are performed primarily to confirm the cause of a disease state. However, immunological testing also is carried out to determine vaccine effectiveness, test an individual's immunological responsiveness, and obtain information relative to disease processes. The selection of appropriate tests depends on the type of information needed, the stage of the disease process, accuracy required, the rapidity with which tests can be carried out, and cost factors. This chapter presents several of the most commonly used diagnostic procedures and their applications.

II. PREPARATION

Chapters 7-10, 13, and 17-20 should be read before continuing. The following terms are important for you to know. Refer to the glossary and the appropriate chapters of the text if you are uncertain of any of them. A pronunciation guide for selected terms is provided as a learning aid.

1. ABO system
2. agglutination (ah-gloo-teh-NAY-shun)
3. AIDS
4. amebic dysentery (ah-ME-bik DIS-en-ter-ee)
5. antibody
6. antigen (AN-tee-jen)
7. antiserum
8. blood typing
9. cold agglutination
10. complement (KOM-plee-ment)
11. complement fixation
12. Coombs (koomz) test
13. countercurrent immunoelectrophoresis
14. cytolysis (sye-TOL-ee-sis)
15. cytomegalovirus (sye-toh-meg-ah-loh-VYE-rus)
16. delayed hypersensitivity
17. diagnostic skin test
18. electrofusion (ee-lek-troh-FEW-shun)
19. ELISA
20. ferritin (FER-ee-tin)
21. fetus (FEE-tus)
22. hemagglutination (he-mah-gloo-teh-NAY-shun)
23. hemagglutination inhibition
24. hemagglutinin (he-mah-GLOO-teh-nin)
25. human immunodeficiency virus (HIV)
26. hybridoma (hi-breh-DOH-mah)

27. immediate hypersensitivity
28. immunodiffusion (im-mu-no-deh-FEW-shun)
29. immunoelectron microscopy
30. immunoelectrophoresis (im-mu-no-eh-lek-troh-foh-RES-sis)
31. immunofluorescence (im-mu-no-floo-RES-ens)
32. immunogen (IM-mu-no-jen)
33. immunoglobulin (im-mu-no-GLOB-u-lin)
34. immunoperoxidase (im-mu-no-peh-ROCK-see-dace)
35. *in utero* (in U-ter-oh)
36. *in vitro* (in VEE-troh)
37. *in vivo* (in VEE-voh)
38. latex agglutination
39. monoclonal (mon-oh-KLOH-nal) antibodies
40. plasmacytoma (plaz-mah-sye-TOH-mah)
41. precipitin (pre-SIP-ee-tin)
42. radial immunodiffusion
43. Rh factors
44. rocket electrophoresis
45. serum
46. single-use diagnostic system
47. Weil-Felix test
48. Western blot
49. wheal and flare

III. PRETEST

Correct answers to all questions can be found at the end of this section. Write your responses in the appropriate space provided.

Completion Questions

Provide the correct term or phrase for the following. Spelling counts.

1. The red blood cell agglutination reaction caused by viruses such as influenza and mumps is known as ______________________.
2. The Weil-Felix procedure uses antigens of the bacterium ______________________.
3. The specific strains of *Proteus* used in the Weil-Felix test for the diagnosis of various rickettsial diseases are (a) ______________________, (b) ______________________, and (c) ______________________.
4. The Widal test can be used in the diagnosis of ______________________.
5. The procedure in which viral hemagglutination is inhibited by specific antibodies against the virus in a reaction is called ______________________.
6. The components of the indicator system in the complement fixation procedure are (a) ______________ and (b) ______________________.
7. The process by which fluorescent dye markers are attached to serum proteins is called ______________.
8. The diagnostic skin test for lymphogranuloma venereum is the ______________________.
9. The material usually employed in the tuberculin skin test is ______________________.
10. An individual with blood type O can be expected to have antibodies against blood types (a) __________, (b) ______________________, and (c) ______________________.

11. Parents having the genotypes AO and AB can have offspring with blood types of (a) ________________ or (b) ______________________________.

12. The compatibility blood test involving the recipient's serum and the donor's cells is called the __________.

Matching

Select the answer from the right-hand side that corresponds to the term or phrase on the left-hand side of the question sheet. An answer may be used more than once. In some cases, more than one answer may be required.

Topic: Diagnostic Testing

______ 13. Positive complement fixation
______ 14. Blood typing
______ 15. Hemagglutination
______ 16. Hemagglutination inhibition
______ 17. Virus neutralization
______ 18. Positive immunodiffusion reaction
______ 19. Negative complement fixation test

a. clumping of red cells by homologous antibodies
b. glowing on exposure to ultraviolet light
c. cloudy red suspension
d. formation of precipitate lines within agar
e. absence of the destructive effects of viruses
f. none of these

True or False

Indicate all true statements with a "T" and all false statements with an "F".

______ 20. The sera of patients with the protozoan disease trypanosomiasis contain cold agglutinins.

______ 21. Viruses such as the agents of influenza and mumps can agglutinate red blood cells.

______ 22. The antibodies involved in the precipitin reaction are precipitinogens.

______ 23. In a positive precipitin test, a visible reaction develops in a region of optimal, immunologically equivalent proportions of antigens and antibodies.

______ 24. Delayed skin test reactions occur within 30 minutes after the introduction of the test material.

______ 25. The Rh′ antigen is generally the main factor in the hemolytic disease of the newborn.

______ 26. A single hybridoma cell generally produces two or more immunoglobulin classes.

______ 27. A positive result in the ELISA procedure is indicated by the formation of precipitin lines within the test agar.

Answers

1. hemagglutination
2. *Proteus*
3a. OX-2
3b. OX-19
3c. OX-K
4. typhoid fever
5. hemagglutination inhibition
6a. sheep red blood cells
6b. hemolysin
7. labeling or conjugation
8. Frei test
9. purified protein derivative (PPD)
10a. A
10b. B
10c. AB
11a. A
11b. B
12. major cross match
13. c
14. a
15. a
16. f
17. e
18. d
19. f
20. T
21. T
22. F
23. T
24. F
25. F
26. F
27. F

IV. CONCEPTS AND TERMINOLOGY

Determining the level of antibodies to a given antigen is valuable in: (1) establishing the identity of a disease agent, (2) charting a patient's recovery from infection, and (3) evaluating the effectiveness of immunization. Because of the specificity of the antigen-antibody reaction, if either the antigen or antibody is known, it is possible to identify and measure the other by using one of a variety of *in vitro* and *in vivo* techniques. Table 19-1 outlines the features of several immunologic procedures and their positive test results. The branch of immunology concerned with the nature and behavior of humoral antibodies is called *serology*.

Production of Antisera

To obtain potent antisera (blood serum that contains antibodies) for use in diagnostic tests, an experimental animal is inoculated with suspensions of a particular antigen. Animals used for this purpose include chickens, mice, horses, rabbits, sheep, and even humans. The immunogen may be introduced intraperitoneally, intravenously, or subcutaneously, usually in a series of inoculations. A blood sample is taken periodically from the laboratory animal to determine the antibody level (trial titrations). The antibodies produced are capable of binding with the antigenic determinant that caused their formation.

TABLE 19-1 Immunologic Procedures Used in Diagnosis and Microbial Identification

Procedure	*Principle Involved*	*Positive Test Results*	*Applications*
Agglutination	Antibody clumps cellular or particulate antigens (insoluble particles coated with antigens, e.g., latex particles, *Staphylococcus* A protein)	Aggregates (clumps) of antigens	Diagnosis of typhus, Rocky Mountain spotted fever (Weil-Felix test), and typhoid fever (Widal test); identification of disease agents such as *Haemophilus influenzae, Neisseria meningitidis,* and *Streptococcus pneumoniae*
Complement fixation	Antigen-antibody complex of test system binds complement, which is thereby unavailable for binding by sheep red blood cells and hemolysin of the indicator system.	Cloudy red suspension	Diagnosis of various bacterial, mycotic, protozoan, viral, and helminth (worm) diseases
Countercurrent immunoelectrophoresis (one form of electro-immunodiffusion)	Antigen and antibody are placed in separate wells and driven toward each other with an electric current (electrophoresis)	Precipitation forms at a point intermediate between the two wells	Antigen and/or antibody semi-quantitative detection
Enzyme-linked immunosorbent assay (ELISA)	Antigen or antibody from specimens trapped by corresponding, specific antibody or antigen coating a solid phase support combines with enzyme-labeled specific antibody. The formed complex reacts with an added enzyme substrate in proportion to the amount of antigen or antibody first bound by the coating antibody or antigen.	Color changes occurring with the addition of enzyme substrate are proportional to either antibody or antigen in specimens	Detection of IgM to rubella and influenza A; identification and detection of herpes simplex viruses types 1 and 2, cytomegalovirus, measles, and hepatitis B viruses
Ferritin-conjugated antibodies	Antibody, to which ferritin (iron-containing) particles are attached, binds various types of antigens.	Presence of localized dark spheres in electron micrographs	Location of bacterial, fungal, viral, and other biological antigens by electron microscopy

TABLE 19-1 Immunologic Procedures Used in Diagnosis and Microbial Identification ***(continued)***

Procedure	*Principle Involved*	*Positive Test Results*	*Applications*
Hemagglutination	Homologous antibody (hemagglutinin) aggregates of red blood cells.	Aggregates of red blood cells	Blood typing
Hemagglutination inhibition (viral)	Antibody inhibits the agglutination or red blood cells by coating hemagglutinating virus.	Formation of a circle of unagglutinated cells	Determination of the immune status toward German measles; identification
Immunodiffusion	Antibody and soluble antigen diffuse toward one another through an agar gel and react where homologous antibody is in proper proportion to homologous antigen.	Lines of precipitate form within the agar	Antigen and/or antibody identification
Immunofluorescent microscopy	Antibody (usually) or antigen is labeled with a fluorescent dye. which fluoresces on exposure to ultraviolet or blue light.	Glowing on exposure to ultraviolet light	Detection of antigen or antibody; identification of microbial pathogens of diseases such as rabies, syphilis, Legionnaire's disease, etc.
Immunoperoxidase	Antibody is labeled (conjugated) with an enzyme, usually horseradish peroxidase, which is detected by color reaction produced upon treatment with a peroxidase substrate.	Color changes occurring with the addition of a peroxidase substrate	Detection and identification of several viruses including cytomegaloviruses and rabies virus
Precipitation	Antibody and soluble antigen react where they are in proper proportion to one another.	Lines of precipitate form	Diagnosis of microbial diseases; detection of antigens
Radioimmunoassay	Antibody or antigen can be labeled with radioactive element, and the resulting complex precipitated and monitored for radioactivity.	Radioactivity counts	Detection of antigen and/or antibody; detection of hepatitis antigen
Rocket electrophoresis (electroimmunodiffusion)	Antibodies to a specific antigen or antigens is incorporated into a solid supporting agar medium. The specimen is added to a small well and allowed to migrate in an electrical field (electrophoresed)	Precipitation pattern forms a rocket or spike shape. The amount of antigen is proportional to the length of the rocket	Quantitation of most antigens
Virus neutralization	Antibody neutralizes infectivity.	Absence of virus destructive effects	Determination of the neutralizing effects of antibody; virus identification and diagnosis
Western blot assay	Proteins of antigen are separated by electrophoresis, transferred to and immobilized on nitrocellulose strips, and then exposed to serum specimens. Antigen-antibody reactions are detected by an added enzyme-linked anti-human immunoglobulin reagent.	Formation of a black precipitate in the regions where enzyme-immunoglobulin reagent is bound	1. Diagnosis of infectious diseases such as AIDS 2. Detection of antibody against different antigenic components

The Diagnostic Significance of Rising Antibody Titers

The titer, or concentration of antibody in serum, fluctuates as a consequence of immunizations and infectious states, both subclinical and full blown. At least two specimens of a patient's serum are necessary to distinguish the antibody production due to an actual ongoing infection from the effects of vaccination or from antibodies produced during a past infection. The first specimen is obtained soon after the onset of the disease and the other, approximately 12 to 14 days later. With the greater antibody activity in the later specimen, identification of the causative agent is possible. If little or no antibody is detected in either specimen, it can be assumed, barring any abnormalities, that an organism other than the one being tested for is the cause of the infection.

Diagnostic Procedures

The serological procedures used for diagnosis and microbial identification (see Table 19-1) are powerful tools not only in the diagnosis of disease states, but also in the identification of microorganisms. They are based on antibodies produced *in vivo* in response to the antigens of microorganisms and other cells. Some of these antigens are *type specific,* limited to a particular species, whereas others are *common group antigens,* antigenic to related groups of microorganisms.

Table 19-2 lists specific diagnostic tests and reagents (test materials) used to detect unknown antibodies or antigens.

Diagnostic Skin Tests

The injection of test antigens just under the skin can be used in diagnosis and to follow the progress of recovery from a disease (see Table 19-3). Diagnostic skin tests include the Frei test for the venereal disease lymphogranuloma venereum and the tuberculin, Mantoux, and patch tests for tuberculosis.

Positive responses to test antigens may be either *immediate* or *delayed.* An immediate reaction develops shortly after exposure to the test antigen and appears as an elevated, flat, pale, swollen area surrounded by a region of redness (wheal and flare reaction). The delayed reaction appears several hours after the introduction of the antigen. The area around the site of inoculation becomes reddened, firm, and swollen within 24 to 48 hours.

Skin tests can also be used to determine an individual's susceptibility to a microbial toxin. Small quantities of toxins are injected into the skin and the reaction observed. Positive responses, those showing susceptibility, are indicated by reddened areas appearing within 24 hours. The Schick test for diphtheria and the Dick test for scarlet fever are examples of this type of procedure.

Monoclonal Antibodies

Somatic cell hybridization or *hybridoma formation* results in the formation of a hybrid cell usually containing two nuclei in a cytoplasm that is the mixture of the two parental cells. Monoclonal antibodies can be obtained by first injecting a mouse with an antigen and then obtaining and chemically fusing the antibody-making cells of its spleen with a cancerous type of mouse cell known as *plasmacytoma.* The hybrid cell so formed produces the single type of antibody molecule of its spleen cell parent and continues to grow and divide as its plasmacytoma cell parent.

The hybridoma technique provides a means by which a constant and uniform source of antibody can be produced. (The immune system usually produces a mixture of antibodies with each type directed against a different feature of the antigen.) Monoclonal antibodies are chemically, physically, and immunologically uniform.

Immunohematology

The major blood types are referred to as the ABO system. In addition to having specific blood group antigens, some humans possess antibodies that react with the erythrocytes from individuals of other blood types, causing them either to agglutinate (isohemagglutination) or to lyse. The immunoglobulins are *isohemagglutinins*

TABLE 19-2 Diagnostic Tests and Reagents

Diagnostic Test	*Diagnostic Reagent(s)*
Agglutination	
Blood typing (hemagglutination)	Known antisera against A, B, and Rh_0(D) blood factors
Cold hemagglutination	Blood cells used at 2° C and 37° C
Viral hemagglutination	Human, chicken, or other suitable red blood cells
Weil-Felix test	Strains of the bacterium *Proteus* OX-2, OX-19, and OX-K
Widal test	Twenty-four-hour culture of *Salmonella typhi*
Passive agglutination (e.g., latex agglutination, bentonite flocculation)	Soluble antigens attached to surface of insoluble particles such as bentonite (a mineral colloid), polystyrene latex spheres, and red blood cells
Hemagglutination-inhibition test	Human, chicken, or other suitable red blood cells, and suspension of suspected virus
Precipitin	
Ring or interface test	Soluble extract of known (test) antigen
Oudin test (single diffusion)	Either known (test) antigen or antibody contained in an agar preparation
Radial immunodiffusion	Test antigen placed in wells of agar-containing antiserum (antibodies)
Two-dimensional immunoelectrophoresis	Specific antibodies against serum globulins
Ouchterlony test (double diffusion)	Either known antigen or antibody (one or the other)
Electroimmunodiffusion	
Counterimmuno-electrophoresis	Either known antigen or antibody
Rocket electrophoresis	Particular antigens, immunoglobulins, or complement components
Western blot assay	Enzyme-linked antihuman immunoglobulin (Ig) reagent, such as a horseradish peroxidase-conjugated antihuman immunoglobulin preparation; hydrogen peroxidase and 3-diamino-benzidine
Complement fixation	
Complement-fixation test	*Test system:* contains commercial antigen and complement *Indicator system:* contains sheep red blood cells and antibody against these cells (hemolysin)
Immunofluorescent techniques	
Fluorescent antibody technique	Specific antibody labeled with fluorescent dye
Immunoelectron microscopy	
Ferritin-conjugated antibody test	High-molecular-weight iron molecule (ferritin) coupled with antibody
Immunogold staining	Different sized gold particles linked to different antibodies
Enzyme reactions	
Immunoperoxidase test	Antibody labeled with an enzyme such as horseradish peroxidase
Enzyme-linked immuno-sorbent assay (ELISA)	Enzyme-labeled antibody and specific enzyme substrate
Virus neutralization	
In vivo virus neutralization test	Dilutions of suspected virus

TABLE 19-3 Representative Examples of Skin Diagnostic Tests

Disease State	*Types of Infective Disease Agent Involved (If Applicable)*	*Nature of Preparation Used*	*Type of Reaction*
Brucellosis	Bacterium	Brucellergin (extract of *Brucella* spp.)	Delayed
Leprosy	Bacterium	Lepromin (extract of lepromatous tissue)	Delayed
Lymphogranuloma venereum	Bacterium	Chorioallantoic membrane (extract from infected chick embryo)	Delayed
Psittacosis	Bacterium	Heat-killed organisms	Delayed
Tuberculosis	Bacterium	Purified Protein Derivative (PPD) or Old Tuberculin (OT)	Delayed
Blastomycosis	Fungus	Concentrated culture filtrate	Delayed
Coccidioidomycosis	Fungus	Coccidioidin (concentrated culture filtrate)	Delayed
Histoplasmosis	Fungus	Histoplasmin (concentrated culture filtrate)	Delayed
Leishmaniasis	Protozoan	Extract of cultured organisms	Delayed
Echinococcosis (sheep tapeworm)	Helminth	Hydatid fluid extract	Delayed
Trichinosis (pork tapeworm)	Helminth	Extract of the causative agent	Immediate
Schistosomiasis	Helminth	Extract of the causative agent	Immediate
Contact dermatitis	Simple chemical compounds	Small quantities of suspected chemicals	Delayed

Note: The information obtained from such tests also can be used in epidemiologic surveys (Chapter 21).

and *isohemolysins,* respectively. The lysis occurs as a consequence of antibody molecules (isohemolysins) sensitizing the blood cells, making them vulnerable to the hemolytic activity of complement.

Most individuals have antibodies against blood factors not present in their blood cells. For instance, a person of blood type O has antibodies against type A (anti-A) and type B (anti-B). Antibody concentrations can be increased as a result of transfusions, or, in women, by bearing children of a different blood type. Table 19-4 lists the recognized blood groups with their respective agglutinogens and agglutinins.

TABLE 19-4 Selected Characteristics of the Major Human Blood Groups

		Presence or Absence of Agglutinins Within Normal Sera	
International Designation	*Agglutinogen Associated with Cells*	*Anti-A or α (alpha)*	*Anti-B or β (beta)*
A	A	—	+
B	B	+	—
AB	AB	—	—
O	O	+	+

The Universal Donor and Recipient

Persons with an AB blood type do not have agglutinins against the A or B factors. Consequently, such individuals can receive blood from donors belonging to any of the four major blood groups. They are therefore often called *universal recipients.*

Individuals with O type blood are commonly referred to as *universal donors.* However, this designation is misleading. Transfusions of O type blood cannot be safely performed in all cases.

The ABO Subgroups

Antigenic variations of the ABO system have been identified. The blood group A has been divided into the major subdivisions A_1 and A_2. The AB group is subdivided into A_1B and A_2B. Subgroups of the B group include B_V, B_3, B_k, B_w, and B_x. A_1 and A_2 are the most frequently encountered subgroups.

Inheritance of the ABO Blood Groups

The ABO blood factors constitute one example of genetic characteristics determined by a multiple allelic series. An *allele* is a gene belonging to a group of alternate genes that occur on a specific region of a chromosome. Inheritance of blood type follows Mendelian principles. An individual's blood type is determined by receiving from each parent one of the four allelic genes: A_1, A_2, B_1, and O.

Table 19-5 shows the possible genotypes (assortments of genes) and phenotypes (demonstrable expressions of the individual's genotype) from each possible parental mating. A and B alleles are dominant over the O gene. Generally, the O blood type appears only in persons lacking both A and B genes, that is, a homozygous O (OO) state must be present. In AB individuals, neither allele is dominant; a state of codominance exists.

TABLE 19-5 Genotypes and Phenotypes Resulting from Selected Parental Matings (Checkerboard System)

Parental		*Offspring Possible*	
Genotypes	*Corresponding Phenotypes*	*Genotypes*	*Corresponding Phenotypes*
AA x AO	A x A	AA, AO	A
AO x AO	A x A	AA, AO, OO	A or O
BB x BO	B x B	BB, BO	B
BO x BO	B x B	BB, BO, OO	B or O
AA x BB	A x B	AB	AB
AO x BO	A x B	AO, BO, AB, OO	A, B, AB, or O
AB x AO	AB x A	AA, AO, BO	A or B
AB x BO	AB x B	AO, AB, BB, BO	A, AB, or B
AB x OO	AB x O	AO, BO	A or B

The genetic composition and the observed blood type of offspring can be determined by the *Punnett square technique,* commonly called the checkerboard system. The possible types of genes of one parent appear across the top of the checkerboard square, and the genetic contributions of the other parent along the side. It is important to remember that each parent contributes one gene for each characteristic. With the parental blood types shown, AO and AB, the possibilities for the offspring's blood type are AA, AB, AO, and BO.

Parent 1 (genotype AB)

Parent 2 (genotype AO)	A	B
A	AA	AB
O	AO	BO

The blood group substances (BGS) A and B also are found in various body fluids and tissue cells, including the kidney, liver, lung, muscles, amniotic fluid, gastric juices, ovarian cyst fluid, perspiration, saliva, semen, tears, and urine.

The Rh-Hr System

The Rh-Hr blood factors are of great clinical significance, second only to the ABO system. Controversies have arisen over the nomenclature for the Rh agglutinogens. Two principal methods are currently used: the Wiener scheme (the original Rh-Hr nomenclature designations) and the Fisher-Race system (combination of the letters C, D, and E). Table 19-6 shows the comparison between these two systems of notation.

Rh Typing

The Rh factors of red cells are detected by agglutination tests using the appropriate antisera, anti-Rh_0, -rh, and -rh″. The classic and clinically most important of the factors is Rh_0(D).

TABLE 19-6 A Comparison of the Designations for the Rh Blood Factors (Wiener) and Rh Agglutinogens (Fisher-Race)

Wiener System	*Fisher-Race System*
Rh_0	D
rh′	C
rh″	E
hr′	c
hr″	e
hr	d

Note: Designations for the more recently discovered Rh factors are symbolized by various combinations of subscript or superscript letters and numerals (e.g., D_u, C_w).

Hemolytic Disease of the Newborn (The Rh Baby)

Hemolytic disease of the newborn, erythroblastosis fetalis, is the blood incompatibility normally encountered when the mother and child have different Rh factors. Usually, this condition develops in babies born to mothers who are negative and fathers who are positive for the Rh_0 factor. The RH_0(D) antigen generally is the causative factor in the hemolytic disease of the newborn. However, all of the other blood antigens are also capable of causing the condition.

Prevention of Rh Isoimmunization

The major threat of isoimmunization occurs at the end of the third trimester of pregnancy and immediately after childbirth. To prevent this from occurring, an immunoglobulin G preparation containing high titers of anti-D (anti-Rh_0) antibody is administered to the mother during the postpartum period to suppress the antigenic stimulus provided by fetal cells. Unsensitized Rh-negative mothers receive this material within 72 hours after a delivery, abortion, or miscarriage if a danger of isoimmunization exists.

Blood Typing and Cross Matching

Two diagnostic procedures, the tube and slide tests, are commonly used to determine blood types. The determination can be made either by testing red cells with standardized anti-A and anti-B sera or by testing the patient's serum with standard, sensitive, known A and B red cells. For reliable results, both test systems should be employed; however, most often, only the first combination is used. Table 19-7 shows the patterns that can occur in blood type determinations.

Cross matching is performed to detect any incompatibility between a recipient's serum and a donor's cells (the major cross match) or between the recipient's cells and the donor's serum (the minor cross match). Cross matching is done after the blood specimens of both the recipient and the donor have been typed as to the ABO, Rh, and other relevant factors. Agglutination or lysis of the red cells in either the donor's or recipient's tests usually is considered to indicate incompatibility.

TABLE 19-7 Reaction Patterns in Blood Type Determinations

	Isoagglutinins	
Blood Type	*Anti-A*	*Anti-B*
A	+	—
B	—	+
AB	+	+
O	—	—

Coombs Tests

Two tests for the detection of incomplete antibodies, the direct and indirect Coombs tests, are instrumental in diagnosing cases of hemolytic anemia and hemolytic disease of the newborn. In the direct procedure, red cells are examined for the presence of antibody on cells (sensitization *in vivo*); in the indirect test, Rh-positive cells sensitized *in vitro* by the antibodies present in the patient's serum are used to detect the Rh antibodies.

V. CHAPTER SELF-TEST

Continue with this section only after you have read Chapters 3 and 17-20 and have completed the Concepts and Terminology section. A score of 80 percent or better is good. If your score is less than 65 percent, reread the chapter.

Correct answers to all questions can be found at the end of this section. Write your responses in the appropriate space provided.

Completion Questions

Provide the correct term or phrase for the following. Spelling counts.

1. Three types of abnormalities that could cause a lack of antibody response are (a) ____________________ ,
   (b) ______________________________ , and (c) ______________________________ .

2. The clumping of red blood cells by corresponding antibodies on a slide is an example of ________ .
3. The somatic antigens of the bacterial genera ________________ are used in diagnosis of various rickettsial infections.
4. The antigens used in the precipitin test are ________________ .
5. The ________________ test is used for the detection of incomplete antibodies.
6. Determination of antibody levels against the thermolabile hemolysin of streptococci, streptolysin O, is known as the ________________ test.
7. Electroimmunodiffusion combines (a) ________________ and (b) ________________ for the separation of proteins in body fluids.
8. Major cross matching involves mixing of the recipient's (a) ________________ and the donor's (b) ________________ .
9. Two tests used in the detection of AIDS virus infection are (a) ________________ and (b) ________________ .

Multiple Choice

Select the best possible answer.

______ 10. The antigen material used in agglutination type of reactions is:
(a) isohemagglutinins; (b) particulate; (c) soluble; (d) agglutinins; (e) none of these.

______ 11. The Weil-Felix test is used as a diagnostic tool in cases of infections caused by:
(a) rickettsia; (b) *Brucella* species; (c) trypanosomes; (d) mycoplasma; (e) all of these.

______ 12. The diagnosis of certain rickettsial diseases involves the use of the somatic antigens of:
(a) *Brucella;* (b) mycoplasma; (c) trypanosomes; (d) *Proteus;* (e) influenza virus.

______ 13. The antigens used in precipitin type reactions are referred to as being:
(a) soluble; (b) cellular; (c) particulate; (d) *a* and *c* only; (e) none of these.

______ 14. In the complement-fixation test, the heating of a patient's serum for 30 minutes at 56°C is to inactivate:
(a) hemolysin; (b) sheep red blood cells; (c) patient's antibody; (d) complement; (e) none of these.

______ 15. The components of a minor cross match include:
(a) donor's cells and serum; (b) donor's serum and recipient's cells; (c) donor's cells and recipient's serum; (d) recipient's cells; (e) recipient's cells and serum.

______ 16. Type IV hypersensitivity skin reactions can be used as a diagnostic aid for:
(a) histoplasmosis; (b) leprosy; (c) tuberculosis; (d) scarlet fever; (e) *a, b,* and *c* only.

______ 17. Which of the following is a component of the indicator system in the complement-fixation test?
(a) complement; (b) patient's serum; (c) hemolysin; (d) antigen preparation; (e) none of these.

______ 18. In a positive complement fixation reaction, complement is fixed by:
(a) the antigen in the test system; (b) the antigen-antibody complex in the test system; (c) hemolysin; (d) the sheep red blood cell-hemolysin complex; (e) none of these.

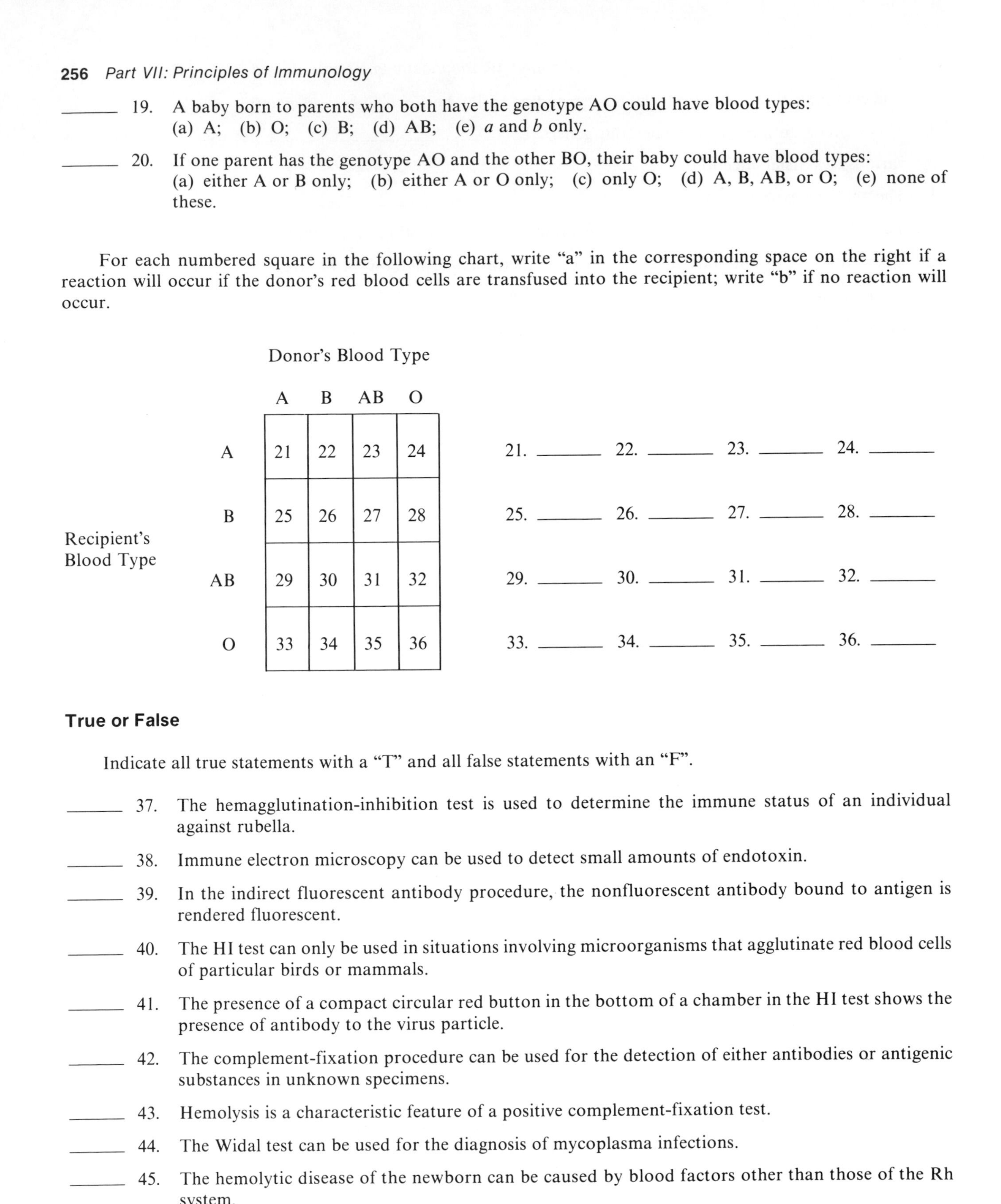

______ 19. A baby born to parents who both have the genotype AO could have blood types: (a) A; (b) O; (c) B; (d) AB; (e) *a* and *b* only.

______ 20. If one parent has the genotype AO and the other BO, their baby could have blood types: (a) either A or B only; (b) either A or O only; (c) only O; (d) A, B, AB, or O; (e) none of these.

For each numbered square in the following chart, write "a" in the corresponding space on the right if a reaction will occur if the donor's red blood cells are transfused into the recipient; write "b" if no reaction will occur.

Donor's Blood Type

Recipient's Blood Type	A	B	AB	O
A	21	22	23	24
B	25	26	27	28
AB	29	30	31	32
O	33	34	35	36

21. ______ 22. ______ 23. ______ 24. ______

25. ______ 26. ______ 27. ______ 28. ______

29. ______ 30. ______ 31. ______ 32. ______

33. ______ 34. ______ 35. ______ 36. ______

True or False

Indicate all true statements with a "T" and all false statements with an "F".

______ 37. The hemagglutination-inhibition test is used to determine the immune status of an individual against rubella.

______ 38. Immune electron microscopy can be used to detect small amounts of endotoxin.

______ 39. In the indirect fluorescent antibody procedure, the nonfluorescent antibody bound to antigen is rendered fluorescent.

______ 40. The HI test can only be used in situations involving microorganisms that agglutinate red blood cells of particular birds or mammals.

______ 41. The presence of a compact circular red button in the bottom of a chamber in the HI test shows the presence of antibody to the virus particle.

______ 42. The complement-fixation procedure can be used for the detection of either antibodies or antigenic substances in unknown specimens.

______ 43. Hemolysis is a characteristic feature of a positive complement-fixation test.

______ 44. The Widal test can be used for the diagnosis of mycoplasma infections.

______ 45. The hemolytic disease of the newborn can be caused by blood factors other than those of the Rh system.

______ 46. In a positive immunoperoxidase test, the peroxidase is joined to the antigen by the antibody in the conjugate.

_______ 47. The immunoperoxidase test results are identified by specific color changes.

_______ 48. Monoclonal antibody (single types) are chemically, physically, and immunologically uniform.

Essay Questions

Answer the following questions on a separate sheet of paper.

49. How can antibody production associated with an actual ongoing infection be distinguished from the effects of immunization or from immunoglobulins associated with a past infection?

50. What is the basis of cold hemagglutination reactions?

Answers

1a. immunosuppressive drugs	8b. cells	21. b	35. a
1b. excessive radiation exposure	9a. ELISA	22. a	36. b
1c. certain congenital defects	9b. Western blot test	23. a	37. T
2. hemagglutination	10. b	24. b	38. F
3. *Proteus*	11. a	25. a	39. T
4. soluble	12. d	26. b	40. T
5. Coombs	13. a	27. a	41. T
6. antistreptolysin	14. d	28. b	42. T
7a. gel diffusion	15. b	29. b	43. F
7b. electrophoresis	16. e	30. b	44. F
8a. serum	17. c	31. b	45. T
	18. b	32. b	46. T
	19. e	33. a	47. T
	20. d	34. a	48. T

49. To distinguish the antibody production associated with an actual ongoing infection from the effects of immunization or from antibodies associated with a past infection, at least two specimens of a patient's serum are necessary. The first is obtained soon after the onset of the disease and the other approximately 12 to 14 days later. The sera from both specimens are tested to determine if a rise in concentration of the suspected antibody has occurred. If the titer has risen, as indicated by a greater immunoglobulin activity in the later specimen, identification of the causative agent is possible. If little or no antibody is detected in either specimen, it can be assumed, barring any abnormalities, that an organism other than the one being tested for is the cause of the infection.

50. The sera of patients with atypical primary pneumonia of mycoplasmal origin and protozoan infections, such as trypanosomiasis, contain antibodies that are capable of agglutinating erythrocytes from these patients at 2° C, but not at 37° C. Such antibodies are called *cold agglutinins,* and the phenomenon is referred to as *cold hemagglutination.* These unusual antibodies are important to diagnosis in that they appear in association with only a few diseases.

VI. ENRICHMENT

ELISA Applications to the Diagnosis of Infectious Diseases of Laboratory Animals

Infectious diseases of animals other than humans, especially those associated with a variety of experimental studies, pose significant problems. Such diseases may cause reduced birth rates, affect physiologic mechanisms important to the interpretation of special studies, and destroy entire animal breeder colonies. Early and accurate detection of disease agents is of major interest and importance in diagnostic laboratory animal medicine. Immunoassays are among the most widely used analytical procedures in such laboratories. While the main application of immunoassays has employed radioisotope-labeled antibodies and antigens (RIA), this system has always presented disadvantages, including reagent instability, potential health hazards, and disposal problems. Alternative techniques have received much attention, especially assay systems that eliminate the disadvantages of radioisotopes, yet maintain comparable specificity and sensitivity.

During the past 20 years or so, the most effective and widely used tests for the diagnosis of infectious diseases in laboratory animals have been assays employing hemagglutination inhibition (HAI), complement fixation (CF), immunofluorescence, and to a lesser extent, electrophoresis, immunodiffusion, and neutralization. A significant advance in diagnostic immunologic methodology occurred around 1970 when enzymes were successfully linked to antibody, with the resultant complex retaining its immunologic and enzymatic activity, thus providing the basis for the development of the enzyme-linked immunosorbent assay (ELISA). ELISA tests have been developed for an increasing number of infectious agents, including mouse hepatitis virus, pneumonia virus of mice, reovirus of mice, and *Mycoplasma pulmonis* (a bacterial respiratory disease agent in laboratory mice and rats).

20

Immunologic Disorders

I. INTRODUCTION

The concept that disease could be caused by immunoglobulins contradicted the pioneering work of investigators such as Bordet, Ehrlich, and others, which showed that these molecules have protective properties. It is now quite apparent from numerous documented studies that various forms of allergies and hypersensitivities are immunoglobulin related. These proteins are not the only initiators of immunologic disorders; several cell-dependent hypersensitivities also have been identified. This chapter describes the mechanisms and examples of different types of hypersensitivities and immunologic disorders including acquired immune deficiency syndrome (AIDS). Attention also is given to the relationship and significance of the major histocompatibility complex to histocompatibility antigens.

II. PREPARATION

Chapters 17-20, 22, 26, and 30 should be read before continuing. The following terms are important for you to know. Refer to the glossary and the appropriate chapters of the text if you are uncertain of any of them. A pronunciation guide for selected terms is provided as a learning aid.

1. acute
2. AIDS
3. allergen (AL-err-jen)
4. allograft (AL-oh-graft)
5. anaphylaxis (an-ah-feh-LAK-sis)
6. antireceptor antibodies
7. atopy (AT-oh-pee)
8. autograft (AW-toh-graft)
9. autoimmune
10. basophil (BAY-soh-fil)
11. B cell
12. biological response modifiers
13. cell-mediated immunity
14. Chediak-Higachi syndrome
15. chronic
16. complement cascade
17. congenital immunodeficiency
18. cytolysis (sye-TOL-ee-sis)
19. desensitization
20. diarrhea (die-ah-REE-ah)
21. Di George syndrome
22. elicitor
23. histamine (HISS-tah-mean)
24. histocompatibility (hiss-toh-kom-pat-ee-BIL-ee-tee)
25. HLA system
26. hypersensitivity
27. hypogammaglobulinemia (hi-poh-gam-ah-glob-u-lin-EE-me-ah)

28. IgE
29. IgG
30. immunodeficiency (im-mu-no-dee-FISH-en-see)
31. immunoglobulin (im-u-no-GLOB-u-lin)
32. immunologic rejection
33. immunosuppression
34. incubation
35. *in vitro*
36. Job's syndrome
37. leukotriene
38. lymph node
39. lymphocyte
40. lymphokine
41. malignancy (mah-LIG-nan-see)
42. mast cell
43. MHC system
44. osteomyelitis (os-tee-oh-my-ee-LIE-tis)
45. penicillin allergy
46. phagocytosis (fag-oh-sye-TOE-sis)
47. pneumonitis (noo-moh-NYE-tis)
48. prostaglandin (pros-tah-GLAN-din)
49. Rh baby
50. serum
51. T cell
52. thymus gland
53. tuberculin
54. wheal and flare

III. PRETEST

Correct answers to all questions can be found at the end of this section. Write your responses in the appropriate space provided.

Completion Questions

Provide the correct term or phrase for the following. Spelling counts.

1. The congenital abnormality resulting in an inability to produce all immunoglobulins is called ________ .
2. Cell-mediated hypersensitivity reactions are mediated by ________________________ .
3. In cases of type I hypersensitivity, chemicals such as histamine and serotonin are released from ________________ cells.
4. Anaphylactic reactions are believed to result from the interaction between the specific allergen and the ________________ .
5. The three steps required to cause anaphylactic responses are (a) ________________ , (b) ________________ , and (c) ________________ .
6. A procedure that renders an atopic individual immunologically unreactive to a particular allergen is called ________________ .
7. The transplant of tissues involving two different species is referred to as a(an) ________________ .
8. A widely applied procedure designed to measure the histocompatibility antigenic composition of donor and recipient tissues is the ________________ .

9. When immunocompetent tissues are transplanted in an immunologically handicapped host, a reaction known as the ______________________________ occurs.

10. The inability of an individual's immune system to recognize normal parts of the body's tissue as self can result in a disorder known as ______________________________ .

Matching

Select the answer from the right-hand side that corresponds to the term or phrase on the left-hand side of the question sheet. An answer may be used more than once. In some cases, more than one answer may be required.

______ 11. Involves IgE and IgG

______ 12. Inhibited by cortisone

______ 13. Anaphylaxis

______ 14. Passive transfer with serum possible

______ 15. T lymphocytes involves

______ 16. Graft rejection

______ 17. Reactions occur after 24 hours

______ 18. Inhibited by antihistaminic drugs

______ 19. Prausnitz-Küstner reaction

______ 20. Atopy

______ 21. Contact dermatitis

______ 22. Antireceptor antibodies

______ 23. Insulin-dependent diabetes

a. immediate types of hypersensitivity

b. delayed (cell-mediated) hypersensitivity

c. *a* and *b*

d. neither *a* nor *b*

True or False

Indicate all true statements with a "T" and all false statements with an "F".

______ 24. An infant with Di George's syndrome is highly resistant to fungi and atypical tubercle bacilli.

______ 25. Electrophoretic examination of sera from an individual with agammaglobulinemia shows the presence of large quantities of IgA and IgM.

______ 26. Secondary immunodeficiencies occur more frequently than primary ones.

27. An allergen is an acquired exaggerated response toward a specific substance that does not produce similar reactions in the majority of previously unexposed members of the same species.

______ 28. Features of anaphylactic shock vary, depending on the species involved.

Answers

1. hypogammaglobulinemia
2. antigen-sensitized T lymphocytes
3. mast
4. cell-mediated anaphylactic antibodies

5a. injection or exposure to an inducer
5b. incubation or sensitization period
5c. exposure to elicitor

6. desensitization
7. xenograft
8. lymphocyte toxicity test
9. graft-versus-host response
10. autoallergy
11. a
12. b
13. a
14. a
15. b
16. b
17. b
18. a
19. a
20. a
21. b
22. d
23. d
24. F
25. F
26. T
27. F
28. T

IV. CONCEPTS AND TERMINOLOGY

To protect the body, the immune system must be able to recognize and destroy molecules that are *foreign,* or *nonself,* while conserving those body components that are *native,* or *self.*

Immunodeficiencies

Immunodeficiency diseases are characterized by the inability of the immune system to perform normally. Four major immune systems normally provide the defense against constant attacks by microbial and helminth agents and factors that have the potential to produce infection and disease. These systems are antibody-regulated (B cell) immunity, cell-mediated (T cell) immunity, phagocytosis, and complement. Each system may work in cooperation with one or more of the others, or it may function independently. Two general categories of immunodeficiencies are recognized: *primary* and *secondary.*

Primary immunodeficiencies usually arise from an inherited failure of one or more of the immune system's components to develop. Examples include (1) bone-marrow stem cell deficiency (neither B- nor T-type lymphocyte components are functional); (2) Di George's syndrome (the thymus gland and T lymphocytes fail to develop); and (3) hypogammaglobulinemia (a sex-linked inability to produce immunoglobulins).

Secondary, or acquired, deficiencies occur much more frequently than the primary immunodeficiencies. The causes of the secondary deficiencies include malnutrition, malignancies (cancers), and extensive exposure to radiation, burns, and drugs that interfere with the lymphatic system. Secondary immunodeficiency also results in an increased susceptibility to opportunistic infection. One example of a severe form of such a state is acquired immune deficiency syndrome (AIDS).

Hypersensitivity

Hypersensitivity is an acquired exaggerated response toward a specific substance that does not produce similar reactions in the majority of previously unexposed members of the same species. The inciting, or inducing, agents are referred to as *allergens.*

Initial exposure to an allergen, immunologically primes, or sensitizes, the individual. Further contact with the same allergen not only can lead to a secondary boosting of the immune response, but also can cause tissue-damaging reactions. Several factors influence the nature and extent of hypersensitive responses, including the portal of entry, the cellular and tissue responses to the allergen, and the individual's genetic makeup.

Categories of Hypersensitivity

Hypersensitivity reactions that developed within a few seconds to 24 hours following allergen exposure and involved circulating antibodies were once classified as immediate. Those that developed after 24 hours but within 48 hours and involved the direct participation of sensitized lymphocytes were known as delayed hypersensitivity reactions. These designations are now understood to have different meanings. Four types of hypersensitivity categories are recognized today: type I (classic immediate), type II (cytotoxic), type III (immune-complex-mediated), and type IV (cell-mediated, or delayed). Types I through III are immunoglobulin-dependent hypersensitivities. Type IV reactions are mediated by antigen-sensitized T lymphocytes rather than by antibody molecules. In addition, specific interactions of sensitized T cells with antigens are involved in the rejection of transplanted tissues and organs and in protection against cancer.

Table 20-1 compares the features of these hypersensitivity states.

The type of hypersensitivity acquired is determined by several factors: (1) the chemical nature of the allergen; (2) the route involved in the sensitization (e.g., inhalation, ingestion, or injection); and (3) the physiological state of the individual. The form of the response depends on the type of contact with the precipitating allergenic dose and on the acquired hypersensitive states.

Type I (Classic Immediate) Hypersensitivity

Examples of type I allergic responses include certain forms of asthma, anaphylactic shock, hay fever (allergic rhinitis), and hives (urticaria). One of the distinguishing features of this hypersensitive state is the rapid liberation of physiologically active chemicals, such as histamine and serotonin, from affected cells. Substances of this kind are normally released from cells as a consequence of antigen and antibody interaction.

Anaphylaxis

Two general types of anaphylactic responses are recognized: systemic, or generalized, and cutaneous. Both states are temporary. If the individual does not die almost at once, recovery usually occurs within an hour.

The allergens or sensitizers that originally cause the hypersensitive state are called *inducers*. The anaphylactic response is brought about by substances similar to the inducers called *elicitors*. Generalized anaphylaxis produces suffocation, respiratory-passage constriction, shock (failure of the peripheral circulation), engorgement of blood vessels, and other respiratory changes. Several active participants, or mediators, are involved in producing the symptoms of anaphylaxis. These include the preformed mediators (regulators) found existing in the granules of mast cells and basophils, histamine and serotonin, the well-established secondary mediators such as the leukotrienes, and various chemical-attracting (chemotactic) factors and hydrolytic enzymes found in mast cell granules. The secondary mediators are produced by stimulated mast cells, basophils, and other cells recruited in the anaphylactic reaction. The release of mediators from activated mast cells or basophils is started by the bridging of two IgE-specific receptors on these cells.

Atopy

This type of hypersensitivity develops in susceptible persons because of repeated accidental absorption of the allergenic substances. Absorption can occur via the mucous membranes of the gastrointestinal or respiratory tracts or the skin. The various atopic allergic disease states are precipitated by later exposure to these allergens.

Upon injection of an appropriate allergen, atopic individuals exhibit itching at the inoculation site followed by a wheal-and-flare response. The Prausnitz-Küstner (P-K) procedure can be used for patients who are extremely sensitive to specific allergens and cannot undergo direct testing.

Treatment for atopy falls into two categories: (1) pharmacologic intervention—the use of medications to block the release or reduce the effects of the chemicals involved in allergic states; and (2) immunotherapy procedures to render an atopic individual immunologically unreactive to a particular allergen, that is, to *desensitize* the individual.

Measurement of IgE levels may require highly sensitive laboratory procedures such as the radioallergosorbent test (RAST).

TABLE 20-1 A Comparison of Different Types of Hypersensitivity

	Immediate (Immunoglobulin Dependent)			*Delayed*
Characteristic	*I (Classic Immediate)*	*II (Cytotoxic)*	*III (Immune Complex Mediated)*	*IV (Cell-Mediated)*
Maximum reaction time for clinical manifestation (response)	30 min	Variable	3-8 hr	24-48 hr
Immunoglobulins involved	IgE and IgG	IgG and IgM	IgE, IgG, and IgM	None
Cells involved	Most cells, basophils	Red blood cells, lymphocytes, eosinophils, neutrophils, and body tissue cells	Body tissue cells	Body tissue cells, T lymphocytes, and macrophages (eosinophils in certain cases)
Reaction mediators (regulators)	Pharmacological substances including histamine, leukotriene D, ECF-A, kinins, and prostaglandins[b]	Complement	Complement, and lysosomal enzymes	Soluble mediators such as MIF and TF[a] (lymphokines)
Appearance of response to intradermal antigens	Wheal and flare		Erythema (redness) and edema (swelling)	Erythema and induration (hardened tissue)
Inhibition by antihistaminic drugs	Yes, in certain situations	No	No	No
Inhibition by cortisone	No, with normal dosages	No	No	Yes
Passive transfer with serum from sensitive donor	Generally yes	Yes	Yes	No
Passive transfer with lymphocytes or lymphocyte extracts from sensitive donor	No	No	No	Yes
Examples of allergic states	Anaphylaxis, asthma, serum sickness, and drug, food, and insect allergies	Transfusion reaction, hemolytic disease of the newborn, and drug-induced allergies	Arthus reaction, serum sickness, and certain autoimmune diseases	Infection allergies, autoimmune disease, graft rejection, contact dermatitis to drugs, and tumor immunity

[a] MIF is a migration inhibitory factor; TF is a transfer factor.

[b] Leukotriene D was known as SRS-A, a slow-reacting substance of anaphylaxis; ECF-A is an eosinophilic chemotactic factor of anaphylaxis.

Type II (Cytotoxic) Hypersensitivity

Examples of type II responses in humans include: (1) blood transfusion reactions, in which blood group antigens in red blood cell membranes are the inciting factors; (2) erythroblastosis fetalis (the Rh baby), in which Rh antigens on fetal or newborn red blood cells are the targets of antibodies formed by the mother; (3) drug-caused blood loss (anemias), in which a drug forms an antigenic complex with the surfaces of blood cells and brings about the production of antibodies that are destructive for the cell-drug complex; and (4) nephrotoxic nephritis, in which antibodies against antigen in the basement membrane of kidney glomeruli combine and produce a cytotoxic reaction.

The Complement Cascade (Pathway)

Two different mechanisms are recognized for the activation of complement—classic and alternate (properdin) pathways). Each pathway contains an initial set of different reactions, but at the midpoint of the system, they share a common reaction sequence. (Refer to the text for specific details.) The common remaining portion of pathway, known as the membrane attack system, requires the attachment of complement components c5b through 9 to cell and viral surfaces.

Control mechanisms exist to limit complement activation; these include various serum proteins. Uncontrolled complement activation or congenital complement deficiency may lead to various disease states.

Type III (Immune-Complex-Mediated) Hypersensitivity

In this form of hypersensitivity, soluble antigen and immunoglobulins combine in the body, which may produce an acute inflammatory reaction. If complement (C) is fixed by this combination, biological molecules called anaphylatoxins are produced. These molecules ultimately cause the release of histamine and the production of other active substances which, in turn, set in motion a chain reaction that damages tissues and intensifies the inflammatory response. Type III responses involve IgG and IgM immunoglobulins and include the Arthus reaction, serum sickness, certain autoimmune reactions, and hypersensitivity pneumonitis.

Type IV (Cell-Mediated or Delayed) Hypersensitivity

Type IV responses do not involve the release of histamine or chemically related substances, nor can this type of sensitivity be transferred by the injection of serum from a sensitive individual to a nonsensitive one, as in the case of the P-K reaction. Passive transfer of type IV responses to nonsensitive recipients can be accomplished by the injection of living lymphoid cells from sensitized donors and by a nonantibody-active transfer factor. While type IV responses are not inhibited by antihistamines, they are inhibited by steroid compounds such as cortisone and hydrocortisone.

Type IV responses depend upon the interaction of T lymphocytes with an antigen. Several cell-mediated immune reactions are beneficial. When cell-mediated responses cause tissue damage, the condition generally is considered to be an allergic one. Such immunologic injury, also known as delayed hypersensitivity, involves T type cells bearing specific receptors on their surfaces that are stimulated by contact with antigens to release factors called lymphokines. Whenever the receptors react with specific antigenic determinants, the T cells become sensitized. This exposure to antigens and the sensitization are necessary for the occurrence of delayed hypersensitivity reactions. Table 20-2 lists several lymphokines.

Lymphocytes and lymphokines are not the only factors involved in delayed hypersensitivity. Under the influence of lymphokines, activated macrophages show increased phagocyte-biosynthetic and microbe-destroying properties. These macrophages are far more effective than normal ones in eliminating microorganisms.

Allergy of Infection

This type IV allergic state brings about an accelerated and exaggerated tissue reaction to certain infectious agents. The most thoroughly studied example of this type of hypersensitive reaction is that associated with tuberculosis.

TABLE 20-2 Representative Lymphokines and Related Functions

Lymphokine	*Designation (If Appropriate)*	*Functions or Activities*
Individual chemotactic factors for eosinophils, lymphocytes, macrophages, and neutrophils	—	Causes chemotaxis (chemical attraction) for specific type of white blood cells
Cloning-inhibition factor	CLIF	Blocks *in vitro* multiplication of certain cell types
Dialyzable transfer factors	TF	Converts normal lymphocytes *in vivo* and *in vitro* to antigen-sensitized lymphocytes
Inhibitor RNA synthesis	IDS	Reversibly inhibits mitosis of lymphocytes
Lymphotoxin	LT	Causes cytotoxicity
Macrophage-activation factor	MAF	Is partly responsible for increased lysosomal activity of macrophage
Macrophage-aggregation factor	MAF	Restricts macrophage movement and induces formation of giant cells
Macrophage-inhibition factor	MIF	Inhibits migration of macrophages
Mitogenic factor	—	Stimulates lymphocyte transformation
Proliferation-inhibition factor	PIF	Blocks *in vitro* multiplication of certain cell types
Skin reactive factor	—	Produces skin inflammation

Contact Dermatitis

This form of hypersensitivity, which includes certain types of drug allergies, is one of the most commonly encountered human allergic diseases. The sensitization and the production of symptoms result simply from contact with various causative compounds, which combine with proteins in the skin.

Two types of responses are generally recognized: contact skin sensitivity and allergic contact dermatitis. In contact skin sensitivity, exposure to the incitant causes local irritation. While these substances by themselves are not allergenic, they can become so by combining with proteins in the skin. In allergic contact dermatitis, both humans and various laboratory animals can be sensitized by skin contact with the incitant or by intradermal injection of the incitant.

Antireceptor Antibodies: A Possible New Type of Hypersensitivity

Certain disease states are associated with the actions of immunoglobulins against the cellular receptors for various hormones or other reaction-controlling substances. Examples of such disease states include insulin-dependent diabetes and myasthenia gravis.

Tissue Transplantation Reactions

Tissue transplants activate both humoral mechanisms (circulating antibodies) and cell-mediated mechanisms of immunity. Transplants of tissues are classified in terms of genetic relationships. Examples include:

1. *Autograft*—the tissue of one individual is involved
2. *Allograft*—tissue to be transplanted is taken from a genetically dissimilar member of the same species
3. *Isograft*—tissue to be transplanted is from a genetically identical individual
4. *Xenograft*—tissue to be transplanted is taken from a different species

Genetic Control

The success of organ or tissue transplants involves a group of genes that control the formation of histocompatibility antigens. This genetic system or human leukocyte antigen (HLA), also referred to as the human major histocompatibility complex (MHC), controls both humoral immune and cellular immune responses. In the human, these, genes as well as others possibly responsible for immunological reactions are located on chromosome 6.

As shown in Table 20-3, the MHC genes can be divided into three classes.

TABLE 20-3 MHC Genes

Class	*Genes*
I	Genes that control the expression of the classical transplantation antigens of all cells of an individual. In humans, these genes are HLA-A, -B, and -C.
II	Genes that control the expression of surface antigens (Ia) involved with the immune response on lymphocytes and macrophages. In humans, these genes are found in the HLA-D region.
III	Genes that control a series of serum proteins that belong to the complement system. These are complement components C4 and C2 and factor B of the properdin system. The specific human genes of this class have not been identified.

The HLA system has several alternate genes (alleles) at each known locus. Closely linked allele combinations on a single chromosome fòrm that unit called the *haplotype.* One HLA haplotype is inherited from each parent.

Determining the suitability of tissue for grafting—histocompatibility testing—is accomplished by the microcytotoxicity assay or the mixed leukocyte (lymphocyte) culture assay. The microcytotoxicity test is a widely applied method designed to measure the histocompatibility antigenic (HLA) composition of donor and recipient. Other applications of HLA typing include establishing paternity and determining associations of the HLA system with certain diseases.

In mice, the set of antigens known as the H-2 system provides by far the strongest barrier to transplantation. The system is under the control of a specific gene complex on the mouse's no. 17 chromosome. Other parts of the H-2 gene complex play an important role in the functioning of the immune system; they include the I region and the S region. A series of genes within the I region known as immune response (Ir) genes controls the manufacture of antibodies. The S region has been shown to control the manufacture of a component of serum complement.

HLA Disease Associations

Two major groups of diseases have shown relationships with the HLA system. One group consists of the rheumetic diseases, which include arthritis of the spine and sacroiliac and axial joints. The second group is characterized by long-lasting inflammation and abnormal immunologic responses. Other HLA associations have been found with pernicious anemia, myasthenia gravis, psoriasis, and juvenile diabetes.

Graft Rejection

When tisue is transplanted from one individual to another, two distinct rejection processes can take place: the host rejection of grafted tissue, that is, the *host-versus-graft reaction,* and the reverse of this reaction, namely, the *graft-versus-host response.* The second type of reaction occurs when immunocompetent tissues are transferred to an immunologically handicapped host.

Means for limiting or totally inhibiting graft rejection include the injection of immunosuppressive drugs, such as corticosteroids or antilymphocytics, whole-body irradiation, or a combination of these measures.

Autoimmune (Autoallergy) Diseases

The body's tolerance mechanism occasionally falters and breaks down, thus creating autoallergic (autoimmune) disease states. In these disorders, self-antigens, or autoantigens, stimulate the production of circulating immunoglobulins (humoral response) or of specifically sensitized lymphocytes (cell-mediated response), either of which will react with the autoantigen.

Several factors may trigger an autoallergic response: (1) the presence of antigens that normally do not circulate in the blood; (2) the presence of an altered antigen, which can develop through exposure to chemicals, physical agents, or microorganisms; (3) the introduction of a foreign antigen similar to an autoantigen; and (4) the occurrence of a mutation in immunocompetent cells that results in a responsiveness to normal autoantigens.

Autoallergic or autoimmune diseases comprise a broad range of disorders and a confusing array of overlapping symptoms, pathological lesions, and immunological properties. Some diseases, such as Hasimoto's thyroiditis, are organ specific. Others, such as the autoimmune hemolytic anemias, are localized but are not organ specific. Still others are nonorgan specific. Systemic lupus erythematosis (SLE) is an autoimmune disorder in which neither lesions nor autoantibodies are limited to any one organ or tissue. Autoimmunity may play an important role in rheumatoid arthritis, a common inflammatory condition of joints, and in aging.

Tissue lesions in autoimmune diseases can result from humoral, cellular, or combined mechanisms. In situations involving a humoral mechanism, antibody-antigen complexes are deposited in the tissue, especially if the immunoglobulin is IgG. The complement cascade is activated, and inflammatory reactions produce a type II hypersensitivity reaction. When antigen is bound to cells, a type II hypersensitivity reaction, cell lysis, occurs. The direct action of cytotoxic T cells and antibody-dependent cell-mediated cytotoxicity (ADCC) of killer cells and macrophages are examples of cell-mediated or type IV hypersensitivity reactions.

Autoimmune disease states have certain shared characteristics: (1) their incidence tends to increase with age; (2) they occur most commonly among persons with generalized immunological deficiencies; (3) they occur most often in females; and (4) any one may produce destructive effects that resemble those associated with the other types of hypersensitivities.

The diagnosis and identification of autoimmune diseases involves the use of laboratory tests for autoantibodies.

V. CHAPTER SELF-TEST

Continue with this section only after you have read Chapters 17-20, 22, 26, and 30. A score of 80 percent or better is good. If your score is less than 65 percent, reread the chapter.

Correct answers to all questions can be found at the end of this section. Write your responses in the appropriate space provided.

Matching

Select the answer from the right-hand side that corresponds to the term or phrase on the left-hand side of the question sheet. An answer may be used more than once. In some cases, more than one answer may be required.

Topic: Hypersensitivity Reactions

______	1. Inhibited by cortisone	a. immediate hypersensitivity type I
______	2. Involves IgG	b. immediate hypersensitivity type II
______	3. Leukotriene D involved	c. cell-mediated hypersensitivity type IV
______	4. Hemolytic disease of the newborn	d. none of these
______	5. Anaphylaxis	
______	6. Inhibited by antihistaminic drugs	

_____ 7. Wheal-and-flare reactions

_____ 8. T lymphocytes

_____ 9. Passive transfer by serum

_____ 10. Arthus reaction

_____ 11. Tumor immunity

_____ 12. Reactions occur within 30 minutes

_____ 13. Prausnitz-Küstner reaction

_____ 14. Di George's syndrome

_____ 15. Insulin-dependent diabetes

_____ 16. Myasthenia gravis

True or False

Indicate all true statements with a "T" and all false statements with an "F".

_____ 17. Primary immunodeficiencies result from an inherited failure of one or more components to develop.

_____ 18. In Di George's syndrome, B cells are lacking.

_____ 19. In actual cases of tuberculosis, the allergic states become established long after the obvious signs of infection are apparent.

_____ 20. An isograft refers to a tissue transplant involving genetically different organisms.

_____ 21. A tissue transplant from one part of an individual's body to another is an example of an isograft.

_____ 22. Histocompatibility antigens differ from one individual to another.

Completion Questions

Provide the correct term or phrase for the following. Spelling counts.

23. Four major immune systems that normally provide protection against infectious disease agents are (a) _____, (b) _____, (c) _____, and (d) _____.

24. Kaposi's sarcoma and opportunistic infections are associated with the severe form of secondary immunodeficiency known as _____.

25. The rapid release of physiologically active chemicals such as histamine from affected cells is characteristic of type _____ allergic responses.

26. In desensitization or hyposensitization procedures, allergenic preparations stimulate the formation of _____.

27. The complement components _____ are involved in the classic pathway.

28. _____ is the specific complement component that must be present for the activation of the alternate complement pathway.

29. Attachment of the complement component ________________ to the activating enzyme formed by the classic or alternate pathways completes the cytolytic principle.

30. The use of antifungal antibiotics for cases of hypersensitivity pneumonitis ________________ recommended.

31. The ________________ controls both the humoral immune and cellular immune responses of individuals.

32. The most effective use of HLA typing has been in the selection of ________________ for tissue and organ transplants.

33. Tissue lesions in autoimmune diseases can result from (a) ________________, (b) ________________, or (c) ________________ mechanisms.

Essay Questions

Answer the following questions on a separate sheet of paper.

34. How are immunodeficiency disorders diagnosed?

35. Is the complement cascade necessary for the formation of lesions in type II reactions?

36. What are the immunoglobulin classes involved in type II hypersensitivity responses?

Answers

1. c
2. *a* and *b*
3. a
4. b
5. a
6. a
7. a
8. c
9. a, b
10. d
11. c
12. a
13. a
14. d
15. d
16. d
17. T
18. F
19. F
20. F
21. F
22. T

23a. antibody-regulated immunity
23b. cell-mediated immunity
23c. phagocytosis
23d. complement

24. acquired immune deficiency syndrome (AIDS)
25. I
26. IgG blocking antibodies
27. C1 through C4
28. C3b
29. C9
30. is not
31. major histocompatibility complex (MHC)
32. donor

33a. humoral
33b. cellular
33c. combined

34. Diagnosis of many immunodeficiency disorders can be made with the aid of tests that screen for the presence of respective immune system components. Such approaches include electrophoresis, radial immunodiffusion, antinuclear and anti-DNA tests, T cell function tests, total lymphocyte counts, the detection of lymphokines, and monoclonal antibody tests.

35. For lesions to occur in most of type II reactions, fixation and activation of the complement cascade system are necessary. One type of complement independent reaction is antibody-dependent cell-mediated cytotoxicity.

36. The antibodies involved in type II hypersensitivity responses are IgG and IgM.

VI. ENRICHMENT

Histamine Production by Bacteria

Large-scale commercial tuna fishing is done primarily by the purse seining method in which enormous nets are used to trap schools of fish. However, the netted fish cannot always be collected immediately but must remain in the ocean for a considerable time until they can be brought on board the fishing vessel and cooled before being frozen and stored. If these delays are prolonged, some postmortem decomposition can occur in the fish. Comparable delays also may happen when unusually large loads of fish must be cooled or when the equipment experiences mechanical failure.

Fresh scombroid fish such as tuna, bonito, and mackerel have virtually no histamine in their muscle. However, after decomposition caused by spoilage, these fish contain substantial amounts of histamine. The histamine is produced by bacteria that decarboxylate histidine, an amino acid that is abundant in scombroid fish muscle. A large and diverse group of bacteria have been reported to be responsible for the histamine found in fish, dairy products, and even some vegetable products. Such organisms include facultatively or obligately anaerobic species such as *Clostridium perfringens, Enterobacter aerogenes, Klebsiella pneumoniae, Proteus mirabilis,* and *Vibrio alginolyticus.* The optimum temperature for bacterial histamine formation is 38° C.

PART VII
MICROBIOLOGY TRIVIA PURSUIT
(Chapters 17 through 20)

Let us test your attention to detail and fact-gathering ability once again. This is the next to last opportunity to do so in this *Study Guide*. This time, there are *five* separate and challenging Parts to this section. Remember, a certain number of questions must be answered in each Part before you proceed up the MICROBIOLOGY TRIVIA PURSUIT trail. The answers and directions for continuing are given at the end of each Part. Try Part 1.

PART 1

Completion Questions

Provide the correct term or phrase for the following. Spelling counts.

1. What is the soluble protein which normally circulates in the plasma, and is converted into fibrin?

   ____

2. The yellow fluid resulting during and after a blood clot shrinks is called ________________ .
3. ______________________ are modified amino acids found in mature neutrophils, and have antimicrobial activity.

4 and 5. Adoptive immunity is a form of (4) ________________ acquired (5) ________________ immunity.

Directions

Check your responses with the *Answers* section, and add up your score. Enter the number of correct answers in the space provided.

Total Correct Answers for Part 1: ____________ .

If your score was 4 or higher, proceed to Part 2.

Answers
1. fibrinogen; 2. serum; 3. defensins; 4. artificially; 5. passive.

PART 2

Very good! Congratulations on the successful completion of Part 1. Now try your hand with Part 2.

Completion Questions

Provide the correct term or phrase for the following. Spelling counts.

6. In a normal differential blood count, which granulocyte appears in the greatest numbers?

   ____

7 and 8. Two specific antibacterial proteins normally found in the respiratory tract are (7) ____________ and (8) ____________.

9. Cationic proteins are rich in the basic amino acid ____________.

10. Identify the English physician shown in Figure VII-1. ____________

(He was responsible for the original vaccination procedure against smallpox, which was introduced near the end of the 1700s.)

Figure VII-1

Directions

Check your responses with the *Answers* section, and add up your score. Enter the number of correct answers in the space provided.

Total Correct Answers for Part 2: ____________.

If your score was 4 or higher, proceed to Part 3.

Answers

6. neutrophils; 7. lysozyme; 8. lactoferrin; 9. arginine; 10. Edward Jenner.

PART 3

Excellent! You are doing very well. Here is a challenging Part 3.

Completion Questions

Provide the correct term or phrase for the following. Spelling counts.

11. The bactericidal property of serum and whole blood is associated with the complex protein known as ______________________________ .
12. In a phagocyte's ingestion activity, a phagocytic vacuole called a ______________________________ is formed.
13. The ability of a substance to stimulate antibody production is referred to as its ______________ .
14. Most antibody can be found in the protein portion of serum or plasma known as ______________ globulin.
15. The particular immunoglobulin that structurally consists of five Y-shaped units similar to IgG is ______________________________ .

Directions

Check your responses with the *Answers* section, and add up your score. Enter the number of correct answers in the space provided.

Total Correct Answers for Part 3: ______________ .

If your score was 5, proceed to Part 4.

Answers
11. complement; 12. phagosome; 13. immunogenicity; 14. gamma; 15. IgM.

PART 4

You have come a long way. Now for the next to last challenge.

Completion Questions

Provide the correct term or phrase for the following. Spelling counts.

16, 17, and 18. The three general categories into which variations in immunoglobulin structure can be placed are (16) ______________________ , (17) ______________________ , and (18) ______________________ .

19. The ______________________________ gland is believed to be the central location for the multiplication of lymphocytes during embryonic development.
20. The T ______________________________ cell regulates the other types of T cells and B cells.

Directions

Check your responses with the *Answers* section, and add up your score. Enter the number of correct answers in the space provided.

Total Correct Answers for Part 4: ______________ .

If your score was 4 or higher, proceed to Part 5.

Answers

16. isotypes; 17. allotypes; 18. idiotypes; 19. thymus; 20. suppressor.

PART 5

Good work. Congratulations on the successful completion of Part 4. Now try your hand with the last Part of this Microbiology Trivia Pursuit.

Completion Questions

Provide the correct term or phrase for the following. Spelling counts.

21 and 22. To distinguish the antibody production associated with an actual ongoing infection from antibody production associated with past infections or the effects of immunization, at lease two specimens of a patient's serum are needed. These are known as (21) ______________ and (22) ______________ specimens.

23. The Rh blood factor indicated in the Wiener System as Rh_0 is designated as ______________ in the Fisher-Race System.

24. The disease of AIDS is an example of a ______________ immunodeficiency.

25. An unknown blood was typed. The results are shown in Figure VII-2. On the left (a), anti-A serum was added to a blood drop, and on the right (b), anti-B serum. On the basis of the results shown, what is the patient's major blood type? ______________

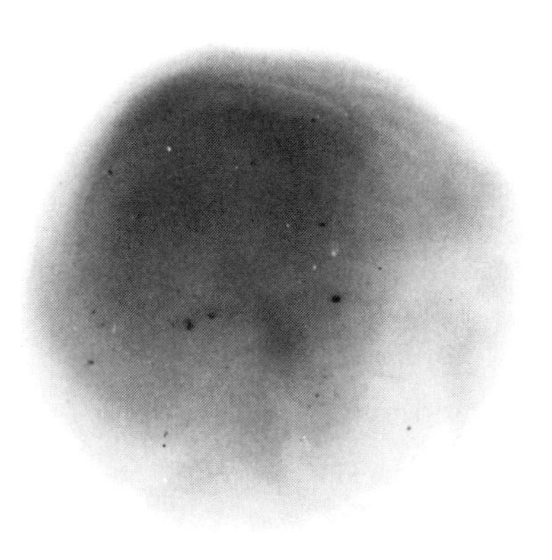

(a) **Figure VII-2** (b)

Directions

Check your responses with the *Answers* section, and add up your score. Enter the number of correct answers in the space provided. Then add all your scores for Parts 1 through 5 together and find your score on the *Performance Score Scale.*

Total Correct Answers for Part 5: ______________ .

Total Correct Answers for Parts 1 through 5: ______________ .

Answers

21. acute; 22. convalescent; 23. D; 24. secondary; 25. B.

PERFORMANCE SCORE SCALE

Number Correct	*Ranking*
15	You should have done better
17	Better
19	Good
21	Excellent
23	Outstanding! Keep it up

Note that each major section of this Study Guide has a *MICROBIOLOGY TRIVIA PURSUIT* section. If you did not do as well as you expected, there is still one more opportunity ahead.

21

An Introduction to Epidemiology and the Identification of Disease Agents

I. INTRODUCTION

Epidemiology is concerned with the distribution, occurrence, and control of disease states in a population. Particular attention is given to known factors—for example, microorganisms—that contribute to the presence of a specific disease in a given group of individuals. The early recognition and treatment of an infectious disease state also is of prime importance to a patient and to all exposed individuals. While the prompt, accurate diagnosis of such diseases is the major responsibility of the physician, a number of other health-care personnel are also involved.

In this chapter, we consider several of the factors contributing to the causes, distribution, and transmission of diseases within a population. Attention also will be given to the approaches used in the isolation and identification of microbial and related disease agents. Such information will be important for Chapters 23 through 31.

II. PREPARATION

Before continuing, you should read Chapters 3, 13, 19, 21, and 22 and the portions of Chapters 3, 7, 8, 9, and 10 dealing with staining reactions and microbial structures and cultivation. The following terms are important for you to know. Refer to the glossary and the appropriate chapters of the text if you are uncertain of any of them. A pronunciation guide for selected terms is provided as a learning aid.

1. abscess (AB-sess)
2. acid fast
3. antibiotic
4. anticoagulant
5. arthropod
6. aseptic
7. autopsy (AW-top-see)
8. bacteremia (back-teer-EE-me-ah)
9. bacteriostatic (back-tee-ree-oh-STAT-ik)
10. biopsy (BY-op-see)
11. CAMP factor
12. capsule
13. carrier
14. chemotherapeutic

15. coagulase (koh-AG-you-lace)
16. communicable
17. disease
18. droplet nucleus
19. ELISA
20. endemic (en-DEM-ik)
21. epidemic (ep-ee-DEM-ik)
22. epidemiology (ep-ee-dee-mee-OL-oh-je)
23. etiology (ee-tee-OL-oh-je)
24. fastidious
25. feces (FEE-sees)
26. fomite
27. genus (GEE-nus)
28. glucose
29. gram-negative
30. gram-positive
31. helminth (HEL-minth)
32. hemolysis (hee-MOL-ee-sis)
33. hyphae (HI-fee)
34. hypothesis
35. immunofluorescence (im-mu-no-floo-RES-ens)
36. immunosuppression
37. infestation (in-fes-TAY-shun)
38. inoculating loop
39. larvae (LAR-vee)
40. medium
41. mite
42. morbidity (mor-BID-ee-tee)
43. mortality
44. multiple test system
45. nematode
46. nosocomial (nos-oh-KOH-me-al)
47. pandemic (pan-DEM-ik)
48. parasite
49. pathogen
50. quality control
51. rapid test system
52. reagent (ree-A-jent)
53. reservoir of infection
54. saprozoonosis
55. sporadic (spo-RAD-ik)
56. sputum (SPEW-tum)
57. sterile
58. surgical asepsis
59. syndrome (SIN-drohm)
60. tick
61. total protected environment
62. vector (VEK-tor)
63. virulence (VIR-u-lence)
64. zoonosis (zoh-oh-NO-sis)

III. PRETEST

Correct answers to all questions can be found at the end of this section. Write your responses in the appropriate space provided.

Completion Questions

Provide the correct term or phrase for the following. Spelling counts.

1. The study of the distribution and causes of disease prevalence in humans is called ________________ .
2. A disease that is constantly present but affects few persons is referred to as being ________________ .

3. The number of individuals having a specific disease per unit of the population within a given time period is known as ______________________________ .
4. The sources of infectious body fluids are referred to as ______________________________ .
5. An individual who harbors infectious disease agents is called a(an) ______________________________ .
6. Diseases that primarily affect lower animals but also can be transmitted to humans by natural means are called ______________________________ .
7. A contaminated inanimate object is called a(an) ______________________________ .
8. Infections acquired in a hospital are known as ______________________________ .
9. The process by which certain infectious disease agents pass into the eggs of infected ticks is called ______ .
10. The reduction of the direct or indirect transmission of pathogens forms the basis of ________________ .
11. A zoonotic disease for which humans are the natural host of the infectious agent is called a(an) __________ ______________________________ .
12. The formation of a green zone around a bacterial colony growing on a blood agar plate is referred to as ______________________________ hemolysis.
13. A pathogenic bacterial species known for its production of acid from mannitol is ________________ .
14. The formation of oxygen bubbles upon the addition of hydrogen peroxide to a culture shows the presence of the enzyme ______________________________ .

Matching

Select the answer from the right-hand side that corresponds to the term or phrase on the left-hand side of the question sheet. An answer may be used more than once. In some cases, more than one answer may be required.

Topic: Means of Disease Transmission

______ 15. Fomites

______ 16. Arthropod bites

______ 17. Feces

______ 18. Injection of blood and blood products

______ 19. Ingestion of contaminated food or water

______ 20. Fingers

______ 21. Flies

a. mechanical means of disease transmission

b. biological means of disease transmission

c. both *a* and *b* apply

Topic: Arthropods and Diseases

______ 22. Yellow fever

______ 23. Polio

______ 24. Trench fever

a. spread by the common house fly

b. spread by mosquitoes

c. spread by ticks

______ 25. Rickettsial pox

______ 26. Hepatitis A infection

______ 27. Acariasis

______ 28. Epidemic typhus

______ 29. Rocky Mountain spotted fever

______ 30. Relapsing fever

______ 31. Chagas' disease

______ 32. Malaria

______ 33. AIDS

d. spread by lice

e. none of these

Topic: Properties of Disease Agents

______ 34. Catalase-positive coccus

______ 35. Ferments mannitol

______ 36. Produces urease

______ 37. Forms pseudomycelia

______ 38. Bacitracin sensitive

______ 39. Acid-fast rod

______ 40. CAMP positive

______ 41. Spore-forming aerobe

______ 42. Coagulase positive

______ 43. Sensitive to optochin

______ 44. Oxidase positive

a. Group A beta hemolytic streptococci

b. *Staphylococcus aureus*

c. *Mycobacterium* species

d. *Candida* species

e. *Neisseria gonorrhoeae*

f. *Streptococcus pneumoniae*

g. Group B streptococci

h. *Cryptococcus neoformans*

i. none of these

True or False

Indicate all true statements with a "T" and all false statements with an "F".

______ 45. Communicable diseases do not interfere with normal body functions.

______ 46. A pandemic is a series of sporadic infections affecting several countries.

______ 47. Human tissues and secretions can serve as reservoirs of infection.

______ 48. A person who harbors infectious disease agents is called a fomite.

______ 49. Various warm-blooded animals are reservoirs for human disease agents.

______ 50. Coccidioidomycosis is an example of a dust-borne infection.

______ 51. What remains of a droplet of bacteria after evaporation is the droplet center.

______ 52. Arthropods do not have complete digestive tracts.

______ 53. The frequent and indiscriminate use of certain antibiotics contributes to the appearance of nosocomial infections.

______ 54. Various types of equipment and practices employed in hospitals have been implicated as sources of disease agents.

______ 55. Obstetrical delivery rooms are considered to be high-risk areas for sources of nosocomial infections.

______ 56. Disinfection and isolation of infected persons are among the approaches used in medical asepsis.

______ 57. The isolation and identification of an unknown pathogen is important to effective disease control.

______ 58. The procedures used for the identification of a pathogen do not vary significantly with the type of organism involved.

______ 59. Blood smears are used in the identification of most helminths.

______ 60. Several fungi can be differentiated on the basis of microscopic examinations.

______ 61. All pathogenic yeasts produce the enzyme urease.

______ 62. Regardless of the type of specimen used for microbial isolation, it must be handled aseptically.

______ 63. A general medium used in the identification of *Staphylococcus aureus* is mannitol salt agar.

______ 64. An important aspect of quality control involves the routine monitoring of equipment performance.

______ 65. Hydrogen peroxide is a common reagent used to test for catalase activity.

______ 66. Quality control refers primarily to a collection of clinical specimens.

Answers

1. epidemiology
2. endemic
3. morbidity
4. portals of exit
5. carrier
6. zoonoses
7. fomite
8. nosocomial
9. transoverial passage or transmission
10. medical asepsis
11. zooanthroponosis
12. alpha hemolysis
13. *Staphylococcus aureus*
14. catalase
15. a
16. b
17. a
18. b
19. a
20. a
21. a
22. b
23. a
24. d
25. e
26. a
27. e
28. d
29. c
30. c or d
31. e
32. b
33. e
34. a
35. b
36. h
37. d
38. a
39. c
40. g
41. i
42. b
43. f
44. e
45. F
46. F
47. T
48. F
49. T
50. T
51. F
52. F
53. T
54. T
55. T
56. T
57. T
58. F
59. F
60. T
61. F
62. T
63. T
64. T
65. T
66. F

IV. CONCEPTS AND TERMINOLOGY

Humans, lower animals, and plants can be successfully parasitized by a variety of microorganisms and by such forms of animal life as worms, mites, and ticks. The transmission of such agents from host to host is essential both to their survival and their biological success as pathogens.

Epidemiology is the study of the distribution and causes of disease prevalent in humans. Such disease states include microbial infections as well as debilitating, noninfectious diseases such as cancer, cardiovascular conditions, congenital defects, diabetes mellitus, emphysema, and vitamin deficiencies such as pellagra, rickets, and scurvy. Epidemiologists collect and analyze data from many individuals to reach conclusions about the nature, distribution, and incidence pattern of a particular disease.

For centuries, the pattern of occurrence among various diseases has differed noticeably. The prevalence of disease states can be described by the terms listed in Table 21-1. Epidemiologists collect and analyze data from many individuals to reach conclusions about the presence or absence of a particular disease in a given group. Their investigations include determining the effectiveness of methods used to control disease, describing the natural course or history of a disease, classifying diseases, and obtaining information necessary for the planning and evaluation of health care.

TABLE 21-1 Terms Used to Describe the Prevalence of Diseases

Term	*Description*	*Examples of Microbial Diseases*
Endemic	A disease that is constantly present but affects relatively few persons	Coccidioidomycosis, leprosy, tuberculosis
Epidemic	An unusual occurrence of a disease involving large segments of a population for a limited length of time	Influenza, poliomyelitis
Pandemic	A series of epidemics affecting several countries or even major portions of the world	Influenza, AIDS
Sporadic	Uncommon, occurring irregularly, and affecting relatively few people	Diphtheria, whooping cough

Infection and Disease

An *infection* is caused by a disease-causing agent or *pathogen* and results from the ability of the agent to invade and multiply in the tissues of a host. A *disease,* on the other hand, is caused by a pathogen's interference with the normal functioning of body systems or organs.

Signs, Symptoms, and Syndromes

Recognition of the presence of disease is based on the obvious presence of objective *signs* or specific recognizable abnormalities known as *symptoms.* Signs and symptoms occurring in a characteristic pattern are referred to as *syndromes.* Syndromes are of value in diagnosing and determining the distribution as well as the cause or *etiology* of diseases.

Methods of Epidemiology

Epidemiological studies may be *descriptive, analytic,* or *experimental.* Descriptive studies determine the rate of occurrence of a disease, the kinds of individuals suffering from it, and where and when it occurs. Analytic studies are used to test hypotheses concerning factors believed to determine susceptibility to disease. Such studies are designed to show whether particular events or environmental conditions are responsible.

Morbidity and Mortality Rates

Some of the principal epidemiological findings used in various reports are frequently expressed in terms of morbidity and mortality rates. *Morbidity* is the number of individuals having the disease per unit of the population—usually 100,000—within a given time period. The *mortality rate* is the number of deaths attributable to a particular disease per unit of the population (usually 1,000) within a given time period.

Reporting Communicable Infectious Diseases

State administrative codes require that actual or even suspected cases of certain communicable infectious diseases be reported to local health authorities. The actual requirements of each state vary. Individuals charged with the responsibility of notifying local health authorities include physicians, coroners, hospital directors, clinics, laboratories, and any persons knowing of a disease's existence.

Sources and Reservoirs of Infection

The sources of infectious disease agents are many and varied. The most disabling and common infections among humans are caused by microorganisms capable of living and reproducing in human tissues. Where these organisms are present, human tissues and secretions serve as a reservoir of infection. The sources of infectious body fluids are referred to as *portals of exit.* They include (1) the gastrointestinal tract, (2) the genitourinary system, (3) the oral region, (4) the respiratory tract, (5) the blood and blood derivatives, and (6) lesions of the skin and other areas.

Individuals who harbor pathogens are called carriers, several types of which are recognized: *healthy carriers* who harbor pathogens without ill effects, *incubatory carriers* who harbor pathogens while undergoing the beginning stages of a disease, and *convalescent carriers* who harbor pathogens during recovery period.

Zoonoses

Diseases that primarily affect lower animals but can also be transmitted to human beings by natural means are referred to as *zoonoses.* Examples of such diseases are listed in Table 21-2. Several animals also serve as sources of parasites that affect human beings.

Categories of Zoonoses

Zoonotic diseases may be grouped in several ways. Generally, they are classified according to the major reservoir of the infectious agent and the mode of transmission of the infectious agent among natural hosts. Zoonoses can also be less formally grouped as to the major human populations at risk for acquiring a zoonotic disease.

Principal Modes of Transfer for Infectious Disease Agents

Infectious disease agents can be transmitted to susceptible individuals in a variety of ways. For example, *mechanical transmission* refers to a pathogen being physically transported by an insect—e.g., cockroaches and flies—from contaminated material such as food or water to other objects. The mechanical means of disease transmission include the *5 Fs:* food, fingers, flies, feces, and fomites.

In *biological transmission,* a portion of the pathogen's life cycle is carried out in the vector. Transmission of the malarial pathogen by the *Anopheles* mosquito is an example of biological transmission. The biological means of transmission include the injection of blood and blood products, warm-blooded animal bites, arthropod bites, and the introduction of arthropod feces into bites and wounds (see Table 21-3).

Arthropods and Disease

Many arthropod species are of medical and economic importance. Their significance lies partially in their ability to produce serious injuries, or even sensitization, as in the case of centipedes, wasps, and spiders. Several arthropods serve either as intermediate hosts for parasites or as vectors for pathogenic microorganisms. It is the latter aspect that concerns us in this section.

Although arthropods vary greatly, they all share several major characteristics: (1) a rigid or semirigid exoskeleton composed of chitin, (2) a complete digestive tract, (3) an open circulatory system (with or without a dorsally situated heart) that forms a body cavity (hemocoel), and (4) excretory, nervous, and respiratory systems.

TABLE 21-2 Representative Zoonoses Produced by Microorganisms

Disease	*Associated Animal(s)*	*Major Mode of Transmission*
Bacterial infections		
Anthrax	Domestic livestock	Direct contact with infected and contaminated soil
Brucellosis (undulant fever)	Domestic livestock	Direct contact with infected tissues; ingestion of milk from diseased animals
Bubonic plague	Rodents	Fleas
Leptospirosis	Dogs, rodents, wild animals	Direct contact with infected tissues and urine
Relapsing fever	Various rodents	Lice and ticks
Rocky Mountain spotted fever	Dogs, rodents	Ticks
Salmonellosis	Dogs, poultry, rats	Ingestion of infected meat; contamination of water
Tularemia	Wild rabbits	Direct contact with infected tissues; deer flies, ticks
Fungus infections		
Several forms of ringworm	Various domestic animals (e.g., cats, dogs)	Direct contact with infected animals and/or fomites
Protozoan infections		
African sleeping sickness (trypanosomiasis)	Humans, wild game animals	Tsetse flies
Chagas' disease	Humans, wild animals	Kissing bugs
Kala azar, leishmaniasis	Cats, dogs, rodents	Sandflies
Toxoplasmosis	Birds, wild rodents, domestic animals (e.g., cats)	Aerosols; possibly contamination of food and water
Viral infections		
Eastern equine encephalitis (EEE)	Birds; horses and related animals	Mosquitoes
Influenza	Humans, swine, horses	Direct contact with droplets
Jungle yellow fever	Various species of monkeys	Mosquitoes
Rabies	Bats, cats, dogs, humans, skunks, wolves, etc.	Bites, contamination of wounds with infectious virus-containing material

Note: Refer to Chapters 25-29 for additional details concerning the infectious diseases listed in this table.

Members of the class Insecta have segmented bodies and jointed legs. Properties of different arthropod classes are summarized in Table 21-4.

The Hospital Environment

The hospital environment is a potential reservoir of infection, for it houses patients with a variety of pathogenic microorganisms and a large number of susceptible individuals. Today, *nosocomial* (hospital-acquired) infections pose serious and far-reaching problems.

Many factors contribute to the problem of nosocomial infections: (1) overcrowding and staff shortages in hospitals; (2) the closing of most communicable disease hospitals; (3) the indiscriminate, frequent, and prolonged use of broad-spectrum antibiotics; (4) the tendency toward longer, more complicated surgical procedures; (5) a false sense of security that has fostered neglect of aseptic techniques; and (6) the use of immunosuppressive agents such as steroids, anticancer drugs, and irradiation.

TABLE 21-3 Modes of Disease Agent Transfer

Category	*Description*	*Diseases Transmitted*	*Control Measures*
Direct contact	Physical contact with infected wounds, lesions, mouth, or contaminated hands. Several disease agents gain access to the body through the nose and throat.	Chicken pox, poliomyelitis, common cold, bacillary dysentery, and streptococcal infections	Washing of hands following blowing the nose, defecating, urinating, or contacting infected individuals
Indirect contact	Ingestion of or contact with food, water, or inanimate objects that are contaminated (*fomites*) or contain infectious body fluids or disease agents.	Amebic dysentery, shigellosis, cholera, typhoid fever, brucellosis, tapeworms	Regular inspection of food-handling personnel and equipment; better sanitary procedures in homes and restaurants
Air and dust	Inhalation of particles bearing microorganisms produced during normal activities of the respiratory tract and the redistribution of accumulated particles of room dust.	Fungus infections, coccidioidomycosis, histoplasmosis	Preventing the use of sprays, use of special floor coverings to trap dust; wearing masks in high-risk situations
Fomites	Inanimate objects or substances capable of absorbing and transferring infectious organisms.	Common cold, influenza, infectious mononucleosis, chicken pox, various diseases of plants	Improved sanitation, proper disposal of fomites, use of chemicals for sterilization
Accidental inoculation	Introduction of infectious agents through accidental injuries.	Tetanus, hepatis B infection	Proper sterilization of contaminated tools, instruments, etc.; practicing laboratory safety
Arthropods	The introduction of infectious agents through arthropod bites or fecal contamination of wounds.	Arbovirus infections, malaria, typhus fever, Rocky Mountain spotted fever	Control of arthropod populations, immunizations, effective treatment of infected individuals

Medical devices can also bring about infections by (1) damaging or invading skin or membrane barriers to infection, (2) supporting the growth of microorganisms and thus serving as reservoirs of disease agents, (3) interfering with host defense mechanisms, and (4) directly infecting individuals when contaminated.

To safeguard the well-being of patients and prevent the transmission of infectious diseases, health-care personnel use the basic measures of medical asepsis and surgical asepsis. *Medical asepsis* employs techniques to reduce the direct or indirect transmission of pathogenic microorganisms—both by reducing their number and hindering their transfer from one person or place to another. The techniques include washing, dusting, disinfection, and isolation.

Surgical asepsis involves those practices that make and keep objects and areas sterile, that is, free from all microorganisms. Aseptic techniques are necessary in all surgical procedures or those involving the body's deeper tissues. Surgical asepsis exacts very high standards of sterility of materials used in the hospital. Any object not known to be sterile—including the outside wrappings of sterile supplies—is assumed to be unsterile. In addition, sterile objects and packages must be kept dry, since moisture can carry bacteria to the sterile area.

One new method developed, the *total protected environment* (TPE), has been shown to reduce significantly the incidence of serious infections in severely compromised patients. The goal of TPE is to eliminate the patient's endogenous microbial flora and to prevent colonization by new microorganisms.

Health care facilities vary greatly in the specific policies they maintain for the isolation and control of communicable diseases. Each institution should put its policies in writing, and complete details of procedures to be used should be available to all staff members.

TABLE 21-4 Properties of Arthropods

Arthropod Class	*Common Example(s)*	*Description*	*Diseases Transmitted or Caused*
Arachnida	Ticks, mites, spiders, etc.	The life cycle of ticks and mites begins with an egg. Six-legged larvae hatch from the eggs and develop into eight-legged nymphs. These forms later become adults. Certain infectious disease agents are known to pass into the eggs of infected ticks (through *transovarian passage* or *transmission*) and into the larvae of mites. Ticks are larger than mites and have leathery bodies on which are found a few short hairs or none at all. Long hairs are present on the membranous bodies of mites. Ticks also have an exposed, teeth-bearing *hypostome* (a rodlike organ at the base of the beak) used for attachment.	Rocky Mountain spotted fever and other rickettsial diseases, tick paralysis, scabies (mite infestation)
Insecta	Lice	Generally, the mouth parts of lice are adapted either for sucking the blood and tissue fluids of mammals or for chewing epitheloid structures in their host's skin. The life cycle of lice is composed of the egg (nit), nymph, and adult stages. Male and female organs are in separate insects. Blood meals can be taken by both the nymph and adult forms. Human lice can be found clinging to hair or clothing.	Epidemic typhus, trench fever, relapsing fever
Hemiptera	Fleas, reduviid bugs, mosquitoes, aphids	Most of the Hemiptera are plant eaters. However, several species apparently have abandoned plant feeding for insects. Probably no other group of arthropods exerts as pronounced an effect on human welfare as these insects do.[a]	Chagas' disease, dog tapeworm, arbovirus infections, various plant diseases

[a] Refer to the text for details concerning the reproduction cycles of the different hemiptera.

Diseases of Plants

The causative agents of plant infections and crop spoilage are for the most part the same types as those responsible for animal infections—bacteria, fungi, viruses, and parasitic nematodes. Fortunately, however, none of the organisms affecting plants has thus far been shown to be capable of producing infections in humans or other animals. Plant-disease agents are transmitted in a wide variety of ways—by arthropods, through contaminated soil and tools, during grafting procedures, by mechanical inoculation (the rubbing of abrasives on the surfaces of leaves), and by seeds and even certain species of fungi and nematodes. Many factors, such as host resistance, temperature, moisture, and virulence of the disease agent, are important to the development of plant diseases. In many ways, the disease process in plants closely resembles that in animals.

Approaches to Identifying Pathogens

The isolation of an unknown pathogen is important to providing adequate treatment and to disease control. Laboratory identification of an organism includes:

1. The collection and transport of appropriate specimens
2. Prompt microscopic examination of specimens (if appropriate)
3. Determination of antibiotic sensitivity
4. Use of specific and nonspecific diagnostic immunological tests

Results of laboratory findings should be communicated to physicians or other appropriate individuals (e.g., public health officials).

The procedures required for the identification of a disease agent vary significantly with the type of organism involved. Such procedures include microscopic examinations, staining, and the use of various laboratory media for isolation and identification purposes.

Collecting and Handling Specimens

The responsibility for determining the type of specimen for clinical identification and diagnosis of a disease agent belongs to the physician. The collection of many different types of specimens may involve physicians, nurses, medical technologists, or other individuals. Examples of specimen types used for the isolation of causative agents of specific bacterial, mycotic, and some worm (helminthic) infections are listed in Table 21-5. Several methods used for vital detection and identification are described in Chapter 10.

TABLE 21-5 Representative Types of Specimens Used for the Laboratory Diagnosis of Certain Bacterial, Mycotic, and Helminthic Infections

Infection	*Autopsy Material*	*Biopsy Material*	*Blood*	*Spinal Fluid*	*Sputum*	*Stool*	*Mucous Membranes or Vesicular Fluid or Skin Scraping*	*Throat Swab*	*Throat Washing*	*Urethral Discharge*	*Urine*	*Hair or Nail Material*
Actinomycosis (lumpy jaw)		+			+							
Boils		+					+					
Diphtheria								+	+			
Enteritis (salmonellosis)		+			+					+		
Gas gangrene	+	+										
Gonorrhea										+		
Meningitis	+		+	+								
Shigellosis						+						
Syphilis		+	+	+								
Tuberculosis	+	+		+	+			+			+	
Urinary tract infection		+								+	+	
Athlete's foot (Tinea pedis)							+					+
Coccidioido-mycosis	+	+	+		+							
Histoplasmosis	+	+	+		+							
Moniliasis (oral thrush)					+		+					
Nocardiosis	+	+			+							
North American blastomycosis	+	+			+							
Ringworm of most skin and hair surfaces							+					+
Vaginal thrush							+					
Tapeworm infections						+						

Materials containing microbial pathogens for identification should be transported rapidly and in a manner that favors the survival of the disease agent. Attention should also be given to public safety and federal health guidelines.

Laboratory Procedures

Several different properties of suspected disease agents that have been isolated are used for identification purposes—staining reactions, colonial appearance, and odors of cultures. A variety of media are used to assist the microbiologist. Several media are used to determine the ability of a particular organism to utilize or enzymatically attack certain substances. Specific chemical products frequently are produced that can also be analyzed for identification purposes. Biochemical tests and media used for identification and study of microorganisms such as bacteria can take several forms. These include:

1. Media in separate Petri plates or test tubes, depending on the nature of the substance (e.g., blood agar would go in a Petri plate and carbohydrate broth in a test tube);
2. Combinations of several test substrates in one tube (e.g., triple sugar iron agar slants);
3. Several miniaturized compartments containing material for separate biochemical tests organized into one plastic strip or container;
4. Paper disks impregnated with biochemical substrates.

Computerized inoculating systems also are in use.

General Identification Procedures

Specimens containing pathogens must be handled aseptically. With such specimens and cultures resulting in inoculated media, staining procedures are routinely used.

Identification Keys

Cellular and colonial appearance, hemolytic reactions, and various biochemical tests are important to the identification of pathogens. The reaction patterns of biochemical tests for known bacterial species are used in the identification of isolated unknown pathogens.

Guidelines for the Identification of Selected Microorganisms

Bacterial pathogens have several distinctive properties that can be demonstrated with the aid of staining procedures, various media, and biochemical tests. Several of these properties are summarized in Table 21-6. Microbial identification and the determination of antibiotic sensitivities can be obtained by automated and computerized systems.

Quality Control Considerations

To assure their high quality, diagnostic procedures are checked frequently. This involves routinely challenging the procedures of laboratories and personnel with unknown specimens, monitoring of equipment performance, and maintaining laboratory equipment properly.

TABLE 21-6 Characteristics of Some Bacterial Pathogens

Pathogen	*Gram/ Acid-Fast Reactions*	*Morphology*	*Medium or Test*	*Distinctive Reaction of Pathogen*
Staphylococcus aureus	+/—	Coccus	Mannitol-salt agar	Acid from mannitol (yellowing of medium)
			Blood agar	Beta hemolysis (clear area around colonies)
			Coagulase	Positive coagulase test (solidification of plasma substrate)
			Catalase	Catalase positive (degradation of hydrogen peroxide into oxygen and water)
Streptococci	+/—	Coccus	Catalase	Catalase negative (no production of oxygen and water from hydrogen peroxide)
			Bacitracin	Inhibition of growth of group A beta hemolytic streptococci
			CAMP	Positive reaction produced by group B streptococci
			Immunodiagnostic tests	Positive results with specific organisms
Streptococcus pneumoniae	+/—	Coccus	Blood agar	Alpha hemolysis (formation of green zone of hemolysis around colonies)
			Optochin disk test	Inhibition of growth around disk
			Bile solubility test	Sensitive (colonies dissolved by test reagent)
Corynebacterium diphtheriae	+/—	Rod	Media containing tellurite salts	Grows well
			Methylene blue staining	Show large darkly staining structures in cells (metachromatic granules)
Neisseria meningitidis	—/—	Coccus	Oxidase test	Oxidase positive (colonies turn dark red when exposed to oxidase reagent)
			Glucose and maltose fermentation test	Produces acid from both carbohydrates
N. gonorrhoeae	—/—	Coccus	Oxidase test	Oxidase positive
			Glucose and maltose fermentation test	Produces acid from glucose only
Clostridia	+/—	Rods	Anaerobic conditions or media	Grow well; spore formers
Bacteroides species	—/—	Rods	Anaerobic condition or media	Grow; nonspore formers
Mycobacteria	—/+	Rods	*Mycobacterium* medium (e.g., egg base)	Characteristic colonies are formed
Nocardia species	+/+	Rods	*Mycobacteria* media	Characteristic colony formation

V. CHAPTER SELF-TEST

Continue with this section only after you have read Chapters 3, 7-10, 13, 19, and 21 and have completed Section IV. A score of 80 percent or better is good. If your score is less than 65 percent, reread the chapter.

Correct answers to all questions can be found at the end of this section. Write your responses in the appropriate space provided.

Matching

Select the answer from the right-hand side that corresponds to the term or phrase on the left-hand side of the question sheet. An answer may be used more than once. In some cases, more than one answer may be required.

Topic: Principles of Disease Transmission

_____ 1. Zoonosis

_____ 2. Epidemic

_____ 3. Sporadic

_____ 4. Pandemic

_____ 5. Endemic

_____ 6. Convalescent carrier

_____ 7. Healthy carrier

a. a disease involving large segments of a population for a limited length of time

b. uncommon, irregular diseases that involve few persons

c. epidemics affecting several countries

d. a disease constantly present but affecting relatively few persons

e. an individual harboring a disease agent during a recovery period

f. an individual incubating a disease state

g. none of these

Topic: Means of Disease Transmission

_____ 8. Malaria

_____ 9. Food poisoning

_____ 10. Plague

_____ 11. Rabies

_____ 12. Rocky Mountain spotted fever

_____ 13. African sleeping sickness

_____ 14. Salmonellosis

_____ 15. Influenza

_____ 16. Cholera

_____ 17. Amebic dysentery

a. spread by mechanical means of disease transmission

b. spread by biological means of disease transmission

c. both *a* and *b*

d. neither *a* nor *b*

Topic: Arthropods and Diseases

_____ 18. Shigellosis

_____ 19. Salmonellosis

_____ 20. African sleeping sickness

a. spread by ticks

b. spread by lice

c. spread by mosquitoes

______ 21. Japanese B encephalitis

______ 22. Western equine encephalitis

______ 23. Malaria

______ 24. Bubonic plague

______ 25. Relapsing fever

______ 26. AIDS

d. spread by common house fly

e. none of these

Topic: Properties of Disease Agents

______ 27. Coagulase positive

______ 28. Inhibited by optochin

______ 29. Ferments mannitol

______ 30. Catalase positive coccus

______ 31. Produces CAMP factor

______ 32. Bacitracin sensitive

______ 33. Spore-forming rod

______ 34. Oxidase positive coccus

______ 35. Acid-fast rods

______ 36. Produces urease

a. *Streptococcus pneumoniae*

b. *Neisseria* species

c. Group A beta hemolytic streptococci

d. pathogenic *Staphylococcus aureus*

e. Group B streptococci

f. *Cryptococcus*

g. none of these

True or False

Indicate all true statements with a "T" and all false statements with an "F".

______ 37. The morbidity rate is the number of deaths caused by a particular disease per unit of the population within a given time period.

______ 38. Bats can transmit rabies to humans.

______ 39. A contaminated animate object is called a fomite.

______ 40. Nosocomial infections are acquired mainly in schools and in the home.

______ 41. Mites do not spread any human diseases.

______ 42. Irradiation and the use of certain anticancer drugs can contribute to the development of nosocomial infections.

______ 43. A disease requiring both an arthropod (such as a mosquito) and a vertebrate is known as a metazoonosis.

______ 44. Medical devices can bring about infection by damaging membrane or skin barriers to infection.

______ 45. Individuals harboring *Staphylococcus aureus* do not pose any particular risk to individuals receiving anticancer drugs or other immunosuppressive agents.

______ 46. There is no need to disinfect materials that were used by a person who has died of an infectious disease.

______ 47. The causative agents of plant infections are not of the same general type that cause animal infections.

______ 48. The disease processes operating in plants quite frequently resemble those taking place in animals.

______ 49. The identification of unknown bacterial pathogens frequently involves the use of schemes or keys based on established properties of known organisms.

______ 50. Currently available commercial multicompartment devices containing a variety of biochemical tests can be effectively used for the identification of unknown bacterial pathogens.

______ 51. Bacteremia refers to the presence of bacteria in urine.

______ 52. Anticoagulants added to media have been known to enhance microbial isolations.

______ 53. Staphylococci produce a positive catalase reaction.

Essay Questions

Answer the following questions on a separate sheet of paper.

54. Distinguish between an infection and a disease.

55. What is a syndrome?

56. Can certain medical devices used for disease diagnosis or treatment increase the risk of nosocomial infection?

57. What is the importance of rapid viral diagnoses?

Answers

1. g	15. a	29. d	43. T
2. a	16. a	30. d	44. T
3. b	17. a	31. e	45. F
4. c	18. d	32. c	46. F
5. d	19. d	33. g	47. F
6. e	20. e	34. b	48. T
7. g	21. c	35. g	49. T
8. b	22. c	36. f	50. T
9. a	23. c	37. F	51. F
10. b	24. c	38. T	52. T
11. c	25. a	39. F	53. T
12. b	26. e	40. F	
13. b	27. d	41. F	
14. a	28. a	42. T	

54. An *infection* is caused by a disease-causing agent or *pathogen* and results from the ability of the agent to invade and multiply in the tissues of a host. A *disease,* on the other hand, is caused by a pathogen's interference with the normal functioning of body systems or organs. The mere presence of microorganisms, however, does not necessarily indicate that an infection or disease is occurring.

55. Groups of signs and symptoms occurring in a characteristic pattern are referred to as *syndromes.* Syndromes are of value in diagnosis and determining the distribution as well as the cause or *etiology* of diseases.

56. Medical devices that are used for disease diagnosis or treatment and are placed in contact with a patient's tissues, either temporarily or semipermanently, represent one of the most important factors in the transmission of nosocomial infections. Interestingly enough, device-related infections appear to be the most preventable of the hospital-acquired infections.

57. The need for rapid viral diagnosis is important to provide specific treatment or to take preventable measures to protect a community.

VI. ENRICHMENT

Specimens for the Laboratory Diagnosis of Viral Infections

Knowledge of the distribution of disease as well as the successful management and control of an infectious disease frequently is dependent on the collection of suitable specimens and the application of appropriate diagnostic laboratory procedures. The selection of appropriate specimens for the laboratory diagnosis of viral infections can be complicated at times. This type of problem occurs because several viruses may have attacked the same target organ, making it necessary for the physician to obtain a number of different specimens to detect the specific virus responsible for the illness. For example, several specimens can be submitted from patients with central nervous system (CNS) disease: cerebrospinal fluid samples (CFS) (enterovirus, mumps virus, and perhaps herpes simplex virus); throat cultures (enterovirus, herpes simplex virus); urine samples (mumps virus); stool or rectal swabs (enterovirus). In addition, blood should be collected as an early (acute) phase specimen if serological diagnostic tests become necessary (AIDS virus, mumps virus, arbovirus, and herpes simplex virus).

While the situation appears to be quite involved, it is important to realize that there are usually several clinical and/or patient history clues at the disposal of a physician. Such clues should enable the physician to select the most appropriate specimens. For example, during the summer, when enteroviral meningitis is prevalent, throat, cerebrospinal fluid, and stool specimens should be the specimens of choice. On the other hand, the development of encephalitis in children after experiencing several mosquito bites in wooded areas endemic for California encephalitis virus would suggest whole-blood specimens for culture and serological testing. A number of specimens used for viral diagnosis in the laboratory together with the associated recommended collection procedure are listed in the following table.

TABLE 21-7 Specimens and Procedures for the Collection of Specimens Used in Viral Diagnosis[a]

Specimen(s)	*Collection Procedure(s)*
Blood	
Culture and serological tests	Specimens should be collected during the early (acute) phase of the infection. For serological testing, two or more specimens are usually necessary, taken 2 to 3 weeks apart. Specimens also are taken for the isolation of the pathogen.
Cerebral spinal fluid or other body fluids	Physicians or other qualified persons should collect fluids by means of a sterile needle and syringe or from a catheter site.
Eye	
Conjunctiva	Specimens are obtained by firmly pressing a sterile swab against the inflamed areas of the inner lining of the eyelid.
Corneal lesions	Corneal scrapings are collected by a qualified physician.
Feces	
Stool	Usually 4 to 6 grams of fresh stool sample are collected.
Swab	A swab is inserted about 3 to 5 centimeters into the rectum to obtain a satisfactory fecal specimen.
Nasopharynx	A flexible wire swab is inserted into the nasopharyngeal area, rotated, and then removed.
Skin and/or genital lesion	Blisterlike lesions (vesicles) are ruptured for specimens. The vesicular surface is scraped to obtain fluid and cells from the base of the lesions. Vesicular fluid also can be removed aseptically with the use of an appropriate sterile syringe and needle.
Sputum	Adults are instructed to produce sputum in response to a deep cough. Sometimes a nebulizer is used to obtain a usable specimen.
Throat	A swab is used to obtain a specimen from any inflamed and swollen areas or visible lesions on the posterior pharynx.
Tissue	Specimens are taken directly from areas next to the affected tissue.
Urine	Specimens should be clean voided, first morning urine. Two or three specimens on successive days are generally needed.

[a] Additional details can be found in an article by M. A. Chernesky, C. G. Ray, and T. F. Smith, *Laboratory Diagnosis of Viral Infections*, Cumitech 15 (Washington, DC: American Society of Microbiology, 1982).

22

Microbial Virulence

I. INTRODUCTION

A parasite is defined as an organism that lives on or within a host and obtains its nutrients from it. The bodies of humans as well as those of lower forms of animal life and plants serve as breeding grounds for a wide variety of microbial parasites. In certain situations, microbes can gain access to specific tissues and establish a disease process. The various properties of microbes that contribute to disease states are discussed in this chapter.

II. PREPARATION

Chapters 7, 10, and 21 through 31 should be read before continuing. The following terms are important for you to know. Refer to the glossary and the appropriate chapters of the text if you are uncertain of any of them. A pronunciation guide for selected terms is provided as a learning aid.

1. AIDS
2. algal toxin
3. animal model
4. antigenic
5. autolysis (aw-TOL-ee-sis)
6. bacteremia (back-ter-EE-me-ah)
7. capsule
8. carcinogenic (kar-si-no-JEN-ick)
9. cell wall
10. coagulase (koh-AG-u-lace)
11. cytopathic effect
12. cytotoxin (sye-toh-TOK-sin)
13. deoxyribonuclease (de-ok-see-rye-bo-NEW-klee-ace)
14. endotoxin (en-doh-TOK-sin)
15. enterotoxin (en-ter-oh-TOK-sin)
16. erythrogenic (ee-rith-roh-GEN-ik)
17. exotoxin (ek-so-TOK-sin)
18. fibrin (FI-brin)
19. human immunodeficiency virus
20. hyaluronidase (hi-ah-lu-RON-ee-dace)
21. immunoglobulin protease
22. infectious disease
23. intoxication
24. invasiveness
25. kinase (KYE-nace)
26. latent (LAY-tent)
27. leukocidin (loo-koh-SI-din)
28. limulus (LIM-u-lus) test
29. mutagenic
30. mycosis (my-KOH-sis)
31. mycotoxin
32. neurotoxin

33. opportunist
34. parasite
35. pathogen
36. phagocytosis
37. phytotoxin (fi-toh-TOK-sin)
38. plasmid (PLAZ-mid)
39. pyemia (pie-EE-me-ah)
40. pyrogenic (PIE-ro-gen-ick)
41. septicemia (sep-tee-SEE-me-ah)
42. siderophore (SID-er-oh-four)
43. thermostability
44. toxin (TOK-sin)
45. toxoid (TOK-soid)
46. transferrin (trans-FER-in)
47. ulcer (UL-ser)
48. virulence (VIR-u-lenz)

III. PRETEST

Correct answers to all questions can be found at the end of this section. Write your responses in the appropriate space provided.

Completion Questions

Provide the correct term or phrase for the following. Spelling counts.

1. The term used for the demonstration of bacteria in blood either by culture or microscopic examination is ______________________________ .
2. The presence of bacteria in the bloodstream and their relation to toxic effects is referred to as ________ .
3. The production of localized collections of pus during a disease state is called ______________________________ .
4. The ability of a parasite to establish itself in a host's tissue and survive is called ______________________ .
5. Diseases such as gangrene, tetanus, and scarlet fever are caused by ______________________________ .
6. The S lysin of certain streptococci is noted for its extreme ______________________________ to heat.
7. Aflatoxins are produced by the organisms in the genus ______________________________ .
8. An interference with the normal functioning of a host's physiochemical process by a microorganism is called a(an) ______________________________ .
9. Two structures that enable bacteria to occupy a site on a host cell's surface without necessarily resulting in tissue invasion are (a) ______________________________ and (b) ______________________________ .

Multiple Choice

Select the best possible answer.

______ 10. Which of the following are necessary for a pathogen to be able to cause disease?
(a) attachment to the surface of a susceptible host; (b) resistance to host defenses; (c) penetration of a susceptible host; (d) secretion of toxins; (e) all of these.

______ 11. A pathogen's virulence depends on which of the following?
(a) toxin production; (b) ability to reproduce in a host; (c) invasive properties; (d) all of these; (e) none of these.

_____ 12. Which of the following are exotoxins?
(a) lipopolysaccharide-protein complexes of gram-negative cell walls; (b) hyaluronidase; (c) coagulase; (d) *b* and *c*; (e) none of these.

True or False

Indicate all true statements with a "T" and all false statements with an "F".

_____ 13. Most microorganisms have disease-producing potential given suitable environmental conditions and host.

_____ 14. The capacity of a pathogen to produce disease is called virulence.

_____ 15. Capsules are found inside certain bacterial cells.

_____ 16. *Klebsiella pneumoniae* is noted for its capsule production.

_____ 17. Coagulase converts fibrin to fibrinogen in forming a clot.

_____ 18. Several parasitic worms produce hyaluronidase.

_____ 19. Hyaluronidase dissolves an essential component of several tissues.

_____ 20. Purified streptokinase can be used to dissolve clots.

_____ 21. Exotoxins are proteins.

_____ 22. Toxoid preparations for immunizations generally are made from endotoxins.

_____ 23. The Limulus assay can be used to detect small amounts of bacterial capsules.

_____ 24. Exotoxins are secretion products of growing bacteria.

_____ 25. Botulism results from the ingestion of a temperate bacteriophage.

_____ 26. Exotoxins are responsible for diseases such as scarlet fever, diphtheria, and leprosy.

_____ 27. Fish tapeworm infections can cause a vitamin B_{12} deficiency.

_____ 28. In bacterial diseases such as leprosy, rickettsial infections, and most viral infections, the causative agents grow within host cells and cause cellular destruction.

_____ 29. Streptolysin S may play a significant role in rheumatic disease.

_____ 30. All viral infections result in death of host cells.

_____ 31. Some neurological diseases may be caused by slow viruses.

_____ 32. Distinctions between opportunists and true pathogens are generally obvious.

Answers

1. bacteremia
2. septicemia
3. pyemia
4. invasiveness
5. exotoxins
6. sensitivity
7. *Aspergillus* species
8. infectious disease

9a. pili
9b. flagella
10. e
11. d
12. d
13. T
14. T
15. F
16. T
17. F
18. T
19. T
20. T
21. T
22. F
23. F
24. T
25. F
26. F
27. T
28. T
29. F
30. F
31. T
32. F

IV. CONCEPTS AND TERMINOLOGY

Infectious Disease and Virulence

An infectious disease is an interference with the normal functioning of an organism's physiochemistry by another organism living within its tissues or on its surfaces.

To cause disease, a pathogen must: (1) attach to the surface of and gain entrance into a susceptible host; (2) multiply in host tissues; (3) resist, or not stimulate, host defenses; and (4) damage the host, often by the secretion of *toxins,* or poisons. An organism's capacity to establish a disease process in an animal or plant host depends on the mechanisms by which it carries out these steps. Table 22-1 lists several mechanisms and disease states in which they can be found.

TABLE 22-1 Some Mechanisms of Action Exhibited by Infectious Disease Agents[a]

Mechanism	*Some Disease States and/or Pathogens Exhibiting the Mechanism*
Allergic reactions (e.g., delayed hypersensitivity)	Deep-seated fungus infections, dermatomycosis (fungus skin diseases), helminthic diseases, leprosy, protozoan infections, syphilis, and tuberculosis
Blood loss and/or utilization of vitamin B_{12}	Hookworm and fish tapeworms
Fusion of cellular and viral membranes; includes the formation of giant cells known as syncytia	Viruses, including herpes simplex viruses, type I and II, measles, parainfluenza, respiratory syncytial disease, and varicella-zoster agent
Genetic integration (incorporation of nucleic acid of a virus or of a plasmid into that of the host)	Botulism, diphtheria, and cancerous states induced by oncogenic viruses
Immunodepression (interference with a host's immune responses)	Lepromatous leprosy, measles, syphilis, tuberculosis, virus-induced cancer, and acquired immune deficiency syndrome (AIDS)[b]
Interference with essential body functions	Anthrax, botulism, cholera, diphtheria, plague, rickettsial infections, salmonellosis, shigellosis, and staphylococcal infections
Interference with phagocytosis	Infections with bacterial pathogens such as anthrax bacilli, meningococci, pneumococci, and group A streptococci; yeasts such as cryptococci; and influenza viruses
Interference with phagocytic killing	Bacterial diseases, including brucellosis, gonorrhea, leprosy, meningococcal meningitis, tuberculosis, and typhoid fever; fungus diseases, including histoplasmosis; and protozoan infections, such as leishmaniasis, pneumocystis pneumonia, and trypanosomiasis
Intracellular growth and cellular destruction	Bacterial diseases, such as brucellosis, leprosy, salmonellosis, shigellosis, tuberculosis, and rickettsial infections; most viral infections
Mechanical blockage of organs and/or associated vessels	Helminth (worm) diseases, including ascariasis, filariasis, and schistosomiasis; fungus diseases, such as aspergillosis and candidiasis; the bacterial disease lymphogranuloma venereum (LGV); and the protozoan disease malaria
Migration through body tissues and/or organs	Helminth diseases, including ascariasis, fasciolopsiasis, hookworm, strongyloidiasis, trichinosis

[a] These include microorganisms as well as helminths (worms). The helminths are discussed in Chapter 31. The other disease states mentioned here are discussed in Chapters 23-30.

[b] Human immunodeficiency virus (HIV) is the causative agent of AIDS. AIDS is discussed in Chapters 20 and 27.

Pathogenicity and Virulence

The ability of an organism to cause disease is called *pathogenicity,* whereas *virulence* refers to the extent of pathogenicity. Virulence depends on certain features of the organism, including its ability to colonize body surfaces, its invasiveness, its ability to spread and reproduce in the face of the host's defensive mechanisms, and its production of toxins harmful to the host. Some pathogens may be both invasive and toxogenic.

The resistance of the host also is an important factor. The relationship of virulence (*V*), numbers of pathogens or dosage (*D*), and the resistant state (*RS*) of the host to the establishment of an infectious disease can be shown by the following formula:

$$\text{Infectious disease} = \frac{V \times D}{RS}$$

Colonization

Certain bacteria have structures such as flagella or pili that enable them to colonize or occupy a site on a host's surface where they can reproduce without necessarily resulting in tissue invasion. The capability to attach to host surfaces increases the probability that a disease state will develop.

Invasiveness

Invasiveness refers to the ability of a parasite not only to survive but to establish itself in the tissues of the host. From a microorganisms's point of view, several obstacles must be overcome as it penetrates deeper into the tissues of the host. Some nontoxic bacterial structures and products contributing to invasiveness are listed in Table 22-2.

Antigenic Variation

Certain pathogens can escape the host's immune system by changing their antigenic surface parts. These include the AIDS virus and certain protozoa.

Microbial Toxin Production

Toxins are categorized either as endotoxins or as exotoxins. Endotoxins are substances that are liberated only after the organism disintegrates by self-destruction or autolysis. Exotoxins are products generally released during the lifetime of an organism and, at times, by autolysis. Occasionally, enzymes released from cells also are considered exotoxins—for example, hyaluronidases; coagulase; fibrinolysins such as streptokinases, which dissolve fibrin clots; proteinases, which dissolve proteins; and lecithinase, which decomposes lipids. Several of the characteristics that distinguish endotoxins from exotoxins are listed in Table 22-3.

Endotoxins are clearly involved in disease states caused by gram-negative bacteria, but their exact role is not understood. Various studies have shown their effects to be nonspecific. Minute concentrations of endotoxins can be detected by the Limulus test, which is based on the fact that an aqueous extract of amoebocytes (lysate) from the blood of the horseshoe crab forms a gel in the presence of very small amounts of endotoxins.

The mechanisms and sites of action have been studied in several of the classic exotoxins, including botulism, diphtheria, and tetanus. In general, they function by destroying specific components of cells or by inhibiting certain cellular activities. Some of these substances work only on specific cell types. Based on their effects, exotoxins can be grouped into several categories such as *cytotoxins, enterotoxins,* and *neurotoxins.*

Properties of bacterial exotoxins are described in Table 22-4.

Various other microorganisms such as algae, cyanobacteria, and fungi produce toxins of importance. The properties of some of these poisonous substances are described in Table 22-5.

TABLE 22-2 Nontoxic Bacterial Structures and Products

Bacterial Structure or Product	*Description*	*Microorganisms Associated with Structure or Product*
Capsule	Outer surface component of cells; consists chemically of polysaccharide, protein, or both. Provides protection against phagocytosis.	*Bacillus anthracis, Haemophilus influenzae, Klebsiella pneumoniae, Neisseria meningitidis, Streptococcus pneumoniae,* and some staphylococci
Cell wall components	Chemical surface components that obstruct phagocytosis and contribute to microbial invasiveness.	Group A streptococci, *Streptococcus pyogenes,* and *Staphylococcus aureus*
Coagulase (staphylocoagulase)	Extracellular enzyme; coagulates human or rabbit plasma if coagulase-reacting factor (CRF) is present. Converts fibrinogen to fibrin. Seven distinct varieties are recognized.	Pathogenic staphylococci
DNAse	Heat-stable, calcium-activated enzyme.	Virulent *Staphylococcus aureus* strains
Hyaluronidase	An enzyme that increases tissue permeability. Hydrolyzes hyaluronic acid, a thick, high-molecular-weight polysaccharide that is an essential component of several cells.	Clostridia, pneumococci, streptococci, and various parasitic worms (helminths)
Immunoglobulin A (IgA) proteases	Enzymes capable of cleaving human IgA molecules.	*Haemophilus influenzae,* and *Streptococcus pneumoniae*
Kinases	Dissolve fibrin from specific mammalian sources.	Several gram-positive bacteria, including *Clostridium perfringens,* hemolytic streptococci, and staphylococci
Siderophores	Low-molecular-weight iron-binding protein enables the organism to acquire available iron from the host	Most pathogenic bacteria
Virulence plasmids	Extrachromosomal DNA, encoded for the production of certain toxins, and surface adherence proteins such as those of pili.	*Escherichia coli,* and *Vibrio cholerae*

TABLE 22-3 A Comparison of Selected Characteristics of Endotoxins and Exotoxins

Characteristic	*Endotoxin*	*Exotoxin*
Chemical composition	Lipopolysaccharide-protein complex	Protein
Source	Cell walls of gram-negative bacteria; released only on autolysis or artificial disruption of cells	Mostly from gram-positive bacteria; excretion products of growing cells or, in some cases, substances released upon autolysis and death
Effects on host	Nonspecific	Generally affects specific tissues
Thermostability	Relatively heat stable (may resist 120° C for 1 hr)	Heat labile; most are inactivated at 60° to 80° C
Toxoid[a] preparation for immunization possible	No	Yes
Bacteria associated with endo- and exotoxin	*Escherichia coli,* and species of *Neisseria, Rickettsia, Salmonella, Shigella, Veillonella*	*Bordetella pertussis, Clostridium botulinum, C. tetani, Corynebacterium diphtheriae, Shigella dysenteriae, Staphylococcus aureus, Streptococcus pyogenes*

[a] Modified protein toxin that is not toxic but still causes the production of antibodies.

TABLE 22-4 Properties of Bacterial Exotoxins

Disease Produced and/or Exotoxin Designation	*Microorganisms Producing Toxin*	*Description of Action*
Botulism	*Clostridium botulinum*	A neurotoxin associated with fatal food poisoning; interferes with the transmission of stimuli to muscles by blocking the release of acetylcholine.
Cholera, choleragen	*Vibrio cholera*	Toxin binds to gangliosides in epithelial target cells, activating adenylate cyclase and resulting in the production of cyclic AMP. The reaction causes excess secretions of chloride bicarbonate and water, leading to diarrhea.
Pseudomembranous colitis, cytotoxin B	*Clostridium difficile*	Causes direct cell destruction and suppression of cyclic adenosine monophosphate.
Diphtheria	*Corynebacterium diphtheriae*	Inhibits protein synthesis and causes destruction of various tissues.
Erythrogenic	Strains of betahemolytic group A streptococci	Exhibits selective action in skin.
Enterotoxin	Coagulase-positive strains of *Staphylococcus aureus*	Protein in nature; poor antigens; resistant to boiling temperatures for approximately 30 minutes; not neutralized by antitoxins prepared with other toxins produced by staphylococci. Humans and monkeys appear to be the only naturally susceptible victims of this toxin. Affected individuals generally experience nausea and vomiting within a few hours after the toxin's ingestion. Fatalities are rare.
Gan gangrene (alpha toxin)	*Clostridium perfringens* and other clostridia	Destroys cell membranes.
Heat-labile (LT) enterotoxin	Enterotoxigenic *Escherichia coli*	Similar to choleragen.
Staphylococcal scalded skin syndrome (SSSS)	*Staphylococcus aureus*	Destroys cellular connections and cell membranes.
Streptolysin O	Strains of beta-hemolytic group A streptococci	Is sensitive to oxygen; toxic for red blood cells.
Streptolysin S	Strains of beta-hemolytic group A streptococci	Is toxic to tissues.
Shigella dysenteriae neurotoxin	*Shigella dysenteriae*	Is toxic to nervous tissue.
Tetanus, tetanospasmin	*Clostridium tetani*	A neurotoxin that exerts destructive effect on the anterior horn cells of the central nervous system.
Toxic shock syndrome, enterotoxin F	*Staphylococcus aureus*	Unknown.
Whooping cough	*Bordetella pertussis*	Causes damage to respiratory tract cilia.

TABLE 22-5 Microbial Toxins

Toxin	*Source*	*Description*
Phytotoxins	Bacteria, including species of *Corynebacterium, Erwinia, Pseudomonas, Xanthomonas*	Plant poisons; cause wilting, loss of color, and extensive destruction to leaves, stems, or flower parts.
Cyanotoxins	Cyanobacteria	Produced during toxic blooms. Some can cause rashes and blisters.
Algae	Dinoflagellates and other algae	Poisonous to fish, waterfowl, mussels, clams, and to humans who eat contaminated shellfish.
Mycotoxins (include aflatoxins)	*Aspergillus flavus* (aflatoxins), and other fungi	Cause food or feed deterioration. Some cause human liver damage and cancer. Certain mycotoxins have anti-microbial and antitumor activities. The distinctive characteristics of poisoning or mycotoxicosis include the following: (1) the disease is not transmissible; (2) drug or antibiotic treatments have little or no effect on the disease; (3) outbreaks in the field are seasonal because certain climatic conditions affect mold develop-ment; (4) the outbreak is usually associated with a specific feed or foodstuff; and (5) examination of the suspected food or feed reveals signs of fungal activity.

Viral Pathogenicity

Viruses cause diseases by breaking through host defenses. As with bacteria, this process depends not only on the strength of the defenses and the microorganism's capacity to counteract them, but also on the number of invaders. A sufficiently large infecting dose can overwhelm the initial defenses of a susceptible host and cause irreparable injury before adequate defenses can be deployed.

That a virus attack produces cellular damage of animal tissues has been known for years. In addition, viral replication can occur in cells without significant damage. It has also become apparent that some slowly developing, persistent diseases that superficially do not appear to be infectious can be caused or triggered by unusually "slow" viruses. Strong evidence has accumulated suggesting that several severe neurological diseases are caused by viral agents.

There is increasing evidence that viruses are more actively responsible for cell damage. One process by which this damage occurs is virus cytotoxic activity. There are two levels at which pathologically important cytotoxic activity can operate, namely, biochemical damage without morphological damage and biochemical damage with morphological damage (cell lysis, fusion, or death). Morphological damage is usually referred to as a cytopathic effect.

Opportunists and True Pathogens

True pathogens are a relatively small fraction of those organisms harbored by obvious infections. *Opportunists* are organisms that have the potential to produce infections if they accidentally gain access to the tissues of the host under special circumstances. An opportunistic microorganism takes advantage of weakened defense mechanisms to cause damage to a "compromised" host. Such microorganisms may or may not be members of a host's normal resident flora, and they may or may not be pathogenic to a normal host.

Animal Models

Experimental methods using laboratory animals are in wide usage to study the various factors associated with the development and control of infectious diseases. Appropriate laboratory animals such as mice, rats, and rabbits are incorporated into experiments designed for various purposes, including: (1) following the course of infection, (2) identifying factors involved in host resistance, (3) uncovering structural and molecular abnormalities, (4) testing the effectiveness of drugs for treatment, (5) determining the effectiveness and safety of vaccines, and (6) identifying the role of virulence factors in disease processes.

V. CHAPTER SELF-TEST

Continue with this section only after you have read Chapters 7, 10, and 21-31 and have completed Section IV. A score of 80 percent or better is good. If your score is less than 65 percent, reread the chapter.

Correct answers to all questions can be found at the end of this section. Write your responses in the appropriate space provided.

Completion Questions

Provide the correct term or phrase for the following. Spelling counts.

1. ______________________ are encoded for toxin production and pili.
2. Most pathogenic staphylococci are noted for their production of the extracellular enzyme __________ .
3. A microbial interference with a host's physiochemical processes is referred to as a(an) __________ .
4. Examples of antiphagocytic bacterial cell surface factors include (a) __________ , (b) __________ , and (c) __________ .
5. ______________________ is an example of a bacterial enzyme capable of dissolving fibrin clots.
6. Bacterial exotoxins, based on their effects on host cells, can be grouped into the following three categories: (a) __________ , (b) __________ , and (c) __________ .
7. The toxin ______________________ is responsible for the effects associated with toxic shock syndrome.
8. Enterotoxins are produced by several bacterial species, including (a) __________ , (b) __________ , (c) __________ , and (d) __________ .
9. ______________________ are organisms that have the potential to produce infections if they accidentally gain access to the tissues of the host under special circumstances.
10. Lower animals are used in experimentally designed studies to (a) __________ , (b) __________ , (c) __________ , (d) __________ , (e) __________ , and (f) __________ .

True or False

Indicate all true statements with a "T" and all false statements with an "F".

______ 11. A large number of pathogens can maintain themselves only within the systems of animals and plants they invade.

______ 12. The capacity of an organism to produce disease is called virulence.

______ 13. Toxins are generally harmful to the microorganisms producing them.

______ 14. Allergic reactions are involved in the mechanisms of action associated with diseases such as leprosy and tuberculosis.

______ 15. Various diseases caused by worms can result in the blockage of body organs.

______ 16. A bacteremia refers to the production of pus by bacteria.

_______ 17. Loss of the ability to form capsules lowers an organism's virulence.

_______ 18. Coagulase prevents the normal coagulation of serum.

_______ 19. Several gram-positive species produce coagulase.

_______ 20. Most exotoxins are produced through the autolysis of bacterial cells.

_______ 21. Chemically, endotoxins are mostly protein.

_______ 22. Toxoids are generally prepared from exotoxins.

_______ 23. Endotoxins are generally obtained from the cell walls of gram-positive organisms.

_______ 24. Botulism results from the action of a specific neurotoxin.

_______ 25. Erythrogenic toxins of streptococci exert a selective action on the skin.

_______ 26. Phytotoxins of certain bacterial species can cause plants to wilt or to lose their coloration.

_______ 27. Most marine cyanobacteria present a serious health problem.

_______ 28. Toxin-producing algae pose a serious threat to human well-being.

_______ 29. Mycotoxicoses are disease states caused by ingestion of foods contaminated by species of mycobacteria.

_______ 30. Mycotoxicoses are often seasonal in occurrence.

_______ 31. Certain mycotoxins such as aflatoxins are carcinogenic.

_______ 32. Some mycotoxins exhibit antimicrobial effects.

_______ 33. Virus infections can result in biochemical and morphological injury to host cells.

_______ 34. An opportunistic microbe takes advantage of weakened host defense mechanisms.

Essay Questions

Answer the following questions on a separate sheet of paper.

35. How do virulence plasmids contribute to microbial pathogenicity?
36. Of what importance is iron to a bacterium?

Answers

1. virulence plasmids
2. coagulase (staphylocoagulase)
3. infectious disease
4a. hyaluronic acid of Group A streptococci
4b. proteins of Group A streptococci
4c. a protein of *Staphylococcus aureus*
5. streptokinase
6a. cytotoxins
6b. enterotoxins
6c. neurotoxins
7. enterotoxin F
8a. *Clostridium difficile*
8b. *Clostridium perfringens*
8c. *Shigella dysenteriae*
8d. *Staphylococcus aureus*
9. opportunists
10a. follow the course of infection
10b. identify host-resistant factors
10c. uncover host abnormalities
10d. determine drug effectiveness
10e. determine vaccine effectiveness and safety
10f. identify virulence factors
11. T
12. T
13. F
14. T
15. T
16. F
17. T
18. F
19. F
20. F
21. F
22. F
23. F
24. T
25. T
26. T
27. F
28. T
29. F
30. T
31. T
32. T
33. T
34. T

35. Virulence plasmids are encoded for various microbial activities, products, or structures. The best known plasmids are those that specify resistance to various antibiotics. Other virulence plasmids are encoded for toxin production and for the formation of surface-adhering structures such as pili.

36. Iron is essential to the metabolism and other activities of a bacterium since it is a vital part of cytochromes, catalase, and nonheme iron-containing electron transport proteins. To be able to grow and to reproduce, a bacterium must stabilize iron in the form of ferric ions and transport them into the cell where they are reduced to the ferrous usable state.

VI. ENRICHMENT

Aflatoxins

Aflatoxin is a collective term that refers to a group of highly poisonous and cancer-causing substances produced by the common molds *Aspergillus flavus* and *A. parasiticus* during their growth on foods including milk, dairy products, and animal feeds. The concern with aflatoxin and the toxigenic aspergilli is twofold. First, there is the potential health hazard of consuming even small amounts of aflatoxin. Second, since toxigenic aspergilli are widely distributed in nature, there is a greater likelihood that they will contaminate various foods and animal feeds.

Major Aflatoxins

Although more than one dozen forms of aflatoxins have been identified, only the five major types—B_1, B_2, G_1, G_2, and M_1—found in milk and dairy products, will be discussed.

Aflatoxins B_1, B_2, G_1, and G_2 are the major forms produced by the molds, with B_1 and G_1 usually produced in largest amounts. The B-aflatoxins are given their designation because they exhibit a bluish color under long-wave ultraviolet light, and the G toxins are given the designation because they produce a greenish-yellow color under the same type of light. The form of aflatoxin produced from aflatoxin B_1 by cows and excreted in milk is called M_1.

Chemically, aflatoxins are substituted coumarins that are relatively small molecules. Aflatoxins are not proteins as are many of the toxins produced by bacteria. Aflatoxins are fairly heat stable, but they can be degraded by strong acidic or alkaline solutions, oxidizing agents, some molds, and a few bacteria.

Milk and Cheese Contamination. Milk becomes contaminated by aflatoxins only when cows consume feed that contains aflatoxin B_1. This is the major and always the most toxic form of aflatoxin. Some of the ingested aflatoxin B_1 is converted to M_1 by the liver of the cow. The resulting aflatoxin is excreted in the milk. Products made from such milk will also contain aflatoxin M_1.

Growth of a toxigenic *Aspergillus* on cheeses also can result in contamination with one or several of the aflatoxins that are formed by the mold. It is possible for cheese to contain M_1, if it is made from contaminated milk, and also B_1 and other forms of aflatoxin, if the same cheese also supports the growth of a toxigenic mold.

Aflatoxin in Moldy Cheese. Various studies have shown that commercial cheese products are not well suited for aflatoxin production. There are several factors that account for this situation, including (1) cheese does not contain the carbohydrate needed by aspergilli for the maximum aflatoxin production, (2) the temperatures commonly used to store cheeses (11° to 13°C) are below the minimum temperatures necessary for aflatoxin production, and (3) the presence of other species of molds can easily outgrow aspergilli, which do not survive well in a competitive environment. Despite the presence of these factors, care must be taken in the handling of cheese and related produced to prevent the development of conditions that favor the growth of toxigenic aspergilli.

23

Microbial Diseases of the Integumentary and Musculoskeletal Systems

I. INTRODUCTION

Skin, hair, nails, and various glands make up the covering of the human body. When the integrity of the skin is impaired as a result of accidental wounds, burns, surgery, or malnutrition, invasion by various types of infectious agents—bacteria, fungi, protozoa, viruses, and worms—can occur. In addition, the skin and related tissues are used by several pathogens in their development and multiplication. These organisms may also invade internal organs and tissues, causing systemic infection. This chapter provides a brief review of skin structure and descriptions of several microbial infections associated with the skin and musculoskeletal system. A Disease Challenge is included at the end of the chapter.

II. PREPARATION

Chapters 4, 7-12, 18-22, and 23 should be read before continuing. The following terms are important for you to know. Refer to the glossary and the appropriate chapters of the text if you are uncertain of any of them. A pronunciation guide for selected terms is provided as a learning aid.

1. abscess (AB-sess)
2. acid fast
3. aerosol (AIR-oh-sol)
4. arthritis (ar-THYRE-tis)
5. bleb
6. booster (BOO-stir)
7. bursitis (ber-SYE-tis)
8. capsule
9. carbuncle
10. crust
11. dermatophyte (DER-mah-toe-fight)
12. epidermis
13. exotoxin
14. fever
15. folliculitis (foh-lick-you-LIE-tis)
16. furuncle
17. gram-negative
18. gram-positive
19. hemolysis (he-MOLL-ee-sis)
20. hypersensitivity
21. incision (in-SIZZ-shun)
22. inflammation

23. *in vitro*
24. laceration (lass-eh-RAY-shun)
25. lepromin (LEP-roh-min) test
26. macule (MAC-ool)
27. maculopapule
28. malaise
29. meningitis (men-in-JYE-tis)
30. mycosis (my-KOH-sis)
31. nasopharyngitis (nay-zoh-far-in-JYE-tis)
32. opportunist
33. osteomyelitis (os-tee-oh-my-ee-LIE-tis)
34. oxidase
35. papule
36. penicillinase (pen-ee-SIL-eh-nase)
37. prophylaxis (pro-feh-LACK-sis)
38. pus
39. pustule
40. ringworm
41. scalded skin syndrome
42. scars
43. septicemia (sep-teh-SEE-me-ah)
44. sequelae (see-KWAL-lah)
45. serology (seh-ROL-oh-je)
46. skin test
47. subcutaneous
48. superficial mycotic infection
49. symptomatic treatment
50. synovial (seh-NO-vee-al) joint
51. thrush
52. systemic infection
53. toxin
54. ulcer
55. vesicle (VES-eh-kel)
56. virion (VYE-ree-on)
57. virulence

III. PRETEST

Correct answers to all questions can be found at the end of this section. Write your responses in the appropriate space provided.

Completion Questions

Provide the correct term or phrase for the following. Spelling counts.

1. The two main structural parts or divisions of the skin are the outer (a) ______________________ and the underneath, the (b) ______________________.
2. Localized inflamed areas of the skin that develop soft centers and eventually discharge pus are ________.
3. Blisterlike elevations appearing on the skin that may contain blood or serum are ______________________.
4. The causative agent of Hansen's disease is ______________________.
5. Species of the genus *Clostridium* can be described with respect to their gram reaction, morphology, and oxygen requirements as (a) ______________________, (b) ______________________, and (c) ______________________, respectively.
6. The causative agent of wound botulism is ______________________.
7. Ringworm of the body's nonhairy skin areas is called ______________________.

Matching

Select the answer from the right-hand side that corresponds to the term or phrase on the left-hand side of the question sheet. An answer may be used more than once. In some cases, more than one answer may be required.

_____ 8. Gas gangrene

_____ 9. Carbuncles

_____ 10. Toxic epidermal necrolysis

_____ 11. Erysipelas

_____ 12. Folliculitis

_____ 13. Gram-positive sporeformer

_____ 14. Tinea capitis

_____ 15. Shingles

_____ 16. Produces a positive reaction on mannitol-salt agar

_____ 17. Green nail syndrome

_____ 18. Leprosy

_____ 19. Scarlet fever

_____ 20. Produces a positive coagulase test

_____ 21. Scalded skin syndrome

a. beta-hemolytic group A streptococci

b. *Bacillus anthracis*

c. *Pseudomonas aeruginosa*

d. *Staphylococcus aureus*

e. *Clostridium perfringens*

f. *Clostridium tetani*

g. *Mycobacterium* species

h. *Trichophyton* species

i. a herpes virus

j. none of these

True or False

Indicate all true statements with a "T" and all false statements with an "F".

_____ 22. Intact skin serves as a natural protective barrier to the majority of pathogens.

_____ 23. Macules are round, circumscribed changes in skin color that are neither elevated nor depressed.

_____ 24. Maduromycosis is a bacterial disease that can occur in a variety of warm-blooded animals.

_____ 25. Both *Staphylococcus aureus* and *Streptococcus pneumoniae* are prominent causes of bacterial arthritis.

_____ 26. Leprosy also is called Hansen's disease.

_____ 27. Bluish-green pus and a grape odor are characteristic features of gas gangrene.

_____ 28. Gas gangrene can occur by the contamination of open wounds by *C. perfringens* spores.

_____ 29. The bacterium known to cause rat leprosy is the same one responsible for human leprosy.

_____ 30. *Staphylococcus aureus* infections are generally limited to the skin.

_____ 31. Lyme disease is caused by a spirochete transmitted by *Ixodes* ticks.

_____ 32. Pustules develop from papules.

_____ 33. The presence of bacteria in the blood is known as septicemia.

_____ 34. The trichophytid reaction indicates an allergy to fungal products.

_____ 35. The same virus is the causative agent of varicella and measles.

Answers

1a. epidermis
1b. dermis
2. furuncle
3. vesicles
4. *Mycobacterium leprae*
5a. gram-positive
5b. rods
5c. anaerobic
6. *Clostridium botulinum*
7. tinea corporis
8. e
9. d
10. d
11. a
12. c
13. b, e, and f
14. h
15. i
16. d
17. c
18. g
19. a
20. d
21. d
22. T
23. T
24. F
25. T
26. T
27. F
28. T
29. T
30. T
31. T
32. T
33. T
34. T
35. F

IV. CONCEPTS AND TERMINOLOGY

The Organization of the Skin

The components of the different layers or regions of the skin are listed in Table 23-1. Functions of the skin include excretion, reception of external stimuli, secretion, and temperature regulation. Intact skin also serves as a natural protective barrier to the majority of infectious disease agents. Factors contributing to this barrier include a low pH, high salinity, unsaturated fatty acids, thick keratinized layers, the presence of an indigenous flora, and a temperature that prevents the growth of certain disease agents.

TABLE 23-1 The Components of the Skin

Layer or Region	*Description*
Epidermis (outer layer)	Contains several layers of skin cells, including regions for new cell and pigment production
Dermis	Located underneath the epidermis; contains dense, irregularly arranged connective tissue and various types of cells, including fibroblasts, histiocytes (phagocytes), and mast cells; also contains blood and lymphatic vessels, nerves, hair follicles, sweat glands, superficial sebaceous glands, and a variable amount of muscle
Subcutaneous	Located below the dermis; attaches the skin to underlying structures. Contains a large number of components, including fat tissue, blood vessels, special nerve endings, nerve trunks, hair follicles, sebaceous glands, and sweat glands.

Skin Lesions

The breaks in the skin that occur during the course of normal living provide opportunities for infectious agents to enter the body. In addition, hair follicles and the openings of secreting glands constitute potential portals of entry for pathogens.

So-called "minor" infections can develop into serious problems if they: (1) spread to involve neighboring tissues, (2) cause bleeding, (3) produce local anemia due to stoppage of circulation, (4) result in edema (swelling), or (5) cause loss of skin.

Table 23-2 provides brief descriptions of skin lesions that result from infection.

Bacterial Diseases

The signs, symptoms, and pathological features of skin infections vary from localized effects, such as those listed in Table 23-2, to extensive penetration of deeper tissues and organs that may result in death. The

TABLE 23-2 Brief Descriptions of Selected Skin Lesions

Type of Lesion	*Description*
Blebs, blisters	Thin-walled, rounded, or irregularly shaped blisters containing serum or a combination of serum and pus
Carbuncle	A deep sore or ulcer lesion of the skin and subcutaneous tissue; usually a hardened border and draining of pus are evident
Crusts (crustae or scabs)	Dried accumulation of blood, pus, or serum combined with cellular and bacterial debris; detachment of thin crusts may leave dry or moist red bases; these generally heal, resulting in a smooth skin surface; scar formations are usually associated with thick crusts covering ulcers
Folliculitis	Inflammation of hair follicles
Furuncles	Localized inflamed regions that develop soft centers and eventually discharge pus
Macules (maculae, spots)	Usually round, circumscribed changes in the color of the skin; the lesion is neither elevated nor depressed; the outline of a spot or macule may be either quite distinct or may blend into the surrounding region
Maculopapules	Slightly raised macules
Papules (papulae or pimples)	Circumscribed, solid, elevated lesions without visible fluid contents; the color, consistency, and size of papules vary
Pustules (pustulae)	Small elevated skin lesions containing pus; they may develop from papules; these lesions also vary in color, size, and contents (pus, blood, or both); pustules may consist of a single cavity or several compartments with fluid
Scars	Newly formed connective tissue that replaces tissue lost through injury or disease; these secondary lesions tend to be pink at first, then they assume a glistening appearance; scars are normal components of the healing process
Ulcers	Rounded or irregularly shaped depressions or excavations; these lesions vary in size
Vesicles (small blisters)	Elevations that may occur irregularly or in groups or rows; they may contain blood, pus, or serum; the color of the lesion depends upon its contents; vesicles may arise from a macule or papule and develop into pustules.

identification of microbial pathogens of the skin frequently depends on their isolation and cultivation and the results of various types of biochemical tests. Skin tests or other immunological procedures may also be useful in diagnosis.

Table 23-3 describes selected bacterial skin infections.

Mycotic Infections

Human skin, nails, and hair are particularly vulnerable to attack by certain pathogenic fungi. Several of these agents are capable of affecting the skin and related tissues that also contain the protein *keratin.* Most of these fungi, called dermatophytes, are found worldwide.

Superficial Mycoses

Fungi that attack mainly the epidermis, hair, nails, and mucosal surfaces are called *superficial fungi.* The diseases caused by such agents include the various forms of ringworm, or *tinea* (from the Latin meaning "growing moth"), and *Candida* infections of mucosal surfaces, such as thrush and vulvovaginitis. These infections are frequently referred to as the *superficial* or *surface mycoses.* Superficial mycoses are further classified on the basis of the location of the effects produced by the causative fungus. Table 23-4 lists several of these infections.

TABLE 23-3 Selected Bacterial Diseases of the Skin

Disease	*Causative Agent*	*Gram Reaction*	*Morphology*	*Transmitters and Sources of Causative Agent*	*General Features*	*Prevention and Control*
Anthrax	*Bacillus anthracis*	+	Rod	Handling of infected animals and their products; inhalation of spores	A reddened, elevated, swollen pimple develops at site of infection; it may lead to bloodstream invasion (septicemia) and tissue death; oral lesions are reddened and swollen; pulmonary infection causes severe lung damage and death.	Immunization (partial control); disposal of infected animal carcasses by cremation or deep burial; chemical or heat disinfection of infected animal products
Boils, carbuncles	*Staphylococcus aureus*	+	Coccus	Humans	Localized swollen areas of tissue destruction in deeper skin layers; may lead to bloodstream invasion; fever and general malaise.	Better nutrition; improved personal hygiene; appropriate treatment
Erysipelas	Beta-hemolytic group A streptococci	+	Coccus	Humans; contamination of wounds	Fever, headache, stinging or itching at a site of infection developing into widespread thickened, reddened areas.	Prompt diagnosis and treatment of wounds
Folliculitis	*Pseudomonas aeruginosa*	—	Rod	Humans, fomites	Fever, swollen lymph nodes, malaise, reddened blisters (some with pus).	Gentamycin ointment, with lymph node involvement, carbenicillin
	S. aureus	+	Coccus	Humans; fomites	Localized swollen areas that develop soft centers, some with pus.	Same as listed earlier for this bacterium
Furuncles	*S. aureus*	+	Coccus	Humans and possibly contaminated clothing	Localized swollen areas that develop soft centers and eventually discharge pus.	Better nutrition; improved personal hygiene; removal of irritating clothing
Gas gangrene[a]	*Clostridium perfringens* and other other clostridia	+	Rod	Contamination of wounds by spores, contaminated materials, or instruments	Usually affects muscle tissue; fever, fast heartbeat, severe pain; infected wounds smell foul, have a discharge and accumulate gas within tissues.	Sterilization of surgical instruments; proper treatment of wounds
Green nail syndrome and toe web infection	*Pseudomonas aeruginosa*	—	Rod	Contamination of wounds; fomites	Greenish discoloration of nail plate; formation of thick, white scaling areas between toes.	Prompt identification of causative agent and appropriate treatment
Impetigo contagiosa	*S. aureus*	+	Coccus	Droplet nuclei; airborne organisms; carriers; direct contact with infected individuals	Crust, scabs; localized pain and fever accompany the disease.	Prompt and appropriate treatment; proper handling and disposal of contaminated objects; treatment of carriers
	Beta-hemolytic group A streptococci	+	Coccus		Less severe form than for *S. aureus*.	

TABLE 23-3 Selected Bacterial Diseases of the Skin *(continued)*

Disease	*Causative Agent*	*Gram Reaction*	*Morphology*	*Transmitters and Sources of Causative Agent*	*General Features*	*Prevention and Control*
Leprosy (Hansen's disease)	*Mycobacterium leprae*	Not done; acid-fast	Rod	Prolonged contact with infected humans; direct contact with infectious material	Four different types of the disease are recognized: *lepromatous*—round, nonelevated patches showing skin color changes (macules); *tuberculoid*—well-defined, reddened or nonpigmented areas, loss of sensations, and nerve destruction; *intermediate*—macules and nerve involvement, lepromin skin test may be positive; *borderline*—infectious form of the disease showing the features of both lepromatous and tuberculoid leprosy.	Prompt and rapid identification of infected individuals; hospitalization when appropriate; disinfection of all materials contaminated by infected persons
Pseudomonas pyoderma	*Pseudomonas aeruginosa*	—	Rod	Same as for green nail syndrome	Eroded and lacerated skin surface, producing a bluish-green pus and grape odor.	Same as for green nail syndrome
Scarlet fever	Beta-hemolytic group A streptococci	+	Coccus	Droplet nuclei; aerosols; contaminated food and water; direct contact with carriers; fomites	Fever, headache, sore throat, vomiting, reddened rash, "strawberry tongue"; peeling of body surface and tongue may occur.	Better sanitation; prompt diagnosis and appropriate treatment; disposal of contaminated objects
Staphylococcal scalded skin syndrome (SSSS)	*S. aureus*	+	Coccus	Humans; fomites	Initially starts with a distinctive, faint, macular, yellow brick-red rash following an eyelid or respiratory infections; skin tenderness involving central portions of face, neck, armpits, and groin; spontaneous skin wrinkling and huge blisters develop followed by skin peeling; skin drying and healing without scarring.	Symptomatic treatment to control fluid and electrolyte balance, semisynthetic penicillins not affected by penicillinase, methicillin, or dicloxacillin (oral therapy)
Tetanus (lockjaw)	*Clostridium tetani*	+	Rod	Fomites associated with infected wounds; contaminated soil	Sudden and violent involuntary contractions of voluntary muscles, convulsions, locking of jaw muscles; fever and pain may be present.	Proper cleansing of wounds; immunization with tetanus toxoid
Wound botulism	*Clostridium botulinum*	+	Rod	Same as for tetanus	Fever, double vision, difficulty in talking and swallowing, neck weakness.	Proper cleansing of wounds

[a] *Staphylococcus aureus* and *Streptococcus pyogenes* functioning together can produce a form of this disease referred to as synergistic gas gangrene.

TABLE 23-4 Representative Superficial Mycoses

Disease	*Causative Agent*	*Source of Infection*	*Geographical Distribution*
Tinea barbae (ringworm of the beard)	*Microsporum canis* (rare), *Trichophyton mentagrophytes, T. rubrum, T. sabouraudi, T. verrucosum, T. violaceum*	Infected animals and children	Worldwide
Tinea capitis (ringworm of the scalp)	*M. audouini, M. canis, M. gypseum, T. mentagrophytes, T. sabouraudi, T. schoenleinii, T. sulfureum, T. tonsurans, T. violaceum*	Infected animals, people, and fomites	Worldwide
Tinea corporis (ringworm of the body)	*M. audouini, M. canis, M. gypseum, T. concentricum, T. mentagrophytes, T. sabouraudi, T. schoenleinii, T. sulfureum, T. tonsurans, T. violaceum*	Infected animals and articles of clothing	Worldwide
Tinea cruris (ringworm of the groin)	*Candida albicans, Epidermophyton floccosum, T. mentagrophytes, T. rubrum*	Infected articles of clothing or athletic supports	Worldwide
Tinea manuum and Tinea pedis (ringworm of the hands and feet)	*C. albicans, E. floccosum, M. canis, T. mentagrophytes, T. rubrum, T. schoenleinii*	Direct contact with fungi in moist environments, including showers, swimming and wading pools	Worldwide
Tinea nigra (ringworm of the palms)	*Cladosporium werneckii*	Infected individuals (possibly)	Worldwide
Tinea unguium (ringworm of the nails)	*C. albicans, E. floccosum, T. mentagrophytes, T. rubrum, T. schoenleinii, T. violaceum*	Infected individuals or regions of the body	Worldwide
Tinea versicolor (branny scaling of the skin involving the body surface)	*Malassezia furfur*	Infected individuals	Worldwide
Black piedra	*Piedraia hortai*	Infected hair (beard, mustache, scalp)	Tropical countries
White piedra	*Trichosporon beigelii*	Infected hair (beard, mustache, scalp)	Temperate and tropical regions

Subcutaneous Mycoses

Subcutaneous fungal infections involve the skin and underlying subcutaneous tissues generally without spreading to the internal organs of the body. Examples of subcutaneous mycoses include chromomycosis, sporotrichosis, and maduromycosis.

Deep-Seated (Systemic) Mycoses

Infections in which the causative agents invade the subepithelial tissues (dermis and deeper regions) are known as deep-seated or systemic mycoses.

Opportunistic Fungi

Some fungi are opportunistic pathogens. They are not normally pathogenic to healthy persons, but under certain conditions, they can produce severe infections. Included among these opportunistic agents are species of *Aspergillus, Candida, Cryptococcus, Geotrichum, Mucor,* and *Rhizopus.*

Diagnosis and Prevention of Mycotic Infections

Detection of fungi in specimens involves techniques for their isolation and cultivation and also may include a direct microscopic examination of tissues. The identification of several fungi is based on the characteristics of hyphae and spores in cultures.

One of the most widely used media for the cultivation of fungi is Sabouraud's dextrose agar. An acid medium (pH approximately 5.6), it permits the fungi to survive while preventing certain bacteria from growing. Media can be made more selective by adding antibiotics that discourage the growth of bacteria and saprophytic fungi. The Gram stain and the periodic acid-Schiff (PAS) stain are also used. Reactions obtained with Wood's light (a form of ultraviolet light) is also of diagnostic value with certain infections.

Preventive measures are directed toward keeping susceptible skin areas as dry as possible and improving sanitation and personal hygiene.

Viral Skin Infections

Several viral infections are either limited to the skin or involve it in the course of disease development. The features of some of the better known viral diseases, including chicken pox, cold sores, measles, smallpox, and warts, are listed in Table 23-5.

The clinical features associated with viral skin diseases vary widely. Yet, certain diseases may possess identical features, creating a problem in diagnosis.

The Musculoskeletal System

The skeleton lies within the soft tissues of the body and performs several important functions that include giving support to soft tissues, providing points of attachment for most of the body muscles, determining the type and extent of body movement, protecting many of the vital organs from physical injury, and, after birth, producing the blood cells found within the circulatory system.

Septic Arthritis

Arthritis, or infection of a joint, by definition means involvement of the space between the joint capsule. In most infections of a joint, the cardinal signs of inflammation are present along with a loss in the range of motion.

The anatomical involvement in arthritis caused by microbial infection is highly dependent on the causative organism. With some pathogens, a specific or definite pattern of joint involvement is common. For example, mycobacteria and fungi tend to infect one joint only, gram-negative bacteria attack unusual joints, and most commonly involve small joints in a symmetrical pattern.

Among the bacterial pathogens associated with arthritis, *Staphylococcus aureus* is the most frequent cause of the disease in all age groups. One of the most common causes of viral arthritis is hepatitis B virus.

Septic Bursitis

The bursae associated with the knee and shoulder are constantly at risk for injury with a resulting inflammation or infection (*bursitis*). *Staphylococcus aureus* is the causative agent in more than 90 percent of these infections.

TABLE 23-5 Representative Viral Diseases of Humans That Affect the Skin

Disease	*Causative Agent*	*Transmitters and Sources of Causative Agents*	*General Features*	*Prevention and Control Measures*
Chicken pox (varicella)	Varicella-zoster virus	Droplets of respiratory secretions; direct or indirect contact with infectious skin lesions of carriers	General red rash leading to vesicles; different forms of rash appear in successive crops and are distributed mainly over the trunk and face; slight fever and itching are commonly experienced.	Better sanitation; avoidance of direct contact with infected individuals
Fever blister (cold sore, herpetic gingio-stomatitis, herpes simplex virus infection)	Herpes simplex virus *(Herpesvirus hominis)*	Direct contact with infected individuals (kissing, hand touching, sexual intercourse); fomites	Localized skin and/or mucous membrane lesions that appear as blisterlike eruptions on the lips, face, ears, etc.	Better sanitation and personal hygiene; avoidance of direct contact with infected individuals
Rubella (German measles)	Rubella virus	Same as for chicken pox	Slight fever, general discomfort, swollen lymph nodes, and macular rash.	Immunization; specific immune globulins for exposed individuals
Measles (morbilli)	Measles virus	Direct contact with infected persons or carriers	Rash appears about 14 days after exposure; fever, cough, muscle pain, general discomfort, photophobia, redness of eyelids, characteristic lesions in mouth, Koplik's spots (red ulcers with bluish-white center on the mucous membrane of inside cheek surfaces).	Immunization
Molluscum contagiosum	Molluscum contagiosum virus (a pox-virus)	Direct contact with fomites; the disease also can be spread by sexual contact	Small, pale, firm, pearlike masses (nodules) appear on the skin; a cheesy substance may be expressed from the center of each lesion; the condition clears within 2 to 12 months without complications or treatment.	Better sanitation and personal hygiene; avoidance of direct contact with infected individuals
Shingles (herpes zoster)	Varicella-zoster virus	Direct contact with infectious respiratory secretions	Blisters along nerve trunk, pain (sometimes extreme), slight fever, and general discomfort.	Better sanitation; avoidance of individuals with chicken pox
Smallpox[a]	*Orthopoxvirus*	Same as for chicken pox	Fever, headache, back pain, aching limbs, and general prostration; small red rash in and around groin areas; rash proceeds from macular to pustule to scab formation.	Immunization
Warts (papilloma)	*Polyomavirus*	Contact with contaminated surfaces and infected individuals	Growths on the back of hands, palms, soles, and other body regions; generally no pain or fever.	Better sanitation; avoidance of contact with contaminated surfaces and infected individuals

[a] According to the World Health Organization, smallpox has been virtually eliminated, through immunization against the virus disease.

V. CHAPTER SELF-TEST

Continue with this section only after you have read Chapters 4, 7-12, 18-22, and 23. A score of 80 percent or better is good. If your score is less than 65 percent, reread the chapter.

Correct answers to all questions can be found at the end of this section. Write your responses in the appropriate space provided.

Multiple Choice

Select the best possible answer.

_______ 1. Small elevated skin lesions containing pus are called:
(a) crusts; (b) vesicles; (c) macules; (d) papules; (e) pustules.

_______ 2. Anthrax infection generally is acquired by:
(a) inhalation of spores; (b) ingestion of spores; (c) trauma; (d) insect bites; (e) all of these.

_______ 3. Common sources of *Mycobacterium leprae* include:
(a) rats; (b) humans; (c) food; (d) water; (e) none of these.

_______ 4. Dermatophytes have a particular affinity for skin and related tissues because of the presence of:
(a) carotene; (b) opportunists; (c) keratin; (d) lipids; (e) all of these.

_______ 5. Fungi specifically attacking the epidermal layers, hair, and nails are referred to as:
(a) superficial mycoses; (b) systemic mycoses; (c) systemic fungi; (d) deep-seated fungi; (e) superficial fungi.

_______ 6. Examples of superficial mycoses include:
(a) thrush; (b) histoplasmosis; (c) tinea capitis; (d) *a* and *b* only; (e) *a* and *c* only.

_______ 7. Ringworm of the groin is also known as:
(a) tinea corporis; (b) tinea cruris; (c) tinea saginata; (d) tinea manuum; (e) none of these.

Matching

Select the answer from the right-hand side that corresponds to the term or phrase on the left-hand side of the question sheet. An answer may be used more than once. In some cases, more than one answer may be required.

Topic: Disease Agents and Diseases

_______ 8. Gram-positive, mannitol-fermenting, coagulase-positive cocci

_______ 9. Acid-fast rods

_______ 10. Gram-positive, spore-forming aerobe

_______ 11. Beta-hemolytic group A streptococci

_______ 12. Deep-seated mycosis

_______ 13. Superficial mycosis

_______ 14. *Clostridium botulinum*

_______ 15. *Microsporum canis*

_______ 16. *Candida* species

_______ 17. Warts

_______ 18. Varicella

a. carbuncles
b. leprosy
c. wound botulism
d. impetigo contagiosa
e. erysipelas
f. scarlet fever
g. tinea pedis
h. oral thrush
i. polyoma virus
j. rubella virus
k. anthrax
l. scalded skin syndrome
m. chicken pox
n. none of these

True or False

Indicate all true statements with a "T" and all false statements with an "F".

_______ 19. Hair follicles can serve as portals of entry for pathogens.

______ 20. The indigenous flora of the skin does not interfere with the growth of skin-invading pathogens.

______ 21. Humans can have anthrax.

______ 22. Pathogenic *Staphylococcus aureus* ferments mannitol and produces a positive coagulase reaction.

______ 23. Robert Koch used *Mycobacterium leprae* to establish Koch's postulates.

______ 24. Fomites can transmit anthrax.

______ 25. Opportunistic fungi are normally pathogenic to healthy persons.

______ 26. Tetanus can only be acquired through a wound caused by a rusty nail or similar material.

______ 27. The identification of several dermatophytes is based on the characteristics of hyphae and spores in culture.

______ 28. The same virus causes varicella and shingles.

______ 29. Warts represent a viral infection.

______ 30. Rheumatic fever can be a consequence of *Staphylococcus aureus* infections.

______ 31. Leprosy is usually acquired during childhood, although exposure and resulting infections occur in adults.

______ 32. Both human and rat leprosy are caused by *Mycobacterium leprae.*

______ 33. *Pseudomonas* species are noted for their production of pigments on cultivation.

______ 34. Toxic epidermal necrosis is typically caused by *Pseudomonas* species.

______ 35. The Group B hemolytic streptococci can be distinguished from other streptococci on the basis of their ability to produce the CAMP factor.

Essay Questions

Answer the following questions on a separate sheet of paper.

36. What damage occurs as a result of leprosy?
37. How can *Clostridium tetani* gain entrance to the body and cause tetanus in a nonimmunized individual?
38. What is a superficial fungus?
39. Define arthritis.

Answers

1. e
2. a
3. b
4. c
5. e
6. e
7. b
8. a and l
9. h
10. k
11. e and f
12. n
13. g
14. c
15. g
16. h
17. i
18. m
19. T
20. F
21. T
22. T
23. F
24. T
25. F
26. F
27. T
28. T
29. T
30. F
31. T
32. F
33. T
34. F
35. T

36. Leprosy affects predominantly the skin and peripheral nerves and results in a high prevalence of deformities involving these tissues in the face, hands, and feet.

37. *Clostridium tetani* is considered, for the most part, a relatively noninvasive organism: foreign objects, including glass and slivers of metal or wood, can introduce spores into the deeper tissues. A newborn infant may develop *tetanus neonatorum* from infection of its severed umbilical cord. Tetanus can affect people of all ages and can develop from abortions, circumcisions, ear-piercing, injections of drugs, and negligent surgical procedures.

38. Fungi that attack mainly the epidermis, hair, nails, and mucosal surfaces are called *superficial fungi.* The diseases caused by such agents include the various forms of ringworm, or tinea, and *Candida* infections of mucosal surfaces, such as thrush and vulvovaginitis. These infections are frequently referred to as the *superficial mycoses* or *surface mycoses.*

39. Arthritis, or infection of a joint, refers to the involvement of the space within the joint capsule. In most infections of a joint, the cardinal signs of inflammation are present along with a loss in the normal range of motion.

VI. DISEASE CHALLENGE

The situation described has been taken from an actual case history. It has been designed to show how clinical, laboratory, microbial, and related information is used in disease diagnosis. A review of treatment and related aspects, if appropriate, is given at the end of the presentation. Answers to questions, laboratory findings, and interpretations are given immediately following a specific question. Test your skills and take the *Disease Challenge.*

Case

An 8-year-old girl came to the emergency room of a City Hospital with tenderness and swelling of the right thumb.

A wound resulted from a cat bite experienced about 24 hours earlier. The cat was owned by the young girl's family and was known to be in good health. The child's immunizations were up to date, but the cat's immunization record was not known.

The child had a fever of 39° C. On physical examination, in addition to her swollen and reddened thumb, she had several draining puncture wounds and pustules on her right hand and lower arm.

At this point, what laboratory specimens should be taken? Blood for general laboratory white blood differential cell counts, the material from the draining puncture wounds and pustules for microscopic examination and the isolation of bacterial pathogens and antibiotic sensitivity determinations.

What microbial disease agents could be suspected? *Staphylococcus aureus, Streptococcus* species, anaerobic bacterial pathogens. While most bites involving domestic animals pose little risk of viral infection of rabies, it could be possible consideration. The local health department should be consulted.

Laboratory Results

White blood cell and differential counts were normal; Gram stains of the drainage specimens showed numerous polymorphonuclear cells but not microorganisms.

On the fourth day, cultures taken from one pustule at the time of admission were reported positive for coagulase-positive *Staphylococcus aureus.* The results of antibiotic sensitivity showed a definite susceptibility to nafcillin.

What type(s) of treatment should be given? The general management of animal bite wounds consists of selective administration of tetanus toxoid, local cleansing of the wound, and antimicrobial therapy for an infected wound. Usually, intravenous penicillin G and oral penicillin V are the drugs of choice for the initial treatment of animal bites. In the case of penicillin-allergic patients, erythromycin or a cephalosporin may be used. With penicillin-resistant staphylococci, oxacillin or nafcillin are recommended.

Treatment

In the hospital, the young patient was given appropriate dosages of nafcillin. Her recovery was uneventful.

VII. ENRICHMENT

Dengue Fever and Dengue Hemorrhagic Fever

Dengue or break-bone fever is a mosquito-borne viral disease with symptoms similar to influenza. The causative agents of dengue fever are flaviviruses, a large group of arthropod-borne viruses that includes both the yellow fever and the St. Louis encephalitis viruses. Four dengue virus serotypes are recognized: types 1, 2, 3, and 4. Infection with any of the four confers long-lasting immunity to that serotype. For three to six weeks after infection with any serotype, an individual is apparently resistant to reinfection with all four but, after this refractory period ends, is again vulnerable to infection with any of the other three serotypes.

Dengue fever is believed by some to have originated in Africa, the ancestral home of *Aëdes aegypti,* the mosquito that transmits it. Indeed, the name "dengue" is Spanish for a Swahili phrase, *ki denga pepo,* meaning a sudden, cramplike seizure caused by an evil spirit. The term was first used for an illness characterized by a rash and painful joints that swept through the Caribbean and into South Atlantic and Gulf port cities of the United States in 1827. Today, a more complete clinical description of the disease exists. Clinical symptoms appear within 2 to 8 days after a bite by an infected mosquito. The onset is sudden, with the principal signs and symptoms being fever, generalized body aching, extremely painful muscles and joints, eye pain, and severe headache. The fever lasts from 3 to 7 days and frequently is followed by a rash, postinfectious fatigue, and severe depression.

Unfortunately, severe complications may result caused by different viral serotypes. One complication is known as dengue hemorrhagic fever, and its most severe form is dengue shock syndrome. Although it may have occurred in North Queensland, Australia, as early as 1897, dengue hemorrhagic fever was first described as a distinct pathologic entity by Filipino pediatricians in 1954. By 1956, the disease was definitely linked to dengue virus infection. Since 1956, over 350,000 hospitalizations and nearly 12,000 deaths have been reported from countries of tropical Asia, particularly the Philippines, Vietnam, and Thailand. The vast majority of cases and deaths have been in children. Dengue hemorrhagic fever is now considered to be endemic in Southeast Asia. Until 1981, only a few possible cases have been reported or recognized in the Western Hemisphere.

The signs and symptoms of dengue hemorrhagic fever include easy bruising, bleeding at blood-taking (venipuncture) sites, the formation of pinpoint-sized hemorrhages in the skin on the extremities, face, armpits, and soft palate. nose, gums or gastrointestinal tract, and a reduction in blood platelets (thrombocytopenia). Victims may go into shock, the duration of which usually is brief. However, victims in shock may either recover or die within 12 to 24 hours.

The first dengue fever epidemic in the United States occurred in Philadelphia in the summer and autumn of 1780. Epidemics of the disease have occurred periodically since this time but have been absent from the continental United States since the last Louisiana epidemic in 1945. Unfortunately, in recent years, dengue fever cases imported to the continental United States have been identified. Moreover, an epidemic occurred in southern Mexico in 1979, and in 1980, transmission of dengue fever within the United States was documented for the first time in 35 years, when 63 cases were diagnosed in Texas. While the number of cases does not appear to be alarmingly high, it is quite apparent that mosquito populations have greatly increased. This situation has come about because federal mosquito eradication programs undertaken in the 1960s have been abandoned. Without suitable control measures, at least 14 states in which *Aëdes aegypti* normally breeds are at risk for epidemic dengue fever and its complications. At present, the most effective way of preventing dengue fever, financial resources permitting, is to eliminate the mosquito, eliminate the places where it breeds, use screens on doors and windows, and take other reasonable precautions to prevent exposure.

24

Oral Microbiology

I. INTRODUCTION

The oral cavity, which includes the teeth, tongue, salivary glands, and the walls of the mouth, serves in the grinding, partial digestion, and swallowing of food, as the sound box for speech, and as a secondary pathway for respiration. This region also provides a diverse environment for colonization by and retention of a variety of microorganisms (*oral microbiota*). Oral infections result from disturbances in the relationship or balance between the oral resident flora and the host response. This imbalance may be caused by an allergic reaction or the result of an immunologic deficiency. This chapter will consider the general organization of the oral cavity and some of the associated microbial diseases (e.g., dental caries and periodontal disease), which account for an enormous proportion of health care problems. A Disease Challenge is included at the end of the chapter.

II. PREPARATION

Chapters 21, 22, and 24 should be read before continuing. The following terms are important for you to know. Refer to the glossary and the appropriate chapters of the text if you are uncertain of any of them. A pronunciation guide for selected terms is provided as a learning aid.

1. abscess (AB-sess)
2. AIDS
3. antibiotic
4. calculus (KAL-ku-lus)
5. *Candida* (KAN-dee-dah)
6. cariogenic (kare-ee-oh-JEN-ik)
7. cementum (see-MEN-tum)
8. central papillary atrophy (PAP-i-lare-ee AT-row-fee)
9. crown
10. *Cytophaga* (sye-TOE-fah-ja)
11. decay
12. dental caries (KARE-eez)
13. dental plaque (plak)
14. dentine (DEN-tin)
15. dry socket
16. enamel
17. facultative
18. focal infection
19. gingiva (jin-JAH-vah)
20. gingivitis (jin-jah-VYE-tis)
21. gonococcal stomatitis (stow-mah-TIE-tis)
22. hair leukoplakia (loo-koh-PLAY-key-ah)
23. incisor (in-SYE-zor)
24. inflammation
25. malaise (MAL-aze)
26. microbial flora

27. molar (MO-lar)
28. mumps
29. oral ecology
30. necrotizing (NEK-roh-tie-zing)
31. osteomyelitis (os-tee-oh-my-ee-LIE-tis)
32. parotitis (par-oh-TIE-tis)
33. pericoronitis (pair-ee-ko-ROH-nye-tis)
34. periodontal disease
35. periodontitis (pair-ee-oh-don-TIE-tis)
36. pharyngitis (far-in-JIE-tis)
37. plaque (plak)
38. pyorrhea (pie-oh-REE-ah)
39. trench mouth
40. Vincent's infection
41. ulceration

III. PRETEST

Correct answers to all questions can be found at the end of this section. Write your responses in the appropriate space provided.

Completion Questions

Provide the correct term or phrase for the following. Spelling counts.

1. The palate is divided into a(an) (a) ______________________ palate at the front of the mouth and a(an) (b) ______________________ palate at the back.
2. The human develops two sets of teeth, the (a) ______________ and the (b) ______________ .
3. The number of permanent teeth in a human is ______________________ .
4. Teeth used for the grinding of food are the (a) ______________ and (b) ______________ .
5. The gummy accumulations found on teeth that contain salivary mucin and bacteria are called ________ .
6. Organisms that can tolerate varying concentrations of oxygen, ranging from very high to very low, are called ______________________ .
7. A localized area of infection is called a(an) ______________________ .
8. The hard, calcified substance that sticks firmly to teeth and is considered to be a major causative factor in periodontal disease is ______________________ .
9. Herpangina is an example of a(an) ______________________ disease.

True or False

Indicate all true statements with a "T" and all false statements with an "F".

______ 10. The oral cavity of a fetus is essentially germ-free.

______ 11. Feeding, contact with other humans, and exposure to new environments provide microorganisms for the microbial oral flora of the newborn.

______ 12. Microorganisms normally present at any one time in an anatomical site exist in balance with others in the region.

_______ 13. Members of the genus *Veillonella* are gram-positive cocci.

_______ 14. Oral *Candida* infections generally follow low dosages of antibiotics.

_______ 15. Bacterial viruses can be found in the oral cavity.

_______ 16. Mixed bacterial infections in deep tooth cavities are not uncommon.

_______ 17. Good oral hygiene and the elimination of infection are important steps in restoring and maintaining good health.

_______ 18. Contamination of a tooth extraction site can be a cause of dry socket.

_______ 19. The signs of periodontal disease appear early in the development of the condition.

_______ 20. The most common disease affecting the soft tissues of the oral cavity is calculus.

_______ 21. Untreated gingivitis can lead to pyorrhea.

_______ 22. Tooth decay is the result of an interaction between host tissues and specific cariogenic microbes.

_______ 23. All bacteria known to cause decay of smooth tooth surfaces are plaque-formers and belong to the genus *Staphylococcus.*

_______ 24. Dental plaque can be effectively eliminated by mechanical means such as brushing and flossing.

_______ 25. Adults who develop mumps may experience severe complications, including infections of the testes and the membrane coverings of the brain.

_______ 26. Members of the genus *Fusobacterium* characteristically have their cells in a picket-fence arrangement.

_______ 27. Necrotizing ulcerative gingivitis also is known as trench mouth.

_______ 28. Complement-controlled damage is thought to be involved in periodontal disease.

_______ 29. Fever blisters in a infected individual can be brought on by several factors, including excessive sun exposure and gastrointestinal upset.

_______ 30. Hairy leukoplakia is commonly found in herpes virus infections.

Answers

1a. hard
1b. soft
2a. deciduous
2b. permanent
3. 32
4a. premolars
4b. molars
5. dental plaque
6. facultative
7. focus of infection
8. calculus
9. viral
10. T
11. T
12. T
13. F
14. F
15. T
16. T
17. T
18. T
19. F
20. F
21. T
22. T
23. F
24. T
25. T
26. F
27. T
28. T
29. T
30. F

IV. CONCEPTS AND TERMINOLOGY

Structure of the Mouth

The oral cavity is situated at the beginning of the gastrointestinal tract. The space is enclosed on the sides by the lips and cheeks, above by the hard and soft palate, and below by the floor of the mouth and tongue. Several of these components are described in Table 24-1.

TABLE 24-1 Components of the Mouth

Component	*Description*	*Function(s)*
Palate	The palate is divided into a hard palate at the front of the mouth and a soft palate at the back. The bones of the hard palate are covered by a thick layer of firm but soft tissue. The soft palate is continuous with the tissues encircling the opening to the pharynx.	The soft palate connects with the passageway from the mouth to the throat.
Tongue	Muscular structure located in the floor of the mouth.	Aids in moving food toward digestive tract and in speech.
Gingiva (gums)	Soft tissue surrounding teeth.	Supportive to teeth.
Salivary glands (submandibular, sublingual, and parotid)	Openings of these glands located near the front and sides of the mouth.	Provide lubricating materials and enzymes for the movement and digestion of food.
Teeth	The human develops two sets of teeth: the deciduous (20) and the permanent (32). Each tooth has three parts: the *crown* (the portion above the gum); the *root* (the structure embedded in the jaw); and the *neck* (the narrower region between the crown and the root). A tooth's crown is coated with *enamel;* the rest of the tooth is covered by a layer of modified bone called the *cementum.* Under the enamel is an ivorylike tissue called *dentine* (the bulk of the tooth), which is quite hard and striated. Within this layer is a cavity, the *pulp chamber,* containing the *dental pulp* (blood vessels, connective tissue, and nerve endings). Four types of teeth are recognized: incisors, canines, premolars, and molars.	Incisors are used for cutting food, canines for tearing food, and premolars and molars for grinding.

The Oral Flora

Because the mouth is warm and moist and is supplied regularly with fresh food, it makes an ideal growth environment for microorganisms. The oral cavity of the fetus is essentially germ-free until the infant passes through the birth canal. Feeding, contact with people, and new environments add many organisms. The assortment of organisms in the oral flora is relatively stable, although subject to change with aging. Although other microorganisms may be introduced, perhaps with food, they are usually transient and seldom take up permanent residence. The organisms present at any one time exist in balance with one another, and any change in this balance may result in disease.

The organisms of the human mouth fall into three groups with regard to their tolerance of, or requirement for, oxygen: strict anaerobes, strict aerobes, and facultatives. This last group includes everything between the two extremes—all those microorganisms that can tolerate some concentration of oxygen, from very high to very low.

Table 24-2 lists several microorganisms that form the oral flora.

The fungi most commonly found in the oral cavity are of the genus *Candida.* Viruses, both human and bacterial, have also been recovered from the oral cavity. The causative organisms of canker sores, herpes simplex infection, and measles can be found in oral lesions during obvious disease.

Nonspecific Infections of the Oral Region

Infections of the face, oral cavity, and neck can be extremely serious, depending upon their location and the microorganisms involved. Specific infections caused by bacteria, fungi, or viruses occur here as well as in other body sites. More commonly, mixed bacterial infections occur in deep cavities in teeth or as a result of tissue injury.

TABLE 24-2 Microorganisms in the Oral Flora[a]

Microorganism	*Gram Reaction*	*Morphology*	*Characteristics and Activities*
Actinomyces	+	Rod to coccoid	Most are facultative anaerobes; organisms can be found between teeth and in gum grooves; certain species are pathogenic.
Bacteroides	—	Rods	Strict anaerobes; some species are pathogenic and are associated with gum disease (necrotizing ulcerative gingivitis).
Borrelia	—	Spirals	Strict anaerobes; some species may cause diseases involving supportive tissues in the mouth.
Branhamella catarrhalis	—	Cocci	Aerobic; parasites of the human mucous membranes.
Candida[b]	+	Large oval cells	Represent only a small percentage of the organisms in the total oral flora; known to cause oral infections in individuals with diabetes or cancer or persons receiving large doses of antibiotics.
Capnocytophaga	—	Rods	Anaerobic rods with tapered ends found within gum region of molars; certain species are believed to be involved with periodontal disease.
Corynebacterium	+	Rods	Aerobes and facultative anaerobes; picket fence arrangement of cells is a common features; some pathogenic species (*C. diphtheriae*) produce exotoxins.
Diphtheroids	+	Rods	Aerobes to microaerophilic; club-shaped cells arranged in patterns resembling Chinese characters; normal inhabitants of the mouth.
Fusobacterium	—	Rods	Strict anaerobes; normally found in mouth and other human cavities.
Lactobacilli	+	Rods	Facultative organisms that produce large amounts of acid from carbohydrates; pathogenicity unusual.
Leptotrichia	—	Rods	Highly anaerobic; found in recesses of crevices between teeth; appear as very thick, long, nonbranching rods with rounded ends.
Mycoplasma	—	Variable shapes	Mostly facultative anaerobes; highly variable in shape (pleomorphic); certain species are parasitic as well as pathogenic.
Neisseria	—	Cocci	Aerobes or facultative anaerobes.
Nocardia	+	Coccoid to rods	Strict aerobes with branching; certain species are pathogenic.
Streptococci	+	Cocci	These organisms make up the largest bacterial group in the oral cavity; streptococci are associated with plaque formation and the production of acids from carbohydrates; alpha-hemolytic streptococci (viridans group) pose danger in cases of tooth extraction and heart valve damage; beta streptococci are noted for diseases such as strep throat and scarlet fever.
Treponema	—	Spirochete	Strict anaerobes; some species normally found in mouth.
Veillonella	—	Cocci	Parasitic anaerobes.

[a] Various protozoa and viruses are also found in the mouth.

[b] Yeast.

Focal Infections

A localized area of infection anywhere on the body is called a *focus of infection.* When organisms or their toxic products spread from this focus to distant tissues, either to form another site of infection or to produce a hypersensitive reaction, the process is known as a *focal infection.* Certain conditions are known to be related to oral foci of infection. Good oral hygiene, for example, is an important part of eliminating and preventing further incidents of infection.

Selected examples of nonspecific oral foci of infection in the periodontium or supporting structures of the teeth are described in Table 24-3.

TABLE 24-3 Features of Nonspecific Oral Foci of Infection

Infection or Condition	*Causes (C) and Contributing Factors (CF)*	*Description*	*Signs and Symptoms*
Dental caries	*CF:* Climate, composition and amount of saliva, hormonal balance, nutritional state, oral hygiene, fluoride level in drinking water, diet, genetic makeup *C:* Interactions between the host tissues and specific caries-producing microorganisms (e.g., *Streptococcus* species)	Loss of calcium salts (decalcification) of inorganic substances of teeth, followed by disintegration of organic portions.	General pain, chronic irritation, headache, and complications resulting in infections of surrounding areas.
Dry socket	*C:* Contamination of the extraction area, excessive injury, rinsing with hot fluids, dislodging of blood clot by vigorous rinsing, lowered host resistance, implanting bacteria or foreign material	Dislodging of a blood clot and exposure of bone from a tooth extraction site.	Foul odor, swollen and inflamed gums, pain, and a mass of dead tissue (slough) along the margin nearest the socket.
Gingivitis	*C:* Improper or inadequate oral hygiene, food impacting between poorly closing teeth or around teeth badly broken from decay	Inflammation of gums.	Swollen, reddened, and bleeding gums; pus formation may occur.
Necrotizing ulcerative gingivitis	*CF:* Fatigue, anxiety, and other forms of stress, debilitating illnesses, such as cancer or diabetes, severe vitamin deficiency diseases, local irritation of gums, calculus, and overhanging gums *C:* Implicated bacteria include *Borrelia vincentii* and *Fusobacterium fusiforme*	This disease is also known as trench mouth or Vincent's disease; destruction of gums and associated tissues.	General pain in gums, slight fever, malaise, ulceration of gums, bleeding, loss of dead tissue, foul odor, and metallic taste; bone involvement may occur in untreated cases.
Osteomyelitis	*C:* Infected pulp, residual infections, severe periodontal disease with extension into the bone, many forms of destructive injury, and specific infections such as actinomycosis, syphilis, and tuberculosis	Inflammation and eventual destruction of bone and surrounding tissues.	Severe pain, elevated temperature, swollen lymph nodes associated with the area, loose teeth, and difficulty in eating.
Pericoronitis	*C:* Contaminated instruments, infections following extractions, and specific bacteria including *Streptococcus* species	Inflammation around the crown of the tooth; the condition may spread to other surrounding tissues.	Face swollen on the side involved, discoloration of tissue, draining of pus from involved area.
Periodontitis (pyorrhea)	*C:* Untreated gingivitis *CF:* Various factors acting together are important considerations, including plaque formation, calculus, allergic responses to bacterial antigens, poor oral hygiene, injury by dental floss, genetic factors, hormonal balance, and poor closure of jaw (malocclusion)	An inflammation of the periodontum, the directly supporting tissue of the tooth.	Inflamed gums, bleeding, loss of bone around the teeth, loose teeth; many cases exhibit few symptoms.

Dental Caries

Dental caries—tooth decay—is the result of an interaction between the host tissues and extremely specific cariogenic (caries-producing) microorganisms that utilize nutrients provided by the host's diet. Considerable controversy exists over the exact mechanism of caries development.

Periodontal Disease

Periodontal disease, a worldwide affliction of humans, appears clinically as an inflammation of the soft tissues around the teeth. Depending on the severity of the disease, the destructive processes may involve both the gums (gingivitis) and the periodontal membrane and the alveolar bone surrounding and supporting the teeth (periodontitis). In advanced stages, destruction of cementum and periodontal membrane accompanied by loss of alveolar bone occurs. One current hypothesis is that periodontal disease is caused not by a particular bacteria species, but rather by certain enzymatic and related activities of organisms in intimate contact with tissues surrounding the teeth. A large body of evidence already indicates that specific microorganisms must first colonize the tooth or epithelial surface as a prerequisite to periodontal disease. Taken as a whole, the *Capnocytophaga* have numerous properties that could place this genus in a central position in the development of periodontal disease.

Dental plaque formation is of great importance, especially since it may well be the initiator of dental decay as well as periodontal disease. Plaque is a mixture of bacteria embedded in an accumulation of saliva and bacterial products sticking to the tooth surface. Dental plaque also contains immunologic agents that have the capacity to enhance or suppress immune responses.

All bacteria known to cause decay on the smooth surfaces of the teeth secrete the complex polysaccharides of plaque, whihch are derived chiefly from sucrose (table sugar). Three polysaccharide-producing streptococci are found in large numbers of humans: *Streptococcus mutans, Streptococcus sanguis,* and *Streptococcus salivarius.* The extracellular polysaccharides that these organisms produce from sucrose enable them to adhere to one another and form colonies on the tooth's surface. Although the streptococci are not the only bacteria known to synthesize polysaccharides, they are the only ones that form plaque.

Immunologic mechanisms are also considered to be major factors in periodontal disease. Host immune responses may be involved in the initiation and development of periodontal disease. These responses may be caused nonspecifically by bacterial plaque or by a specific microorganism or by a combination of microbial antigens.

Bacterial Infections

Various bacterial pathogens affect the oral cavity at some time during their development cycles. Representative agents are listed in Table 24-4.

Mycotic Infections

Fungus pathogens also can affect the oral cavity either superficially or as a consequence of systemic disease. One common mycotic infection is caused by the yeast *Candida albicans.* This and other mycotic diseases are briefly characterized in Table 24-5.

Viral Infections

Representative viral infections of the oral cavity are listed in Table 24-6.

TABLE 24-4 Bacterial Infections Involving the Oral Cavity

Disease Entity	*Causative Agent*	*Gram Reaction*	*Symptoms and Clinical Appearance*
Actinomycosis	*Actinomyces israelii, A. bovis*	+	Facial swelling, usually involving the jaw, with hardened and draining sinuses.
Gonococcal stomatitis	*Neisseria gonorrhoae*	—	Fire-red, worn mucous lining of the mouth, sometimes covered by yellow or white patches; destruction of tongue surface also may occur.
Leprosy	*Mycobacterium leprae*	a	Tumorlike masses of tissue involving the oral lining, tongue, lips, or palate.
Syphilis	*Treponema pallidum*	a	Primary oral chancres, secondary mucous patches, tertiary gummas; chronic, spreading, inflammatory, destructive lesions.
Tuberculosis	*Mycobacterium tuberculosis*	a	Ulcerated punched-out lesions on mucosa or tongue; primary oral tuberculosis lesions are rare; fatigue, malaise, night sweats, productive cough when respiratory system involved.
Yaws	*Treponema pertenue*	a	Oral lesions "daughter yaws" secondary to skin lesions; affecting mucous membranes.

[a] Gram stain reactions are not significant.

TABLE 24-5 Mycotic (Fungus) Infections Involving the Oral Cavity

Disease Entity	*Causative Agent*	*Symptoms and Clinical Appearance*
Candidiasis (moniliasis, thrush)	*Candida albicans*	Superficial lesions of skin or mucous membranes in skin folds or creases, often involves inside surfaces of lips; oral lesions are soft, grayish white, and strip off, leaving raw, bleeding surfaces; may accompany denture sore mouth.[a]
Coccidioidomycosis	*Coccidioides immitis*	Approximately 60% show no symptoms; 40% have symptoms of influenza; pulmonary lesions, skin, and oral mucosal granulomatous lesions.
Cryptococcosis (torulosis)	*Cryptococcus neoformans*	Fever, cough, and pleural pain following inhalation of dusts containing the organism; oral lesions are usually seen in systemically debilitated individuals as ulcerations or sores.
Geotrichosis	*Geotrichum candidum* (and other *Geotrichum* species)	White patches on oral mucosa; may develop pulmonary and intestinal lesions.
Histoplasmosis	*Histoplasma capsulatum*	Lesions of the skin and involvement of reticuloendothelial system; nodules may occur in the mouth or throat and involve the respiratory tract.
Mucormycosis	*Absidia* *Rhizopus* *Mucor*	Purulent nasal discharge; inflammatory, gangrenous mucosa; progressive systemic involvement with fatal outcome if untreated; rarely involves the oral region.
North American blastomycosis	*Blastomyces dermatitidis*	Hard or wavy swellings often with sores, drainage.
South American blastomycosis	*Blastomyces brasiliensis (Paracoccidioides brasiliensis)*	Fever, rales, and cough; chronic granulomatous lesions on the skin or mucous membranes and in various organs; oral lesions common.
Sporotrichosis	*Sporotrichum schenckii*	Primary nodule often with ulceration; infection spreads along lymph channels and involves lymph nodes; may have oral nodules with ulceration.

[a] This infection has become an important consideration in the diagnosis of AIDS.

TABLE 24-6 Viral Infections Involving the Oral Cavity

Disease Entity	*Causative Agent*	*Symptoms and Clinical Appearance*
Chicken pox (varicella)	Varicella-herpes zoster virus	Vesicles with a surrounding erythematous zone appear on the mucosa and soon ulcerate.
Hairy leukopenia	Human immunodeficiency virus (HIV)	Slightly raised, usually white areas on the surfaces of the tongue; some redness and swelling may occur; can be found with some AIDS patients.
Hand, foot, and mouth disease	Coxsackie virus Group A, types 6, 10, and 16	Vesicles and ulcerations involving buccal mucosa, tongue, gingiva, and lips; ulcers are painful and interfere with eating; lesions also present on hands and feet.
Herpangina (aphthous pharyngitis)	Coxsackie virus Group A, types 2, 4, 5, 6, 8, and 10	Sudden onset of high fever, headache, and sore throat, accompanied by papules, vesicles, and later ulcers on the pillars of the fauces, the uvula, and soft palate.
Herpes simplex (fever blister)	Herpes simplex virus type 1	Primary lesions usually in oropharyngeal mucosa as multiple, very small vesicles, which rupture and ulcerate; a bright red zone is present around the periphery; fever, malaise, anorexia, and lymphadenopathy are present; recurrent lesions of mucosa on lips common.
Herpes zoster (shingles)	Varicella-herpes zoster virus	Vesicles form along the distribution of a sensory nerve and soon ulcerate.
Hoof-and-mouth disease	Foot-and-mouth disease virus	Vesicles and ulcerations involving lips, tongue, palate, and mucosa; heal within two weeks.
Measles (rubeola)	Paramyxovirus	Koplik's spots appear on buccal mucosa as bluish-white spots with a reddish surrounding zone, followed in a few days by a diffuse rash, fever, and catarrhal inflammation.
Molluscum contagiosum	Molluscovirus	Usually on skin of face; may rarely involve the intraoral tissues with slightly elevated lesions showing a superficial purulent discharge.
Mumps	Mumps virus	Painful swollen salivary glands, usually the parotid.

V. CHAPTER SELF-TEST

Continue with this section only after you have read Chapters 21, 22, and 24. A score of 80 percent or better is good. If your score is less than 65 percent, reread the chapter.

Correct answers to all questions can be found at the end of this section. Write your responses in the appropriate space provided.

Completion Questions

Provide the correct term or phrase for the following. Spelling counts.

1. A tooth's crown is coated with ______________________________.
2. The type of tooth used for tearing food is the ______________________________.
3. The major portion of a tooth consists of the ivorylike tissue ______________________________.
4. Dental plaque consists of (a) ______________________________ and (b) ______________________________.

Matching

Select the answer from the right-hand side that corresponds to the term or phrase on the left-hand side of the question sheet. An answer may be used more than once. In some cases, more than one answer may be required.

Topic: Diseases of the Oral Cavity

______ 5. Herpangina

______ 6. Varicella

______ 7. Shingles

______ 8. Dental plaque

______ 9. Hand, foot, and mouth disease

______ 10. Ludwig's angina

______ 11. Mumps

______ 12. Moniliasis

______ 13. Fever blisters

______ 14. Histoplasmosis

______ 15. Molluscum contagiosum

______ 16. Hairy leukoplakia

______ 17. Central papillary atrophy

______ 18. Gonococcal stomatitis

a. *Streptococcus salivarius*

b. beta-hemolytic streptococci

c. *Candida albicans*

d. coxsackie viruses

e. herpes simplex virus

f. *Staphylococcus aureus*

g. *Treponema pallidum*

h. *Neisseria gonorrhoeae*

i. human immunodeficiency virus

j. none of these

True or False

Indicate all true statements with a "T" and all false statements with an "F".

______ 19. The teeth used for cutting foods are the molars.

______ 20. Most oral infections result from disturbances in the relationship of the microbial flora acquired at the time of birth to the tissues of the mouth.

______ 21. The number of teeth in the permanent set is 38.

______ 22. Deciduous teeth are permanent.

______ 23. The material found under a tooth's enamel is called dental plaque.

______ 24. The contents of a tooth's pulp chamber include blood vessels, connective tissue, and nerve endings.

______ 25. Aerobic bacteria are the predominant organisms in older plaques.

______ 26. The oral cavity of a fetus essentially and normally contains anaerobic bacteria.

______ 27. Examples of anaerobic bacterial genera that are members of the oral flora include *Bacteroides, Fusobacterium,* and *Borrelia.*

______ 28. A small percentage of the oral flora of certain humans normally contain *Candida.*

______ 29. Herpes simplex virus infections often remain localized.

______ 30. In adults, herpes simplex virus infections can lead to severe complications such as orchitis.

______ 31. The virus of fever blisters can be stimulated to cause infections by excessive exposure to sun, gastrointestinal upsets, and certain psychological factors.

______ 32. The most commonly involved salivary gland in mumps is the sublingual.

______ 33. The assortment of microorganisms in the oral flora varies considerably throughout an individual's lifetime.

______ 34. Pericoronitis is an inflammation of a tooth's crown.

______ 35. Organisms found in the oral flora that are aerobic cocci include members of the genera *Branhamella, Cytophaga, Neisseria,* and *Lactobacillus.*

Essay Questions

Answer the following questions on a separate sheet of paper.

36. Define gingivitis and describe its possible cause.

37. What microorganism is believed to be the principal causative agent of enamel caries? In addition, explain how the microorganism causes its effects.

Answers

1. enamel	9. d	18. h	27. T
2. canine	10. b	19. F	28. T
3. dentine	11. j	20. F	29. T
4a. salivary mucin	12. c	21. F	30. F
4b. bacteria	13. e	22. F	31. T
5. d	14. j	23. F	32. F
6. j	15. j	24. F	33. F
7. j	16. i	25. T	34. T
8. a	17. h	26. F	35. F

36. Gingivitis is an inflammation of the gingiva, the portion of the oral mucous membrane that surrounds a tooth. The inflammation is probably the result of an exaggerated response to large amounts of dental plaque. The condition is the most common disease affecting the soft tissues of the mouth. In mild cases of gingivitis, an increase in the number of polymorphonuclear leukocytes and T lymphocytes occurs.

37. Three polysaccharide-producing streptococci are found in large numbers in humans, *Streptococcus mutans, Streptococcus sanguis,* and *Streptococcus salivarius.* The extracellular polysaccharides that these organisms produce from sucrose enable them to adhere to one another and thus form colonies on the tooth's surface. Although the streptococci are not the only bacteria known to synthesize polysaccharides, they are the major group of organisms that initiate plaque formation. Various experiments have shown that the principal causative microorganism of enamel caries of *S. mutans.* Although other organisms may play a role, *S. mutans* produces acids that are capable of dissolving the enamel surfaces.

VI. DISEASE CHALLENGE

The situation described has been taken from an actual case history. A review of treatment and epidemiological aspects is provided at the end of the presentation. Answers to questions, laboratory findings, and interpretations are given immediately following a specific question. Test your skills and take the *Disease Challenge.*

Case

A 45-year-old male was admitted with a temperature of 40° C (102° F), a painful, swollen neck with reddening (erythema) extending from the submandibular area on the right side of the neck to the shoulder, and persistent contraction of muscles used for chewing (trismus). Chest x-ray films and a panoramic film of the teeth were found to be normal. However, a limited examination of the oral region revealed severe periodontitis and extreme distention of the area under the chin, suggesting pus accumulation. The total white blood cell count was elevated above normal values.

On questioning, the patient indicated that four days earlier he had experienced sudden pain while eating and thought that he had cracked a lower right molar when he bit down on a steak bone. Over the next two days, swelling was noted in the right side, which subsided only to reappear under the chin. The swelling continued and became worse, making swallowing, talking, and even breathing difficult.

At this point, what type(s) of laboratory specimens should be taken? An incision and drainage of the swollen area would be necessary to obtain a specimen. Pus (drainage material) would be the specimen of choice for bacteriological culture and study.

Laboratory Results

Cultures of the pus grew alpha- and beta-hemolytic streptococci.

Probable Diagnosis

Ludwig's angina.

Treatment

The patient, under general anesthesia, underwent an incision and drainage procedure of the involved area and was treated with intravenous penicillin. Approximately 24 hours later, he improved clinically. The redness began to fade and disappear complete within 24 hours. Body temperature returned to normal, and healing of the drained area was uneventful. One week later, a dental examination showed the presence of a split first molar on the lower right side.

VII. ENRICHMENT

Dental Caries and *Streptococcus mutans*

Dental caries (from the Latin *caries*) as it is known today is characteristic of populations consuming a modern "Western-type" diet, highly refined and rich in fermentable carbohydrates. The importance of microbes in the causation of caries was first demonstrated by Underwood and Miller in London in 1881, and the idea was developed by Miller, an American working in Robert Koch's laboratory, as the chemical-parasitic hypothesis. Miller's hypothesis states that bacteria close to the tooth surface, in what is now called *dental plaque,* ferment dietary carbohydrates to produce acids. These acids then dissolve the apatite crystals, of which enamel is mostly composed.

Despite the sound, theoretical basis for the mechanism of carious attack, it was not until the discovery of antibiotics and experiments with germ-free animals that the importance of bacteria was conclusively demonstrated. The source of difficulty is not far to seek: the mouth is a warm, moist, nutrient-rich environment ideally suited for bacterial growth, and it supports a large variety of different organisms. Samples of dental plaque can commonly yield representatives of 20 or more different genera, and the increased efforts of taxonomists in recent years have resulted in the recognition of many novel species and subspecies within each genus.

For many years, lactobacilli were considered prime suspects because of their acid-producing potential, but over the last decade, there has been increasing evidence for the role of *Streptococcus mutans* as a major etiological agent. *S. mutans* was first isolated in 1924. Results from various epidemiological studies have provided strong indications of the importance of this bacterium, particularly in the early stage of dental caries. However, it is not certain that *S. mutans* is responsible for all caries in humans.

S. mutans is highly cariogenic in experimental animals, but other species of streptococci, actinomyces, and lactobacilli also can cause the disease in rodents, but less efficiently. Laboratory-reared rats, mice, and hamsters have all been use as experimental models at various times, but germ-free (gnotobiotic) rats have been found to be the most functional because of the possibility of discounting interference from bacteria other than *S. mutans.* There are, however, substantial differences in tooth structure, salivary composition, and immune experience between such rats and humans. Subhuman primates, although expensive and troublesome to house, provide a more valid model. Monkeys (*Macaca fascicularis*) were first shown to be suitable models in 1966 by Cohen and Bowen at the laboratories of the Royal College of Surgeons of England at Downe in Kent, next door to Charles Darwin's old home. Macaques have a primary and permanent tooth structure resembling that of humans and need little persuasion to consume a sugar-rich diet. Introduction of such a diet results in a dramatic increase in the number of *S. mutans* in plaques and is followed by the development of caries closely resembling the human forms of the disease. Rhesus monkeys (*M. mulatta*) were also found to be useful models.

With the availability of laboratory animals capable of producing a disease state similar to that found in humans, in the late 1960s researchers began their search for a purified vaccine that could provide protection against caries. Certain components of *S. mutans* have received a great deal of attention, namely, the extracellular glucosyl-transferase enzymes. These enzymes synthesize both a water-soluble *dextran* and an insoluble branched polymer called *mutan.* The mutan confers a stickiness on *S. mutans* and enables it to stick to hard surfaces. In studies with rats, antibodies developed against glucosyl-transferases have been found to reduce caries on unprotected smooth tooth surfaces. However, in studies with monkeys, no protection occurred.

25

Microbial Infections of the Respiratory Tract

I. INTRODUCTION

The human respiratory system provides a means of entry and a multiplication site for a fairly large and diverse number of disease-causing microorganisms. Several of the microbial diseases that involve the respiratory tract present serious problems, in that they are more difficult to control and are readily spread within populations. The control and prevention of respiratory diseases caused by bacteria, fungi, and viruses require early and accurate identification, prompt treatment, and the use of immunization if available. This chapter will describe microbial diseases, together with the approaches to diagnosis and control. A Disease Challenge is included at the end of the chapter.

II. PREPARATION

Chapters 8, 10, 12, 13, 17-22, and 25 should be read before continuing. The following terms are important for you to know. Refer to the glossary and the appropriate chapters of the text if you are uncertain of any of them. A pronunciation guide for selected terms is provided as a learning aid.

1. abscess (AB-sess)
2. acid-fast
3. aerosol (AIR-oh-sol)
4. AIDS
5. alveolus (al-VEE-oh-lus)
6. antibiotic resistant
7. antitoxin
8. arthrospore (are-THROW-spore)
9. bacitracin (bas-eh-TRAY-sin)
10. bacteriophage
11. BCG
12. bile
13. carriers
14. Centers for Disease Control
15. coagglutination
16. coccobacillus (kok-oh-bah-SIL-us)
17. croup (croop)
18. cyanosis (sigh-ah-NO-sis)
19. droplet nucleus
20. epiglottiditis (ep-i-glot-tid-EYE-tis)
21. Eustachian (you-STAY-she-an) tube
22. fibrosis (fye-BROH-sis)
23. Gram stain
24. Guillain-Barré syndrome

25. hypersensitivity
26. immunization
27. immunosuppression
28. interferon
29. intradermal
30. *in vitro*
31. Legionnaire's disease
32. macrophage
33. malaise
34. *Mycoplasma*
35. nasopharynx (nay-soh-FAR-ingks)
36. normal flora
37. optochin
38. otitis media (oh-TIE-tis ME-dee-ah)
39. pathogen
40. polymorphonuclear leukocyte
41. *Pneumocystis carinii* (new-moh-SIS-tis kar-I-nee-eye)
42. PPD
43. Reye's syndrome
44. Schick test
45. sinuses
46. sinusitis (sye-nus-EYE-tis)
47. sputum
48. trophozoite (trof-oh-ZOH-te)
49. toxin
50. zoonosis

III. PRETEST

Correct answers to all questions can be found at the end of this section. Write your responses in the appropriate space provided.

Completion Questions

Provide the correct term or phrase for the following. Spelling counts.

1. The tubelike structure that serves as a passageway for both food and air from the mouth to the esophagus is called the ______________________.
2. The Eustachian tubes equalize the air pressure in the middle ear with (a) ______________________ and serve as channels for (b) ______________________ drainage.
3. Each lung of the human respiratory system is divided into ______________________.
4. Gaseous exchange (external respiration) in the respiratory system occurs in the walls of the __________.
5. The letters in the DPT vaccine stand for (a) ________________, (b) ________________, and (c) ______________________.
6. The skin-testing procedure used to determine susceptibility to diphtheria is called the ______________ test.
7. The causative agent of psittacosis is ______________________.
8. Whooping cough is caused by the bacterium ______________________.
9. Psittacosis is an example of a(an) ______________________.
10. Immunization against tuberculosis involves the use of the vaccine containing ______________________.

11. The healing of tuberculosis lesions may occur in several ways, including (a) ______________________,
    (b) ______________________, and (c) ______________________.
12. The letters PPD stand for (a) ______________________ and are used for (b) ______________________.

True or False

Indicate all true statements with a "T" and all false statements with an "F".

______ 13. The respiratory system provides a means for the spreading of microorganisms.

______ 14. The normal flora of the human respiratory tract does not contain pathogens or potential pathogens.

______ 15. At birth, the throat, windpipe, and bronchi are usually sterile.

______ 16. The human respiratory tract provides both a suitable portal of entry and a multiplication site for various microorganisms.

______ 17. The communicability of an infectious disease can be influenced by the specific location of the disease agent in the body.

______ 18. Isolation of infected persons is not an effective control measure for respiratory tract infections.

______ 19. Blood and sputum are typical types of specimens used for the diagnosis of respiratory tract infections.

______ 20. Only strains of *Corynebacterium diphtheriae* infected with specific bacteriophages produce diphtheria-causing exotoxins.

______ 21. Most instances of diphtheria result from direct contact with infectious droplets.

______ 22. Antibiotics are given to prevent and treat viral respiratory infections.

______ 23. Differentiation among the various serological types of *Streptococcus pneumoniae* is based on the existence of immunologically distinct polysaccharide capsules.

______ 24. The tuberculin skin test reveals previous infection but does not prove the presence of an active disease state.

______ 25. Confirmation of an active case of tuberculosis is done primarily by skin-testing methods.

______ 26. The introduction of PPD into a sensitized individual does not cause any observable reaction at the site of injection.

______ 27. Once an individual exhibits a positive tuberculin skin test, additional skin testing is of little value.

______ 28. Histoplasmosis is an acute viral infection associated with histiocytes.

______ 29. *Pneumocystis* infections principally are secondary bacterial disease states that follow influenza infections.

______ 30. *Pneumocystis* infections primarily affect individuals with lowered resistance.

______ 31. Relatively few microorganisms are introduced into the environment during normal breathing, even by an infected individual.

______ 32. The majority of URIs result in severe respiratory illnesses.

______ 33. The etiologic agent of sinusitis includes species of *Bacterioides* and *Aspergillus.*

______ 34. Toxin-producing *Corynebacterium diphtheriae* produces severe throat inflammations.

______ 35. The Frei test can be used to determine susceptibility to diphtheria.

Answers

1. pharynx
2a. atmospheric pressure
2b. middle ear
3. lobes
4. alveoli
5a. diphtheria (toxoid)
5b. pertussis (killed *Bordetella pertussis*)
5c. tetanus (toxoid)
6. Schick test
7. *Chlamydia psittaci*
8. *Bordetella pertussis*
9. zoonosis
10. Bacillus of Calmette and Guérin BCG)
11a. calcification
11b. fibrosis or scarring
11c. resolution
12a. Purified Protein Derivative
12b. tuberculous skin testing
13. T
14. F
15. T
16. T
17. T
18. F
19. T
20. T
21. T
22. F
23. T
24. T
25. F
26. F
27. T
28. F
29. F
30. T
31. T
32. F
33. T
34. T
35. F

IV. CONCEPTS AND TERMINOLOGY

Structures of the Respiratory Tract

The human respiratory tract starts at the nose, passes through the various parts of the respiratory tree, and ends in the air sacs, of alveoli. This entire system is adapted to making air containing oxygen available to the circulatory system, and by it, to the entire body. The respiratory tract is a frequent portal of entry for various microorganisms. It also provides a means of transmitting microbes to other individuals during speaking, coughing, and sneezing, when droplets of microbe-containing secretions are released. Specific components of the system are listed in Table 25-1.

Normal Flora and Microbial Infections of the Respiratory Tract

At birth, the throat, windpipe, and bronchi are sterile. However, within 24 hours after birth, these sites become colonized by streptococci and other bacteria.

Several microorganisms (e.g., streptococci) can cause respiratory tract infections. Many of the resulting diseases are quite common and are among the most serious and debilitating of any that affect humans.

Various microbial pathogens—bacteria, fungi, and viruses—find suitable avenues for entry and sites for multiplication in the respiratory tract.

The various secretions of the respiratory system can transmit infectious agents. The communicability of respiratory infectious diseases is influenced by several factors, including:

1. The survival of respiratory pathogens on fomites or in the air
2. The number of microorganisms inhaled
3. The duration of contact
4. The anatomical site involved in the localization of the infectious agents

Control and Preventative Measures

Many of the measures used to control respiratory tract infections are similar, if not identical, to those for other types of diseases. The procedures are largely determined by the characteristics of the pathogen or the disease it causes. Control of respiratory tract infections may include: (1) isolation of infected persons; (2) concurrent disinfection or sterilization of contaminated equipment, such as mouthpieces, thermometers, rubber tubing, and any and all contaminated articles, such as eating utensils and dishes; (3) the use and proper disposal of gowns following contact with infectious individuals; (4) the disinfection of rooms or other facilities used by infected persons; (5) the washing and disinfecting of hands before and after contact with patients or with any articles handled by patients, such as blankets, dishes, laundry, or pillows; (6) immunizing susceptible members of a population (a procedure followed in cases of influenza and diphtheria); and (7) prompt and effective diagnosis and treatment of infected persons, when possible.

TABLE 25-1 Components of the Respiratory System

Component	*Description*	*Function(s)*
Pharynx	Tubelike structure associated with the pharyngeal *tonsils* (adenoids) and the lingual tonsils, both lymphoid tissues. A total of seven openings lead to or from the throat—one from the mouth, two from the nose, one to the esophagus, and two to the Eustachian tubes.	Passageway for food and air
Eustachian (auditory) tube	Connects the middle ear with the throat.	Equalizes the air pressure in the middle ear with atmospheric pressure and also serves as a channel for middle ear drainage
Trachea (windpipe)	A thin tube averaging about 2.5 cm in diameter. It passes from the voice box, or larynx, into the thoracic (chest) cavity, where it divides into the two primary *bronchi.* The trachea is lined with ciliated epithelial cells.	Passageway for air
Lungs	Large, soft, and spongy structures, each of which is divided into *lobes.* The right lung is composed of three lobes, and the left lung, two.	Respiration
Respiratory pleura (membranes)	Two membranes form a closed sac that envelops the lungs. The inner membrane, the *visceral pleura,* covers the outer lung surface. The outer membrane layer, the *parietal pleura,* lines the inner surface of the chest wall. The potential cavity between these two membranes is called the *pleural cavity.* A serumlike fluid found in this region enables the two pleurae to glide over one another during respiratory movements.	Protective coverings
Bronchial tube	Formed from both bronchi dividing and subdividing into smaller tubelike structures called *bronchioles.* The smallest of these, approximately 0.5 mm, are known as *terminal bronchioles.* These divide and give rise to *respiratory bronchioles,* which in turn branch into several *alveolar ducts.* Alveolar sacs develop from the ducts, the walls of which form the *alveoli.*	Gaseous exchange (external respiration) occurs in the alveoli

Diagnosis

Respiratory tract infections can often be diagnosed on the basis of x-rays, the patient's history, and a physical examination of the patient. Confirmation is made by microscopic examination of sputum, blood, and other specimens; isolation of pathogens from specimens; serological tests; animal inoculations; and other laboratory procedures.

Skin tests are also used as diagnostic aids in bacterial diseases such as diphtheria, tuberculosis, and certain fungal infections, including coccidioidomycosis and histoplasmosis.

The TB skin test, an extremely important epidemiological tool, has been especially important from the standpoint of differential diagnosis. The basis of the tuberculin reaction is the development during the course of an infection of a specific delayed hypersensitivity to certain products of *M. tuberculosis* and related mycobacteria. These products are contained in culture extracts and are referred to as *tuberculins.* Two types of tuberculin, Old Tuberculin (OT) and Purified Protein Derivative (PPD), are widely used for skin testing. Three procedures are currently used in the tuberculin test: (1) intradermal injection (mantoux test), (2) jet injection, and (3) multiple puncture. The intradermal injection serves as the standard for comparison with all other tests and provides more accurate control of dosage. The other two methods are utilized for epidemiological survey and screening. The individual's sensitivity may develop one month or more after infection, usually remaining for several years or for an entire lifetime.

The tuberculin test reveals previous infection but does not prove the presence of an active disease state. Confirmation of an active case of tuberculosis is done with x-ray examination and isolation of the bacteria.

Representative Microbial Diseases

Several infections of the respiratory tract are of great concern because of the ease with which they are transmitted and contracted and the difficulty in eradicating their disease agents. A number of these microbial pathogens and the diseases they cause are described in Tables 25-2 through 25-5.

In the treatment of various viral respiratory diseases, antibiotics are given primarily to prevent secondary bacterial infections, even in the absence of clear evidence that these drugs reduce or eliminate the effects of the causative agents. Additional aspects of treatment are directed toward relieving the patient's symptoms.

Upper Respiratory Infections

The upper respiratory region, which includes the middle ear, the small air cells behind the ears (the mastoids), the sinuses, and the nasal corner of the eyes, are exposed to a variety of pathogens when air is inhaled. These organisms can establish infections in these sites (see Table 25-2 through 25-4) and spread to other regions of the respiratory tract.

TABLE 25-2 Features of Selected Bacterial Upper Respiratory Tract Infections

Disease	*Causative Agent(s)*	*Gram Reaction*	*Morphology*	*General Features of the Disease*
Diphtheria	*Corynebacterium diphtheriae* (Klebs-Loeffler bacillus)	+	Rods appearing in aggregates resembling Xs, Ys, and Chinese letters	Symptoms appear suddenly and include fever, sore throat, general discomfort, formation of diphtheritic pseudo-membrane on tonsils, throat, or in the nasal cavity; complications such as heart and kidney failure can develop.
Middle ear infection (otitis media)	*Haemophilus influenzae,* type b	—	Rod	Pain first limited to ear, then spread to other head regions; other symptoms include varying degrees of deafness, dizziness, noises, feelings of revolving in space or of surroundings rotating, pus formation, and difficulty in swallowing.
	Staphylococcus aureus	+	Coccus	
	Streptococcus pneumoniae	+	Coccus	
Sinus infection (sinusitis)	*Bacteroides* species	—	Rod	Accumulation of mucus in the sinuses; pain, headache, general discomfort, difficulty in breathing.
	Haemophilus influenzae, type b	—	Rod	
	Streptococcus pneumoniae	+	Streptococcus	
Streptococcal infection (strep throat)	*Streptococcus pyogenes*	+	Streptococcus	Chills, fever, headache, nausea or vomiting, rapid pulse, reddened and swollen throat and uvula, and presence of a gray or yellow-white material covering the throat; complications include middle ear infections, kidney infections, and blood poisoning.

TABLE 25-3 Viral Agents and Commonly Associated Upper Respiratory Tract Infections

Respiratory Tract Infection	*Virus by Generic Designation*	*General Features of the Disease*
Common cold	*Coronavirus* *Rhinovirus*	Cough, watery nasal discharge, head cold (coryza), headache, nasal obstruction, sneezing, sore throat
Croup (acute laryngo-tracheobronchitis)	*Adenovirus* *Orthomyxovirus* *Paramyxovirus* Respiratory syncytial virus	Symptoms range from mild to severe and can be grouped into the following types: *type 1*—cough, hoarseness, harsh breathing, and high-pitched sound (stridor); *type 2*—fever, toxic effects, vomiting; *type 3*—convulsions, bluish coloration (cyanosis), dehydration, and restlessness
Minor respiratory illnesses	Adenoviruses Echoviruses Paramyxoviruses Reoviruses	Fever, sort throat, swollen lymph nodes in neck, persistent cough

Diseases of the Ear

Various microorganisms can cause ear infections. Individuals with nasopharyngeal (nose and throat) diseases are often predisposed to ear infections. This is especially true in cases of adenoid growth, Eustachian tube obstructions, and various forms of middle ear and sinus infections. Treatment must be directed not only toward the ear disease, but also against the predisposing condition. Table 25-4 lists some of the diseases associated with such microorganisms.

Lower Respiratory Infections

The lungs inhale many pathogenic microorganisms that normally are eliminated efficiently by the host's defense. However, several diseases of the lower respiratory tract (see Table 25-5) are life-threatening if not treated quickly and adequately.

TABLE 25-4 Representative Microbial Diseases of the Ear

Disease	*Causative Agent(s)*	*Gram Reaction*	*Morphology (if applicable)*
Boils	*Staphylococcus aureus*	+	Coccus
Inflammation of the eardrum	Mixed infections involving hemolytic streptococci and viruses	+	Cocci
Inflammation of the outer ear	*Escherichia coli* *Proteus* spp. *Pseudomonas* spp. Hemolytic streptococci *S. aureus* Mixed infections	— — — +	Rod Rods Rods Cocci
Otitis media (middle ear infection)	*Streptococcus pneumoniae* Beta-hemolytic streptococci *S. aureus* *Haemophilus influenzae*, type b	+ + + —	Cocci Cocci Coccus Coccobacillus
Mycotic infection of the external ear and ear canal	*Aspergillus niger* *Candida albicans*	Not useful +	Mold Yeast
Throat abscess	Beta-hemolytic streptococci *S. aureus*	+ +	Cocci Coccus

TABLE 25-5 Features of Selected Microbial Lower Respiratory Tract Infections

Disease	*Causative Agent*	*Gram Reaction*	*Morphology*	*General Features of the Disease*
Bacterial				
Atypical primary pneumonia	*Mycoplasma pneumoniae* (Eaton's agent)	a	Pleomorphic	Cough, chills, headache, sore throat, general discomfort, thick sputum with pus.
Legionnaire's disease	*Legionella pneumophila*	—	Rod	Chills, rapidly rising fever, abdominal pains, slight headache, muscle aches, nonproductive cough; complications leading to respiratory failure can occur.
Pneumonia	*Klebsiella pneumoniae* (Friedlander's bacillus)	—	Rod	Fever, chest pains, thick reddish-brown sputum.
	Staphylococcus aureus	+	Coccus	High fever, blue coloration (cyanosis), frequent cough, pus-containing discharge from nose, eyes, and rapid breathing;[b] a complication of certain viral infections such as measles.
	Streptococcus pneumoniae	+	Coccus	Sudden onset of symptoms, including severe chills and shaking, high fever, chest pain, thick rust-colored sputum, dry cough, and vomiting.
	Streptococcus pyogenes (group A, beta-hemolytic)	+	Coccus	Chills, cough, difficulty in breathing, fever, general discomfort; complications include infections of the central nervous system and kidneys.
Pontiac fever	*Legionella pneumophila*	+	Rod	A mild, self-limited illness.
Psittacosis	*Chlamydia psittaci*	—	Rod	Sudden onset of symptoms, including cough, difficulty in breathing, fever, pain, and headache.
Q fever	*Coxiella burnetii*	—	Rod	Sudden onset of symptoms, including dry cough, fever, headache, and general stiffness.
Tuberculosis	*Mycobacterium tuberculosis*[d]	c	Rod	Wide variety of symptoms, including fever, general discomfort, weight loss, productive lesion, formation of tubercle (nodule in lung tissue); skin test eventually becomes positive.
Whooping cough (pertussis)	*Bordetella pertussis* (Bordet-Gengou bacillus)	—	Rod	Disease occurs in three stages: *catarrhal*—persistent dry cough, slight fever, poor appetite, excessive mucous secretions, tearing, and vomiting; *paroxysmal*—coughing attacks referred to as "whooping";[e] a production of thick, stringy mucous masses; *convalescent*—coughing attacks decrease in severity.

TABLE 25-5 Features of Selected Microbial Lower Respiratory Tract Infections *(continued)*

Disease	*Causative Agent*	*Gram Reaction*	*Morphology*	*General Features of the Disease*
Fungus				
Coccidioido-mycosis (San Joaquin Valley fever)	*Coccidioides immitis*	f		General flulike symptoms including chills, cough, fever, malaise, chest pain, and a pus-containing sputum in the case of pneumonia.[g]
Histo-plasmosis	*Histoplasma capsulatum*	f		Either no detectable illness or mild effects; small calcified growths appear in several body organs upon recovery.
North American blastomycosis (Gilchrist's disease	*Blastomyces dermatitidis*	f		Symptoms are usually quite mild and self-healing and include cough, fever, and general discomfort; complications and spread to other body regions can occur.
Virus	*Influenza Virus Type or Subtype*	f		
Influenza[h]	A (classic) A_1 (A prime) A_2 (Asian) B C			Uncomplicated influenza symptoms include backache, chills, fever, headache, general discomfort, nasal congestion, cough, dry and sore throat, loss of appetite, nausea, and vomiting; complications include pneumonia.
Protozoan				
Pneumonia	*Pneumocystis carinii*			Course of the disease ranges from 4 to 6 weeks; symptoms include cough, bluish coloration of the skin, fever, rapid breathing, and lung consolidation.[i]

[a] Gram reactions are of little value with this pathogen.

[b] This is a more severe form of pneumonia than that caused by *S. pneumoniae.*

[c] The Gram stain reaction is not used diagnostically. The acid-fast staining procedure is used instead.

[d] Increased recognition of the prevalence of pulmonary disease caused by mycobacteria other than *M. tuberculosis* has focused on the importance of differentiating among these mycobacterial pathogens (e.g., *M. kansasii, M. avium-intracellulare* complex).

[e] The characteristic "whoop" is caused by rapidly inhaling air, which passes quickly over the vocal cords.

[f] Gram reactions are not applicable.

[g] The spread of the disease throughout the body, disseminated form, usually starts from the respiratory system.

[h] Influenza as well as other infections and certain immunizations may be associated with two specific complications: Guillain-Barré and Reye's syndromes.

[i] A major cause of death in AIDS.

V. CHAPTER SELF-TEST

Continue with this section only after you have read Chapters 8, 10, 12, 13, 17-22, and 25. A score of 80 percent or better is good. If your score is less than 65 percent, reread the chapter.

Correct answers to all questions can be found at the end of this section. Write your responses in the appropriate space provided.

Matching

Select the answer from the right-hand side that corresponds to the term or phrase on the left-hand side of the question sheet. An answer may be used more than once. In some cases, more than one answer may be required.

Topic: Respiratory Diseases

______ 1. Croup

______ 2. Associated with toxin-caused disease

______ 3. Causes middle ear infections

______ 4. Histoplasmosis

______ 5. Causes strep throat

______ 6. Common cold virus

______ 7. Major cause of lobar pneumonia

______ 8. Legionnaire's disease

______ 9. A typical primary pneumonia

______ 10. Psittacosis

______ 11. North American blastomycosis

______ 12. Pontiac fever

______ 13. Q-fever

______ 14. A major cause of death among AIDS patients

______ 15. Whooping cough

______ 16. Tuberculosis

a. *Haemophilus influenzae,* type b

b. Rhinoviruses

c. *Staphylococcus aureus*

d. *Corynebacterium diphtheriae*

e. *Streptococcus pyogenes*

f. *Adenovirus*

g. *Coxiella burnetii*

h. *Histoplasma capsulatum*

i. *Streptococcus pneumoniae*

j. *Chlamydia* species

k. *Legionella pueumophila*

l. *Pneumocystis carinii*

m. none of these

True or False

Indicate all true statements with a "T" and all false statements with an "F".

______ 17. Diphtheria is caused by a specific endotoxin.

______ 18. *Corynebacterium diphtheriae* forms spores.

______ 19. The immunity resulting from a severe case of diphtheria is not long-lasting.

______ 20. In the treatment of various viral respiratory diseases, antibiotics are given primarily to prevent secondary bacterial infections.

______ 21. Vitamin C is necessary for interferon formation.

______ 22. Croup commonly occurs in adults between the ages of 25 and 40.

______ 23. *Streptococcus pneumoniae* causes only lobar pneumonia.

______ 24. *Streptococcus pneumoniae* is a capsule former.

______ 25. *Mycoplasma pneumoniae* is the causative agent of tuberculosis.

______ 26. The Gram stain is of major significance in the diagnosis of tuberculosis.

______ 27. Individuals who have recovered from pneumococcal pneumonia can be convalescent carriers.

______ 28. Q-fever is an example of zoonotic rickettsial infection.

______ 29. Scarring accompanies the healing of most tuberculosis lesions.

______ 30. The depositing of calcium within semisolid centers of older tuberculous lesions is called fibrosis.

______ 31. Q-fever characteristically is transmitted by lice.

______ 32. Overcrowding, poor ventilation, and malnutrition favor the establishment of *Mycobacterium tuberculosis* infection.

______ 33. In tuberculosis, pulmonary lesions develop in the most aerated (oxygenated) regions of the lungs.

______ 34. The tuberculin skin test can be used to screen groups for tuberculosis infections.

______ 35. Coccidioidomycosis is a fungal infection that characteristically localizes within macrophages and reticuloendothelial cells.

Essay Questions

Answer the following questions on a separate sheet of paper.

36. At birth, which parts of the respiratory tract are sterile?
37. What is croup?
38. What types of diagnostic materials are used for tuberculosis skin testing?
39. Distinguish between Guillain-Barré syndrome and Reye's syndrome.

Answers

1. f	10. j	19. F	28. T
2. d	11. k	20. T	29. T
3. a	12. k	21. F	30. F
4. h	13. g	22. F	31. F
5. e	14. l	23. F	32. T
6. b	15. m	24. T	33. T
7. i	16. m	25. F	34. T
8. k	17. F	26. F	35. F
9. k	18. F	27. T	

36. At birth, the throat, trachea (windpipe), and bronchi are sterile. However, within 24 hours after birth, these sites become colonized by streptococci and other bacteria. In the adult, the respiratory tract below the level of the epiglottis is normally sterile.

37. Croup, or acute laryngotracheobronchitis, is an acute infectious disease of children under 3 years of age. Males are affected more than are females. Croup may be mild or severe in its effects.

This disease should be distinguished from epiglottis from the standpoint of treatment. Epiglottis is caused almost entirely by the bacterium *Haemophilus influenzae* and can be controlled with antibiotic therapy.

38. Two types of tuberculin, Old Tuberculin (OT) and Purified Protein Derivative (PPD), are widely used for skin testing. The first material was originally described by Robert Koch and incorporates the heat sterilization of an *M. tuberculosis* culture. The active component in Old Tuberculin is a protein noted for its heat stability and retention of specificity for several years.

    The second preparation, PPD, is a slightly more refined testing substance than OT. It is preferred because its strength lends itself to standardization of dosages; skin tests performed with the same dose are comparable. PPD contains an active protein obtained from filtrates of autoclaved tubercle bacilli cultures.

39. Complications found with influenza include Guillain-Barré syndrome (GBS) and Reye's syndrome (RS). The Guillain-Barré condition is a rapidly developing inflammation of several nerves resulting in a spreading muscular weakness of the extremities and possible paralysis.

    Reye's syndrome (RS) is another long-recognized complication of viral and bacterial infections and immunizations. Epidemiological evidence clearly shows a direct relationship between the condition and the viruses of chicken pox and influenza B. Severe vomiting is typical of the onset of Reye's syndrome. Individuals with RS experience a rapid accumulation of fluid in the brain (cerebral edema) without inflammation. While various organs such as the liver are affected during the course of the condition, death results from increased intracranial pressure.

VI. DISEASE CHALLENGE

The situation described has been taken from an actual case history. A review of treatment and epidemiological aspects is provided at the end of the presentation. Answers to questions, laboratory findings, and interpretations are given immediately following a specific question. Test your skills and take the *Disease Challenge.*

Case

A 38-year-old female who worked as a maid was admitted with a persistent pneumonia and cough. She had no fever, but experienced great difficulty in talking because of frequent coughing. Listening to the lungs (auscultation) revealed abnormal sounds (rales) with wheezes during inhalation and exhalation. A chest x-ray showed the right lower lobe of the lung to be heavily infiltrated. The patient's case history indicated that she had been well until five weeks earlier when a runny nose and breathing difficulty developed. Initially, her cough was nonproductive, but soon she began to produce a green, foul-smelling sputum. She visited a physician three weeks earlier and was given ampicillin. The antibiotic treatment was not effective, and the condition continued without improvement. In addition, the patient's PPD skin test was negative and there was no history of contact with a cat, dog, or bird.

At this point, what disease(s) could be suspected? Legionellosis, *Mycoplasma* pneumonia, histoplasmosis, adenovirus infection, or possibly tuberculosis. Since there was no history of animal contact, Q fever and psittacosis would not be likely candidates.

What type(s) of laboratory specimens should be taken? Sputum for bacteriologic and virus cultures, and acute (initial) and convalescent (during recovery period) serum samples to determine changes in antibody levels (titers) to *Legionella, Mycoplasma,* and adenovirus. In addition, serum samples would be necessary to perform the cold agglutinin test, which is of value in mycoplasma infections.

Laboratory Results

Sputum cultures were negative for all bacterial agents suspected. However, a significant increase in antibody levels was observed only with mycoplasma. A significant cold aglutinin response was also noted. (In a positive cold agglutinin test, clumping of red blood cells by the patient's serum in the cold occurs.)

Probable Diagnosis

Mycoplasma pneumonia. The combination of the clinical features, antibody response, cold agglutinin test results, and the ineffectiveness of ampicillin clearly point to the diagnosis.

Treatment

The patient was given a tetracycline antibiotic. Her condition improved remarkably in three days. The cough disappeared, and a follow-up x-ray four days later showed complete clearing.

VII. ENRICHMENT

The Mycobacteria

Despite dramatic improvements in tuberculosis control, *Mycoplasma tuberculosis* remains the most common cause of pulmonary and other forms of mycobacterial infections. It appears that as the number of tuberculosis cases decreases, the relative numbers of disease cases caused by mycobacteria other than *M. tuberculosis* increases. Organisms in this group are referred to by the acronym MOTT, which is derived from the term "mycobacteria other than tuberculosis." The following table briefly presents the properties of the more clinically significant or frequently isolated *Mycobacterium* species. *M. leprae* is not considered.

TABLE 25-6 Clinically Significant Mycobacteria

Mycobacterium Species	*Disease(s) Caused*	*Cultural and Related Characteristics*	*Other Properties*
M. avium-intracellulare	Pulmonary infections	Widely available in the environment.	Infections are very difficult to treat. In addition, this species has been the cause of death in some acquired immune deficiency syndrome (AIDS) victims.
M. bovis	Tuberculosis	A slow-growing organism that may have colony characteristics similar to *M. tuberculosis.*	Occasionally some strains (Bacillus of Calmette and Guérin) are used in immunotherapy for certain forms of cancer.
M. fortuitum-chelonei complex	Skin and corneal infections, pulmonary infections in immunosuppressed patients, as well as in patients following cardiac surgery.	Grows on routine culture media used in a clinical laboratory.	This species can be confused with diphtheroids if an acid-fast stain is not performed.
M. haemophilium	Skin infections in immune suppressed hosts	Grows well on blood agar or egg-based culture medium containing ferric ammonium citrate. Incubated at 30° to 32° C.	This species requires ferric ions for good growth.
M. kansasii	Pulmonary infections	This organism is the most commonly found photochromogenic *Mycobacterium.*[a]	This species occurs mainly in older white males with chronic bronchitis and emphysema.
M. malmoense	Pulmonary infections	A slow-growing and nonpigmented organism.	This species has been found in Austria, Germany, Sweden, and the United States.
M. marium	Skin infections generally associated with contaminated swimming pools	Successful isolation of this organism requires incubation at 30° to 32° C.	This organism also has been found in tropical fish aquariums.

TABLE 25-6 Clinically Significant Mycobacteria *(continued)*

Mycobacterium Species	*Disease(s) Caused*	*Cultural and Related Characteristics*	*Other Properties*
M. scrofulaceum	Seldom causes disease	*M. scrofulaceum* is a scotochromogenic organism.[b]	This organism has been isolated from the cervical lymph nodes of children.
M. simiae	Possible lung infections	This organism is photochromogenic[c] at 24° C and scotochromogenic at 35° C.	This organism may be associated with environmental contamination of water supplies. It also may be transmitted from monkeys to humans.
M. szulgai	Lung disease and infections of tendon sheaths	This organism has a growth temperature range of 24° to 35° C. It is photochromogenic[c] at 24° C and scotochromogenic at 35° C.[b]	
M. tuberculosis	Tuberculosis and other types of pulmonary infections	This is a slow-growing organism and requires a temperature of 35° to 57° C. It is readily identified by various biochemical tests and photoreactivity.	This organism is highly contagious to persons having continuing exposure.
M. ulcerans	Chronic, progressive skin ulcers in patients living in tropical climates	This organism is a slow grower and requires an incubation temperature of 32° C.	This organism and its infections are rarely seen in the United States.
M. xenopi	Skin infections	This organism is yellow pigmented and grows best at 42° to 45° C.	This organism has been implicated in a number of outbreaks of skin infection associated with hot water supply systems.

[a] This organism is the most common *Mycobacterium* species other than *M. tuberculosis* isolated from patients with pulmonary infections.

[b] Scotochromogens develop pigment when grown in the dark.

[c] Photochromogens develop a yellow pigment when they are grown in the dark and then exposed to light.

26

Microbial Diseases of the Gastrointestinal Tract

I. INTRODUCTION

Several varieties of microorganisms use the gastrointestinal tract as a site of entry into the body. They are transported into the system by food, water, fingers, or other objects introduced into the mouth. Most organisms are unable to overcome the barriers created by the gastric juices of the stomach and the microbial intestinal flora. Some of those that do penetrate cannot establish themselves and are eliminated from the system.

This chapter describes the diseases and effects caused by bacterial, viral, and protozoan pathogens that invade the gastrointestinal tract. Consideration also is given to the intoxication caused by the ingestion of microbial toxins. A Disease Challenge is included at the end of the chapter.

II. PREPARATION

Chapters 3, 4, 7-10, 11-13, 17-22, and 26 should be read before continuing. The following terms are important for you to know. Refer to the glossary and the appropriate chapters of the text if you are uncertain of any of them. A pronunciation guide for selected terms is provided as a learning aid.

1. amebiasis (am-eh-BYE-ah-sis)
2. acute diarrheal disease
3. aminoglycoside (ah-me-no-GLY-koh-side)
4. anemia (ah-KNEE-me-ah)
5. antigen (AN-tee-jen)
6. aseptic (a-SEP-tik) meningitis
7. capsid (KAP-sid)
8. carrier
9. catalase
10. cirrhosis (sye-ROH-sis)
11. coagulase (koh-AG-you-lace)
12. colitis (koh-LIE-tis)
13. complement (KOM-plee-ment) fixation
14. cryptosporidiosis
15. cytomegaly (sye-toh-MEG-ah-lee)
16. cyst (sist)
17. dark-field microscopy
18. dehydration
19. delta virus
20. differential and selective media
21. diarrhea (die-ah-REE-ah)
22. dysentery (DIS-en-ter-ee)
23. electrolyte
24. ELISA
25. exotoxin
26. feces (FEE-sees)

27. fermentation
28. fomite
29. gastroenteritis (gas-troh-en-teh-RYE-tis)
30. Gram stain
31. hemagglutination (he-mah-gloo-teh-NAY-shun)
32. hepatitis (hep-ah-TIE-tis)
33. immunofluorescence (im-you-no-floo-RES-ens)
34. immunoglobulin (im-you-no-GLOB-you-lin)
35. IMViC reactions
36. infective intoxication
37. inflammation (in-flah-MAY-shun)
38. jaundice (JAWN-dis)
39. malaise
40. microbiota
41. mucosa
42. nasopharynx
43. neurotoxin
44. pasteurization
45. pediatric gastroenteritis
46. picornavirus
47. rotavirus
48. septicemia (sep-tee-SEE-me-ah)
49. sphincter
50. spirochete (SPY-row-keet)
51. titer (TIE-ter)
52. trophozoite (trof-oh-ZOH-ite)
53. vaccine (VAK-seen)
54. vibrion (VIB-ree-on)
55. villi (VIL-i)
56. virucide (VIE-ruh-side)
57. zoonosis (zoh-NO-sis)

III. PRETEST

Correct answers to all questions can be found at the end of this section. Write your responses in the appropriate space provided.

Completion Questions

Provide the correct term or phrase for the following. Spelling counts.

1. In the human digestive system, most of the enzymatic digestion and absorption of nutrients take place in the ______________________________ .
2. The ______________________________ processes, stores, and eliminates food materials remaining after digestion and absorption.
3. Bile is a product of the ______________________________ .
4. Bile functions to ______________________________ fats.
5. Three ways by which intestinal microflora may cause diarrhea are (a) ______________________________ , (b) ______________________________ , and (c) ______________________________ .

Matching

Select the answer from the right-hand side that corresponds to the term or phrase on the left-hand side of the question sheet. An answer may be used more than once. In some cases, more than one answer may be required.

Topic: GI Diseases

______ 6.	Cryptosporidiosis	a.	*Salmonella typhi*
______ 7.	Traveler's diarrhea	b.	*Vibrio cholerae*
______ 8.	Shigellosis	c.	*Escherichia coli*
______ 9.	Typhoid fever	d.	*Clostridium botulinum*
______ 10.	Asiatic cholera	e.	*Staphylococcus aureus*
______ 11.	Weil's disease	f.	*Proteus vulgaris*
______ 12.	Amoebiasis	g.	*Bacillus cereus*
______ 13.	Botulism	h.	*Leptospira*
______ 14.	Viral hepatitis	i.	*Entamoeba histolytica*
______ 15.	Pseudo-membranous colitis	j.	*Cryptosporidium* species
______ 16.	Bacillary dysentery	k.	*Brucella suis*
______ 17.	Brucellosis	l.	*Clostridium difficile*
______ 18.	Infant epidemic diarrhea	m.	none of these

True or False

Indicate all true statements with a "T" and all false statements with an "F".

______ 19. Polysaccharides, peptones, and fatty acids are absorbed in the small intestine.

______ 20. Liver function tests can be used to evaluate various forms of treatment in liver disease.

______ 21. Many of the microbes picked up by a baby as it passes through the birth canal disappear soon after birth.

______ 22. *Escherichia coli* is a chief inhabitant of most gastrointestinal ecosystems.

______ 23. Species of the genus *Bacillus* are found throughout the human gastrointestinal system.

______ 24. The causative agent of bacillary dysentery is *Entamoeba histolytica.*

______ 25. Leptospira are gram-negative vibrios.

______ 26. The cholera vaccines used today consist of a *Vibrio cholerae* toxoid.

______ 27. Brucellosis can be acquired through the ingestion of unpasteurized dairy products obtained from infected goats.

______ 28. Typhoid fever is considered to be a zoonosis.

______ 29. Leptospirosis is acquired through contact with water contaminated by urine from infected animals.

______ 30. The human is the sole reservoir of infection for shigellosis

______ 31. HBsAg designates the inner core of the hepatitis B virus.

______ 32. *Vibrio vulnificus* is a cause of bloodstream and wound infections.

Answers

1. small intestine
2. large intestine
3. liver
4. emulsify

5a. change dietary foodstuffs or host secretions into substances that affect gut fluid movement

5b. penetrate the intestinal lining and damage the bowel wall

5c. produce exotoxins that cause an emptying of large amounts of water and electrolytes into the intestine without damaging the mucosa

6. j
7. c
8. k
9. a
10. b
11. h
12. i
13. d
14. k
15. l
16. m
17. k
18. c
19. F
20. T
21. T
22. F
23. F
24. F
25. F
26. F
27. T
28. F
29. T
30. T
31. F
32. T

IV. CONCEPTS AND TERMINOLOGY

Microbial diseases of the gastrointestinal tract usually result from the ingestion of food or water containing pathogenic microorganisms or their toxins.

Parts of the Gastrointestinal Tract

The digestive system supplies food and water to the body's internal environment. Once inside the body, basic substances such as amino acids, simple sugars, and fatty acids are carried to the cells via the circulatory system. Table 26-1 lists the components of the gastrointestinal system.

TABLE 26-1 Components of the Gastrointestinal System

Component	*Description*	*Function(s)*
Stomach	Located on the left side of the body just under the lower ribs. The inner layer mucous membrane of the stomach, the *mucosa*, contains millions of glands that secrete mucus and gastric juice. Hydrochloric acid secreted by the stomach keeps the pH of the stomach low and kills many microorganisms that are ingested.	Digestion.
Small intestine	Divided into three portions or loops: the *duodenum*, the *jejunum*, and the *ileum*. Its enormous, highly absorptive surface area is composed of millions of small, fingerlike structures called *villi*. Each of these units contains capillaries and a *lacteal*, a small lymph vessel.	Enzymatic digestion; absorption of amino acids, simple sugars, fatty acids, and glycerol.
Large intestine	Divided into the following regions: *ascending colon*, *transverse colon*, *descending colon*, *rectum*, and *anus*. The "blind sac" at the beginning of the large intestine is called the *caecum*. The *appendix*, a small, fingerlike projection, is located at the tip of the caecum.	Primarily involved with processing, storing, and eliminating of food material and water remaining after digestion and absorption. The waste products of digestion, the feces, are released through the anus.
Liver	One of the largest organs or glands found in the body; located in the upper portion of the abdominal cavity just beneath the diaphragm.	Bile production, detoxification of toxic substances, removal of foreign matter.

Gastrointestinal Microbial Ecology

The structure of the gastrointestinal tract determines the localization of the microbiota and, to some extent, its composition. Microbial habitats may exist in any area from the esophagus to the anus. Each habitat provides a different kind of environmental or nutritional challenge.

The gastrointestinal tract is sterile in the normal fetus up to the time of birth. During normal birth, the baby picks up microorganisms from portions of the mother's reproductive tract and from any other source to which it is exposed. *Escherichia coli* are normally outnumbered by other organisms (e.g., strict anaerobes).

Table 26-2 lists several organisms isolated from the human gastrointestinal system.

TABLE 26-2 Microorganisms Isolated from Various Regions of the Human Gastrointestinal System[a]

	Gastrointestinal Region		
Microbial Type	*Stomach*	*Small Intestine*	*Large Intestine*
Actinobacillus	+	+	—
Bacillus species	—	—	+
Bacteroides	+	+	+
Bifidobacteria	+	+	—
Candida[b]	+	—	—
Clostridia	+	+	—
Coliforms	+	+	+
Lactobacilli	+	+	+
Peptostreptococcus	+	—	—
Staphylococcus	+	+	—
Streptococcus	+	+	+
Torulopsis[b]	+	—	—
Veillonella	+	+	—

[a]The microorganisms isolated and the numbers found vary to some extent, depending on the diet, environmental factors, methods used for isolation, and geographical location of the subjects.

[b] Yeast.

Bacterial Diseases

Acute diarrheal disease results in a disturbance of intestinal functions and, once started, may produce dehydration and the passage of liquid stools. The specific bacterial pathogens that can cause such conditions are shown in Table 26-3 along with other agents of gastrointestinal disease.

Intestinal microflora may induce diarrhea in three ways: (1) they may change dietary foodstuffs or host secretions into substances that affect gut fluid movement; (2) they may penetrate the intestinal lining and damage the bowel wall; or (3) they may produce exotoxins that cause large amounts of water and electrolytes to empty into the intestines without damaging the mucosa.

Food Poisoning (Intoxications)

Food poisoning as discussed here refers to the symptoms resulting from the consumption of food or drink contaminated by pathogenic bacteria or their toxic products.

TABLE 26-3 Bacterial Diseases of the Gastrointestinal Tract

Disease	*Causative Agents*	*Gram Reaction*	*Morphology*	*Transmitters and Sources of Causative Agents*	*General Features*	*Prevention and Control*
Asiatic cholera	*Vibrio cholera*	—	Vibrio	Ingestion of sewage-contaminated food or water; direct contact with an infected person's feces; houseflies; carriers	The disease ranges from mild to severe and symptoms include large amounts of mucus in stools (rice-water stools), sudden loss of water and electrolytes and dehydration; collapse, shock, and death can occur without treatment.	Immunization; purified water supplies; improved sanitation; prompt treatment; education programs
Brucellosis (Malta fever, undulant fever)	*Brucella abortus* *B. melitensis* *B. suis*	—	Coccobacilli	Contact with the tissues or excretions of infected animals; ingestion of contaminated meat or unpasteurized dairy products	Variable symptoms may include general discomfort, weakness, muscle aches and pains, elevated temperature late in the day, falling during the night, enlarged lymph nodes, spleen, and liver involvement; disease may become chronic; residual tissue damage can occur.	Elimination of reservoirs of disease agents; immunization of natural reservoirs (e.g., cattle, hogs, goats—this is not effective for humans); routine pasteurization of dairy products
Gastro-enteritis	*Campylobacter fetus* subspecies *jejuni*	—	Curved rods	Ingestion of contaminated food and water	Disease symptoms range from muted to severe and include abdominal pain, diarrhea, fever, and bloody stools.	Improved sanitation; education programs
	Vibrio parahaemolyticus	—	Vibrio	Consumption of contaminated seafood or exposure to marine environments	Symptoms include abdominal cramps, nausea, vomiting, watery diarrhea, and in severe cases, dehydration.	Ample cooking of seafoods; improved sanitation; education programs
	V. vulnificus[a]	—	Vibrio	Same as noted for *V. parahaemolyticus*	Malaise, chills, diarrhea, vomiting, and rash occur in some cases.	Ample cooking of foods; improved sanitation; education programs
	Yersinia enterocolitica	—	Cocco-bacillus	Ingestion of contaminated food or water, contact with diseased animals	Symptoms appear suddenly and include abdominal pain, diarrhea, dehydration, nausea, chills, vomiting, jaundice, possible convulsions, and weight loss; complications do not usually develop.	Improved sanitation; treatment of diseased animals
Pseudo-membranous colitis (PMC)	*Clostridium difficile*	+	Rod	A side effect associated with the administration of certain antibiotics; disturbance of gastrointestinal ecological balance	Symptoms appear suddenly and include abdominal pain, diarrhea, dizziness, fever, headache, nausea, vomiting, and poor appetite.	Restoration of fluid and electrolyte balance; antibiotics; treatment with chloramphenicol

TABLE 26-3 Bacterial Diseases of the Gastrointestinal Tract *(continued)*

Disease	*Causative Agents*	*Gram Reaction*	*Morphology*	*Transmitters and Sources of Causative Agents*	*General Features*	*Prevention and Control*
Salmonellosis (gastro-enteritis)	*Salmonella typhimurium* and other *Salmonella* spp.	—	Rods	Consumption of contaminated food or water	Symptoms appear suddenly and include abdominal pain, diarrhea, dizziness, fever, headache, nausea, vomiting, and poor appetite.	Proper cooking of meat and meat products; poultry refrigeration and covering of prepared foods; protection of food from contamination by mice, rats, or flies and related insects; the periodic inspection of food handlers; proper sanitation
Shigellosis (bacillary dysentery)	*Shigella boydii, S. dysenteriae, S. flexneri, S. sonnei,* and other *Shigella* spp.	—	Rods	Ingestion of contaminated food or water; the causative agents can be spread by feces, fingers, flies, or food	Symptoms appear suddenly and include abdominal pain, diarrhea, high fever, general discomfort, stools containing mucus, blood, and pus ("red currant jelly" appearance), rectal burning, and dehydration; complications include massive bleeding and perforation of the large intestine.	Prompt diagnosis and appropriate treatment; elimination of flies; proper sanitary disposal of excreta; protection of food and water supplies; education programs
Traveler's diarrhea	Enterotoxin-producing *E. coli* strains and species of *Salmonella* and *Shigella*	—	Rods	Same as for shigellosis	Symptoms appear suddenly and include abdominal pain, diarrhea, dehydration, nausea, chills, vomiting, jaundice, possible convulsions, and weight loss; complications do not usually develop.	Same as for shigellosis
Typhoid fever	*Salmonella typhi*	—	Rods	Same as for shigellosis	Symptoms appear gradually and include abdominal distention, constipation, rising fever, headache, loss of appetite, nausea, vomiting, diarrhea, and appearance of a rash (rose spots) on abdomen; complications include inflammation of gall bladder, perforation of small intestine, intestinal bleeding, and pneumonia.	Immunization; prompt treatment of infected persons and carriers; improved sanitation; pasteurization of appropriate foods; education programs
Weil's disease (spirochetal jaundice, leptospirosis)	*Leptospira interrogans*[a]	Usually not done	Spiral	Direct contact with infectious urine from diseased animals; ingestion of contaminated food or water	Symptoms appear suddenly and include lack of appetite, chest pains, head cold, difficulty in swallowing, swollen lymph nodes, fever, vomiting, and jaundice; complications are severe involvement of the skin, central nervous system, kidneys, and liver.	Reducing human contact with contaminated water or urine from infected animals; immunization of dogs; wearing protective clothing while working with objects or materials that might be contaminated

[a] This organism infects other tissues.

The agents of common bacterial food poisoning are well established. Among the conditions they produce are botulism, perfringens poisoning, salmonellosis, and staphylococcal poisoning (see Table 26-4). Salmonellosis is an example of an active infection (infectious food poisoning); the other three states are examples of poisoning or bacterial intoxications. In poisoning, the toxins alone produce the symptoms.

TABLE 26-4 A Comparison of Common Bacterial Food Poisonings (Intoxication)

Condition	*Causative Agent*	*Gram Reaction*	*Morphology*	*Incubation Period*	*General Features*
Botulism	*Clostridium botulinum*	+	Rod	12-96 hours hours	Difficulty in speaking, double vision, inability to swallow, nausea, vomiting, and paralysis of urinary bladder and all voluntary muscles; death caused by stoppage of heart action and/or breathing may occur.
Perfringens poisoning	*Clostridium perfringens*	+	Rods	Within 18 hours	Abdominal cramps, chills, bluish coloration of the skin, diarrhea, headache, nausea, and vomiting.
Staphylococcal intoxication	*Staphylococcus aureus*	+	Coccus	1-6 hours	Severity of symptoms depends on amount of enterotoxin ingested and include abdominal cramps, chills, bluish coloration of the skin, diarrhea, headache, nausea, and vomiting.

Viral Infections

Various viral pathogens may infect portions of the gastrointestinal system. Viral infections do not necessarily produce gastrointestinal disturbances, but may cause reactions in other parts of the body, such as the nervous and respiratory systems.

Certain viruses that invade the gastrointestinal system use it for replication only. However, they may be involved in bacterial and protozoan infections. Table 26-5 lists the main groups of pathogens known to invade the GI system.

Parasitic Protozoan Infections

Some common protozoan infections that occur throughout the world are listed in Table 26-6.

TABLE 26-5 Viral Diseases of the Gastrointestinal Tract

Disease	*Transmitters and Sources of Disease Agents*	*General Features*	*Prevention and Control*
Cytomegalovirus inclusion disease	Congenital; aerosol; direct contact with fomites	Symptoms of infected newborns include hepatitis, jaundice, increased size of liver and spleen, decreased number of blood platelets, and loss of sight; post-natally infected individuals may show no symptoms or may develop pneumonia; death can occur.[a]	Careful evaluation and monitoring of drug therapy for unrelated conditions

TABLE 26-5 Viral Diseases of the Gastrointestinal Tract *(continued)*

Disease	*Transmitters and Sources of Disease Agents*	*General Features*	*Prevention and Control*
Hepatitis A virus disease (infectious hepatitis)	Consumption of contaminated food or water; fomites; direct contact with infectious feces; aerosol	Symptoms show a wide range and include abdominal discomfort, muscular pains, jaundice, dark urine, light-colored stools, fever, chills, and sore throat	Improved sanitation; sterilization of objects contaminated by infected persons; injections of specific immunoglobulins for exposed persons; identification and treatment of carriers; education programs
Hepatitis B virus disease (serum hepatitis)	Direct contact with contaminated syringes and needles; aerosol; mosquitoes; transfusions with contaminated blood or products	Symptoms are similar to those of hepatitis A disease; however, hepatitis B symptoms tend to be more severe.	Same as for hepatitis A infection; also, sterilization of any and all surgical type instruments and all disposable items
Non-A, non-B hepatitis	Similar to those listed for hepatitis B	Similar to those listed for hepatitis B.	Same as for hepatitis A; alpha interferon has been successful in reducing liver injury.
Infant diarrhea (Rotavirus gastroenteritis)	Consumption of fecal contaminated food or water; fomites	Abrupt onset of vomiting and diarrhea, fever, moderate dehydration, and normal white blood cell count.	Improved sanitation; breast feeding of newborns

[a] Several other infectious diseases may produce identical symptoms. These include the TORCH group (toxoplasmosis, rubella, cytomegalovirus, herpesvirus) and syphilis.

V. CHAPTER SELF-TEST

Continue with this section only after you have read Chapters 3, 4, 7-10, 11-13, 17-22, and 26. A score of 80 percent or better is good. If your score is less than 65 percent, reread the chapter.

Correct answers to all questions can be found at the end of this section. Write your responses in the appropriate space provided.

Completion Questions

Provide the correct term or phrase for the following. Spelling counts.

1. The pH of the stomach is kept low due to the ______________________ secreted by cells lining the organ.
2. Blockage of the bile duct will cause a yellowing of body tissues called ______________________.
3. The formation of nonfunctional scar tissue in the liver over a period of time results in a condition known as ______________________.
4. Traveler's diarrhea may be caused by specific enterotoxin-producing strains of ______________________.

TABLE 26-6 Protozoan Infections of the Gastrointestinal System

Disease	*Causative Agent*	*Transmitters and Sources of Causative Agents*	*General Features*	*Prevention and Control*
Amoebiasis (amoebic dysentery)	*Entamoeba histolytica*	Contaminated water supplies, flies, infected food handlers, and person-to-person contact; poor sanitary conditions provide opportunities for repeated exposure	Abdominal cramps, flatulence, diarrhea, feces containing blood and mucus, weight loss, and general fatigue; complications such as invasion of the liver (amoebic hepatitis) may develop.	Improved sanitation; prompt diagnosis and treatment; purified water supplies; education programs regarding personal hygienic habits and dangers of contaminated food or water; identification and treatment of carriers
Balantidiasis	*Balantidium coli*	Ingestion of contaminated food or water	Intense abdominal pain, diarrhea, loss of weight, and vomiting; rapid destruction of tissue and death may occur.	Improved sanitation; prompt diagnosis and treatment; education programs regarding dangers of contaminated food and water
Cryptosporidiosis	*Cryptosporidium* species	Ingestion of contaminated food or water	Mild abdominal cramps, low-grade fever, nausea, loss of appetite, frequent watery, frothing stools, followed by constipation.	Improved sanitation; prompt diagnosis; education programs regarding dangers of contaminated food and water

Matching

Select the answer from the right-hand side that corresponds to the term or phrase on the left-hand side of the question sheet. An answer may be used more than once. In some cases, more than one answer may be required.

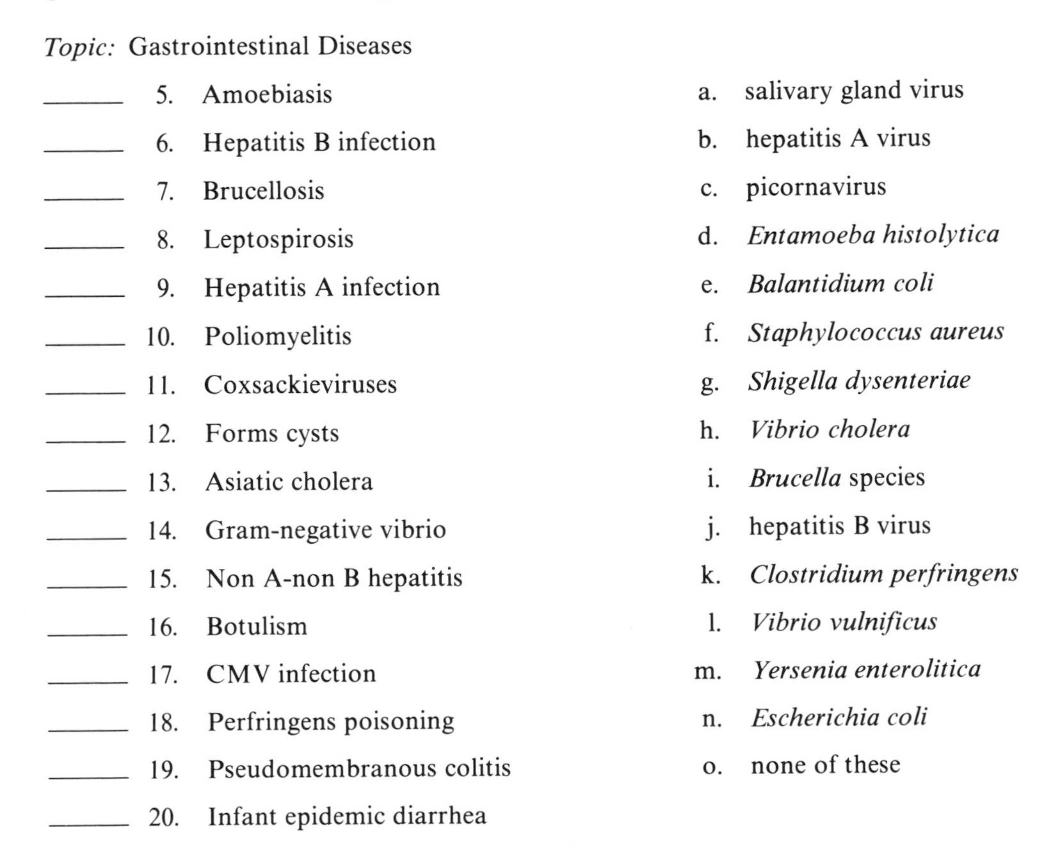

Topic: Gastrointestinal Diseases

______ 5. Amoebiasis		a. salivary gland virus
______ 6. Hepatitis B infection		b. hepatitis A virus
______ 7. Brucellosis		c. picornavirus
______ 8. Leptospirosis		d. *Entamoeba histolytica*
______ 9. Hepatitis A infection		e. *Balantidium coli*
______ 10. Poliomyelitis		f. *Staphylococcus aureus*
______ 11. Coxsackieviruses		g. *Shigella dysenteriae*
______ 12. Forms cysts		h. *Vibrio cholera*
______ 13. Asiatic cholera		i. *Brucella* species
______ 14. Gram-negative vibrio		j. hepatitis B virus
______ 15. Non A-non B hepatitis		k. *Clostridium perfringens*
______ 16. Botulism		l. *Vibrio vulnificus*
______ 17. CMV infection		m. *Yersenia enterolitica*
______ 18. Perfringens poisoning		n. *Escherichia coli*
______ 19. Pseudomembranous colitis		o. none of these
______ 20. Infant epidemic diarrhea		

True or False

Indicate all true statements with a "T" and all false statements with an "F".

______ 21. The gall bladder produces gall or bile.

______ 22. The gastrointestinal tract is sterile in the normal fetus up to the time of birth.

______ 23. The routine pasteurization of dairy products is an effective means of preventing brucellosis.

______ 24. Staphylococcal food poisoning is a communicable disease.

______ 25. Jaundice is a symptom found in most cases of bacterial food poisoning.

______ 26. Cytomegaly refers to an increase in the size of salivary glands.

______ 27. Cytomegalovirus inclusion disease is limited to older persons with leukemia and other cancerous states.

______ 28. Some gastrointestinal tract diseases result from the ingestion of microbial toxin.

______ 29. Humans, apes, and dogs are the only known natural hosts for viral hepatitis.

______ 30. The Vi antigen is a characteristic result of picornavirus infections.

______ 31. Epstein-Barr virus and cytomegalovirus can produce liver damage.

______ 32. Individuals recovering from bacillary dysentery and typhoid fever may become carriers.

Essay Questions

Answer the following questions on a separate sheet of paper.

33. How can *Vibrio parahaemolyticus* infections be prevented?

34. What is the mechanism of action associated with exotoxins involving gastrointestinal bacterial diseases (infective intoxications)?

35. How does infant botulism differ from food-borne and wound forms of botulism?

Answers

1. hydrochloric acid	9. b	17. o	25. F
2. jaundice	10. c	18. k	26. F
3. cirrhosis	11. c	19. o	27. F
4. *Escherichia coli*	12. d, e	20. n	28. T
5. d	13. h	21. F	29. F
6. j	14. h	22. T	30. F
7. i	15. o	23. T	31. T
8. l	16. o	24. F	32. T

33. Preventing *V. parahaemolyticus* infection is simple, regardless of the bacterial content, when seafoods are cooked. Such foods will be safe if they are (1) cooked at high enough temperatures to sterilize them, (2) protected from contamination after cooking, and (3) held until eaten at temperatures too low or too high to permit vibrio reproduction. One commonly encountered problem is that seafoods considered cooked by traditional standards are not sterile. In addition, when seafoods are eaten raw, it is difficult to prevent vibrio infection. It is impossible to guarantee that any raw seafood is totally safe.

34. Infective intoxications are caused by exotoxins produced during microbial growth in the intestinal tract. The toxins act directly on the intestinal tract (primarily in the small intestine) by either an alteration in intestinal function or by causing structural damage to the intestinal lining mucosa.

35. The critical and frightening difference between the infant and adult forms of botulism is that children appear to become ill without having ingested toxin-containing foods. This is because spores have been ingested. Once in the intestine, *C. botulinum* spores germinate into vegetative cells and then multiply and produce botulinal toxin *in vivo.* For as yet unknown reasons, the intestinal tracts of only some infants are susceptible to *C. botulinum* infection. Most infants, older children, and adults regularly ingest *C. botulinum* spores without ill effects. All cases of infant botulism recognized to date have been caused by either type A or type B botulinal toxin. Honey, house dust, and soil are considered to be potential environmental sources of spores.

VI. DISEASE CHALLENGE

The situation described here has been taken from an actual case history. A review of treatment and epidemiological aspects is provided at the end of the presentation. Answers to questions, laboratory findings, and interpretations are given immediately following a specific question. Test your skills and take the *Disease Challenge.*

Case

A newborn boy was well until he was 50 hours old, when blood was noted in a stool. The baby had a normal temperature, was feeding well on formula, and had no abnormal physical signs. An x-ray of the abdomen was normal.

At this point, what type(s) of laboratory specimens should be taken? Blood and stool specimens should be taken for bacterial culture.

Laboratory Results

With appropriate media and incubation conditions, *Campylobacter* ss. *jejuni* was isolated.

Probable Diagnosis

Campylobacter enteritis.

Epidemiology

The mother of the child was questioned as to whether she had had diarrhea or any other symptoms of gastrointestinal disease. She denied having had recent diarrhea, but a stool culture yielded *Campylobacter* ss. *jejuni,* identical to the properties of the organism isolated from her son's stool.

Treatment

The child was treated orally with an electrolyte solution, followed by dilute formula feedings and a course of erythromycin. Blood in the newborn's stools disappeared after 24 hours.

VII. ENRICHMENT

Diarrheal Illness

Annually, more than one-quarter billion people travel from one country to another, thus making tourism, especially in developing countries, economically significant. U.S. travelers to Mexico alone number at least three million annually. Unfortunately, many a traveler succumbs to diarrheal illness, a condition that has plagued travelers and the military for centuries. Diarrheal illness has affected all groups of people, has given rise to a variety of hypotheses concerning cause, treatment, and prevention, and has achieved worldwide fame by its various euphemisms. Examples of the descriptive references coined by individuals experiencing intestinal agonies include gypsy tummy, Casablanca crud, Aden gut, Barsa belly, GI trots, Turkey trot, Hong Kong dog, Delhi belly, Aztec two-step, Montezuma's revenge, and turista. Most recently, a disease acquired by travelers to the Soviet Union caused by *Giardia* has been called "the Trotskys."

Epidemiologic Considerations

Several factors are important determinants in the development of travelers' diarrhea. These include the destination of the trip, the origin of the traveler, the purpose of travel, and the style of eating practiced by the traveler. Advancing age of individuals has been found to result in a lower incidence of the condition. It is not known whether this apparent protection is due to immunity resulting from frequent attacks or differences in eating habits.

Various studies have demonstrated a rather poor correlation between diarrhea and any particular food or drink. People who took great care by avoiding tap water, unpeeled fruits, and fresh vegetables were no better protected than were those who were relatively indiscriminate. One study did show, however, that travelers who consumed fresh salads were at somewhat greater risk for diarrhea. In recent years, diarrheal epidemics in certain developing countries have been traced to commercially bottled beverages, so that even these items cannot be considered safe havens for the thirsty traveler.

Causative Agents

With improvements in laboratory technology in several parts of the world, it has been possible to identify the microorganisms associated with diarrheal illness. Although many cases still lack an identification of a causative agent, the number of microbial pathogens is fairly large. The following list contains most pathogens known to cause diarrheal illness:

Entamoeba histolytica

Enteroviruses

Invasive *Escherichia coli*

Toxigenic *Escherichia coli*

Giardia lamblia

Parvoviruses

Reoviruses

Salmonella species

Shigella species

Vibrio parahaemolyticus

It is important to note that in some cases of the disease, more than one potential pathogen may be found in the patient's stool. This finding makes it difficult to assign a causative role to a specific organism.

27

Microbial Infections of the Cardiovascular and Lymphatic Systems

I. INTRODUCTION

The circulatory system consists of the heart and a series of vessels, arteries, arterioles, capillaries, small venules, and veins through which blood is carried to different regions of the body, collected, and returned to the heart. Consideration is given to the general organization and structures of this system.

Microbial diseases of the circulatory system account for a large number of infections and deaths in the world. Such diseases include AIDS, malaria, rheumatic heart disease, and subacute bacterial endocarditis. General descriptions of these and several other disease states are presented in this chapter. Procedures and laboratory tests used in diagnosis and approaches to treatment and control are briefly described. A Disease Challenge is included at the end of the chapter.

II. PREPARATION

Chapters 9, 10, 11-13, 17-22, 26, and 27 should be read before continuing. The following terms are important for you to know. Refer to the glossary and the appropriate chapters of the text if you are uncertain of any of them. A pronunciation guide for selected terms is provided as a learning aid.

1. AIDS
2. ARC
3. artificially acquired active immunity
4. AZT
5. bacteremia (back-teer-EE-me-ah)
6. bubo (BOO-boh)
7. bubonic plague
8. capillary
9. catalase
10. complement fixation
11. cyst (sist)
12. cytoplasm
13. ELISA
14. endemic
15. endocarditis
16. endocardium
17. epidemic
18. etiology (ee-tee-OL-oh-jee)
19. feces (FEE-sees)
20. fever

21. fluorescent antibody test
22. flagellate
23. Guillain-Barrē syndrome
24. hemoglobin
25. HIV
26. infectious mononucleosis
27. inflammation (in-fluh-MAY-shun)
28. kala agar
29. Kaposi's sarcoma
30. Kawasaki syndrome
31. louse
32. malaria
33. mannitol
34. mite
35. nausea (NAW-zee-ah)
36. pneumonic plague
37. reservoir
38. reverse transcriptase
39. rheumatic fever
40. rheumatic heart disease
41. rickettsioses
42. Rocky Mountain spotted fever
43. schizogony (skeh-ZOG-oh-knee)
44. sporogony (spo-ROG-oh-knee)
45. sporozoite
46. toxin
47. T lymphocytes
48. tricuspid valve
49. trophozoite (trof-oh-ZOH-ite)
50. typhus fever
51. vector
52. ventricle
53. zoonosis (zoh-NO-sis)
54. zygote (ZYE-goat)

III. PRETEST

Correct answers to all questions can be found at the end of this section. Write your responses in the appropriate space provided.

Completion Questions

Provide the correct term or phrase for the following. Spelling counts.

1. The major artery leading from the heart to other arteries and smaller blood vessels is the ____________ .
2. The major portion of the heart is composed of ____________ muscle.
3. The exterior of the heart is covered with an epithelial lining called the ____________ .
4. The two upper chambers of the human heart are the (a) ____________ , and the two lower chambers, thich walled and larger, are the (b) ____________ .
5. Blood vessels can be subdivided into the following three main categories: (a) ____________ , (b) ____________ , and (c) ____________ .
6. The activities or functions of the heart can be altered by the following three ways: (a) ____________ , (b) ____________ , and (c) ____________ .
7. Two specific infectious diseases of the heart are (a) ____________ and (b) ____________ .

8. The heart valve most commonly involved in rheumatic heart disease is the ________________ valve.
9. Rheumatic heart disease, sort throats, and certain kidney diseases are caused by bacteria known as ________________.
10. Enlarged lymph nodes known as buboes are characteristic of the bacterial disease ________________.
11. The arthropod vector for epidemic typhus is the ________________.
12. The causative agents of human malaria belong to the genus of ________________.
13. The infective units of malaria introduced through a mosquito bite are called ________________.
14. The ________________ gender of the mosquito transmits malaria.
15. Infectious mononucleosis and Burkitt's lymphoma are caused by ________________.

Matching

Select the answer from the right-hand side that corresponds to the term or phrase on the left-hand side of the question sheet. An answer may be used more than once. In some cases, more than one answer may be required.

______ 16. Plague
______ 17. Typhoid fever
______ 18. Rickettsial pox
______ 19. Relapsing fever
______ 20. Rheumatic fever
______ 21. Kala azar
______ 22. Tularemia
______ 23. Rocky Mountain spotted fever
______ 24. Epidemic typhus
______ 25. Malaria
______ 26. AIDS
______ 27. Infectious mononucleosis
______ 28. Pneumonic plague
______ 29. PAIDS

a. group A beta-hemolytic streptococci
b. *Rickettsia prowazeki*
c. *Rickettsia rickettsii*
d. *Yersinia pestis*
e. *Salmonella typhi*
f. *Rattus rattus*
g. *Borrelia recurrentis*
h. *Phlebotomus* species
i. *Leishmania* species
j. *Plasmodium* species
k. HIV
l. EBV
m. none of these

True or False

Indicate all true statements with a "T" and all false statements with an "F".

______ 30. Various microbial toxins can be spread to different areas of the body by the circulatory system.
______ 31. The circulatory system aids in the control and elimination of foreign organisms.
______ 32. The right side of the heart receives freshly oxygenated blood from the lungs.
______ 33. The pericardial sac contains a small amount of fluid to lubricate the heart.

_____ 34. One of the major functions of arteries is to carry blood from the heart to capillary systems.

_____ 35. Rheumatic heart disease is restricted to the young of lower-income groups in cold, wet countries.

_____ 36. Antibiotics such as chloramphenicol and tetracyclines are effective in treating most rickettsial diseases.

_____ 37. Bubonic plague is caused by the rat flea.

Answers

1. aorta
2. striated or cardiac
3. pericardium
4a. atria
4b. ventricles
5a. arteries
5b. veins
5c. capillaries
6a. endocarditis
6b. insufficient nourishment of heart tissue
6c. overextension of the heart resulting in exhaustion
7a. rheumatic heart disease
7b. infective endocarditis
8. mitral
9. group A hemolytic streptococci
10. plague
11. body or head louse
12. *Plasmodium*
13. sporozoites
14. female
15. Epstein-Barr virus
16. d
17. e
18. k
19. g
20. a
21. i
22. k
23. c
24. b
25. j
26. k
27. l
28. d
29. k
30. T
31. T
32. F
33. T
34. T
35. F
36. T
37. F

IV. CONCEPTS AND TERMINOLOGY

The development of many infectious diseases follows a consistent pattern. Once pathogens enter the body and establish a local site of infection, they may spread to other regions of the body by means of the circulatory system, forming secondary sites of infection. Depending on the properties of the disease agent, the circulatory system may spread toxins released by the pathogen. Certain microbial disease agents also attack and destroy tissues of the cardiovascular system.

Components of the Circulatory System

The heart and blood vessels form a closed system, so that blood is always contained within this series of vessels and never runs free. The general properties of the heart and blood vessels are listed in Table 27-1.

Diseases of the Heart

Heart disease is the leading cause of death in most developed countries of the world. The activities of the heart can be impaired in at least three ways:

1. Endocarditis, including infection and damage of the heart's valves and associated tissues
2. Insufficient nourishment for cardiac muscle due to a narrowing of the arteries of the heart sometimes associated with a blood clot (thrombus) within these vessels
3. Overexertion of the heart, resulting in exhaustion

Any one of these conditions can cause enlargement of the heart, one of the most important signs of heart disease.

The heart can suffer from infection just as any other part of the body. Of the various infectious diseases of the heart, rheumatic heart disease and infective endocarditis are the most common. General features of these disease states are given in Table 27-2.

TABLE 27-1 Properties of Circulatory System Components

Component	*General Organization and Composition*	*Function(s)*
Blood	Consists of the formed element red and white blood and plasma, the liquid portion; total amount of blood in the human adult is 10-12 pints.	Carries nutrients and hormones, removes cellular waste products, assists in the regulation of body temperature, and aids in the control and elimination of foreign organisms; flows under pressure in a one-way path through the vessels (arteries, veins, and capillaries) and heart directly to the tissues and organs.
Heart	Consists mainly of striated muscle called *cardiac muscle* or *myocardium.* The inner surface of the heart is covered with endothelial tissue called endocardium; the outside is covered with an epithelial tissue called *pericardium,* which forms the *pericardial sac* within which the heart is located. The human heart consists of four separate chambers—the two upper chambers are the *atria;* the two larger, lower chambers are the thick-walled, highly muscular *ventricles.*	The heart functions as a pump, providing the force necessary to maintain adequate blood flow throughout the body. The organ's right side receives blood from the body and pumps it to the lungs, where carbon dioxide in the blood is exchanged for oxygen, and the left side receives the freshly oxygenated blood from the lungs and pumps it to other organs and tissues.
Heart valves	The valves between the atria and the ventricles are the *atrioventricular valves.* The valve on the right side is called the *tricuspid valve* because it has three flaps; the valve on the left is the *bicuspid,* or *mitral,* valve because it has two flaps.	Allow blood to flow in only one direction, preventing its backward flow.
	Semilunar valves are located between the ventricles and their attached blood vessels.	Prevent backward blood flow into the ventricles.
Blood vessels	Blood vessels are divided into three main categories: *arteries, veins,* and *capillaries.* The arteries have the thickest, strongest walls, made up of three layers. The inner layer, or endothelium, lines the vessels; the middle layer contains smooth muscles and elastic tissues; the outer layer, also elastic, is made of various supporting tissues.	Arteries conduct blood from the heart to the capillary network where the exchange of material between blood and tissue cells occurs. The veins collect blood from the capillaries and return it to the heart for a new cycle. Capillaries provide nutrients and oxygen to cells and collect cellular metabolic waste products.

TABLE 27-2 Microbial Diseases of the Heart

Disease State	*Cause(s)*	*General Features*	*Prevention and Control*
Rheumatic heart disease	One or more episodes of group A hemolytic strepto-coccal-caused rheumatic fever. Antibodies formed against the streptococci inflame the mitral and other heart valves.	Scarring and deformation of valves result from abnormal growths of connective tissue on valves (*fibrosis*). Valve damage results in narrowing of valve opening (stenosis) or failure of valve to close completely (valvular insuf-ficiency); universally distributed among all climates, races, social strata, and ages.	Preventative measures include prompt diagnosis and treat-ment with appropriate antibiotics to prevent formation of antibodies against streptococcal antigens. Control involves antibiotic treatment and surgical replacement of damaged valves with artificial devices to restore normal functioning.

TABLE 27-2 Microbial Diseases of the Heart *(continued)*

Disease State	*Cause(s)*	*General Features*	*Prevention and Control*
Infective endocarditis	Bacteria lodged in heart valves scarred by rheumatic fever; congenital malformations; various microbial infections; hardening of the arteries; complications resulting from dental procedures allowing bacteria to enter the bloodstream.	Inflammatory condition of the heart accompanied by prolonged fever, changing heart murmur, and the growth of bacterial *vegetations* (accumulations of tangled masses of fibrin, platelets, blood fragments, and bacteria) on heart valves.	Prompt diagnosis and treatment.

Other Microbial Diseases of the Circulatory System

Microbial diseases of the blood, blood vessels, and other elements of the circulatory system are described in Table 27-3.

TABLE 27-3 Microbial Diseases of the Circulatory System

Disease	*Causative Agent*	*Gram Reaction*	*Morphology*	*Means of Transmission*	*General Features*	*Prevention and Control*
Bacterial						
Plague	*Yersinia pestis*	—	Rod	Infected rat fleas belonging to genera of *Nosopsyllus* and *Xenopsylla*	Two forms in humans: (1) bubonic, an infection causing enlarged regional lymph nodes (buboes), fever, chills, headache, and exhaustion; (2) pneumonic, an infection of respiratory system accompanied by coughing, chest pains, difficulty in breathing, and bluish coloration of skin (cyanosis)	Elimination of infected rodents and fleas; prompt diagnosis and treatment; immunization with killed or attenuated preparations
Relapsing fever	*Borrelia recurrentis*	Not done	Spirochetes	Infected ticks (*Ornithodorus*) and infected lice *Pediculus humanus* var. *capitis*	Sudden fever; period followed by continued fever attacks (relapses) with each one milder than previous ones	Elimination of infected arthropods; prompt diagnosis and treatment[a]
Rickettsialpox	*Rickettsia akari*	—	Rod	Infected mites (*Allodermanyssus sanguineus*)	Formation of a small, red, hard blister at site of mite bite, followed by sudden appearance of backache, chills, fever (104° F), and rash	Elimination of infected rodents and mites; prompt diagnosis and treatment[a]
Rocky Mountain spotted fever	*R. rickettsii*	—	Rod	Infected ticks (*Amblyomma* spp., *Dermocentor* spp., *Rhipicephalus sanguineus*)	Backache, chills, fever, and rash, which develops on ankles, forehead, and wrists and spreads to trunk	Elimination of infected ticks; prompt diagnosis and treatment; immunization with killed organism

TABLE 27-3 Microbial Diseases of the Circulatory System *(continued)*

Disease	*Causative Agent*	*Gram Reaction*	*Morphology*	*Means of Transmission*	*General Features*	*Prevention and Control*
Scrub typhus	*R. tsutsugamushi*	—	Rod	Infected mites (*Trombicula* spp.)	A localized sore site of mite bite followed by backache, chills, fever (102° to 105° F), and rash; deafness and mental disturbances can be complications	Elimination of infected rodents and mites; prompt diagnosis and treatment[a]
Trench fever	*Rochalimeae quintana*	—	Rod	Infected human body lice (*Pediculus corporis*)	Chills, exhaustion, dizziness, fever (102° to 103° F), headache, pain in shins and thighs, and rash	Same as for scrub typhus
Tularemia	*Francisella tularensis*	—	Rod	Handling infected animal carcasses; water contaminated by infected animals or their remains; infected flies or ticks; inhalation of infectious aerosols	Fever, headache, and general malaise; an ulcerating papule can occur at primary site of entry on skin or mucous membranes	Elimination of infected rodents and arthropods; appropriate disposal of diseased animal remains; public education; prompt diagnosis and treatment
Typhoid fever	*Salmonella typhi*	—	Rod	Ingestion of contaminated food or water; flies; fomites; carriers	Abdominal swelling, constipation, loss of appetite, nausea, vomiting, fever (104° F), diarrhea with possible blood in stools, rash on abdomen (rose spots); complications include gall bladder infection	Improved sanitation; treatment of carriers; prompt diagnosis and treatment of infected persons; immunization with killed organisms; education programs
Typhus fever (endemic)	*R. typhi*	—	Rod	Infected fleas (*Xenopsylla cheopsis*) introduce infectious feces into flea bites	Symptoms similar to other typhus fevers	Elimination of infected rodents and fleas; prompt diagnosis and treatment
Typhus fever (epidemic)	*R. prowazekii*	—	Rod	Infected human body lice (*Pediculus)*	Symptoms similar to other typhus fevers; the rash begins on the trunk and spreads to the arms and legs	Elimination of infected rodents and fleas; prompt diagnosis and treatment
Protozoan						
Kala azar (dum-dum fever)	*Leishmania donivani*	a	a	Infected sandflies (*Phlebotomus* spp.)	Fever, bleeding from gums, lips, and nose; enlarged liver and spleen; pneumonia can be a complication	Elimination of arthropod vectors; use of prophylactic devices (e.g., insect repellants, sandfly nets); prompt diagnosis and treatment[a]
Malignant tertian malaria	*Plasmodium falciparum*	a	a	Infected mosquitoes (*Anopheles* spp.)	Headache, nausea, abdominal pain, vomiting, chills, prolonged fever stage, little sweating, convulsions, and coma; enlarged liver and spleen, heart failure, and bloody urine are complications	Elimination of infected arthropods and breeding sites; prompt diagnosis and treatment[a]

TABLE 27-3 Microbial Diseases of the Circulatory System *(continued)*

Disease	*Causative Agent*	*Gram Reaction*	*Morphology*	*Means of Transmission*	*General Features*	*Prevention and Control*
Quartan malaria	*P. malariae*	a	a	Same as for other forms of malaria	A relatively mild form of the disease with some chills, fever, headache, nausea, and abdominal pain	Same as for other forms of malaria
Tertian malaria (ovale)	*P. ovale* (rare)	a	a	Same as for other forms of malaria	Periods of shaking, chills, fever, headache, nausea, muscular pain, and vomiting	Same as for other forms of malaria
Tertian malaria (vivax or benign)	*P. vivax*	a	a	Same as for other forms of malaria	Periods of shaking chills (cold stage), followed by fever (104° to 106° F or 40° to 41° C) with profuse sweating, nausea, vomiting, headache, and muscular pains	Same as for other forms of malaria
Viral						
Acquired immune deficiency syndrome (AIDS)	Human immunodeficiency virus (HIV)	a	a	Sexual activities with infected persons, contaminated syringes and/or needles, placental transmission	Refer to Table 27-6 in the text	Avoidance of means of transmission; ATZ, possibly interferons; treatment of opportunistic infections; symptomatic
AIDS-related complex (ARC)	HIV	a	a	Same as for AIDS	Symptoms include weight loss greater than 10%, night sweats, diarrhea, skin rash, fatigue, oral candidiasis (see Color Photograph 111d), swollen lymph glands, and long-lasting fever (37.8° C)	Treatment of opportunistic infections; symptomatic
Infectious mononucleosis (IM)[b] and Burkitt's lymphoma	Epstein-Barr virus (EBV)	a	a	Direct oral contact (e.g., kissing); fomites	Swollen lymph nodes, increase in abnormal leukocytes, fever, general discomfort	Improved sanitation; education programs; prompt diagnosis and treatment; bed rest and adequate nutrition
Kawasaki syndrome	Unknown	a	a	Unknown	Fever (37.8° to 40° C) for an average of 11 days, inflammation of conjunctiva (both eyes), red rash on palms and feet, enlargement of lymph nodes in the neck area; surface skin loss on finger and toe tips; hardened areas of hands and feet	Supportive to relieve symptoms and prevent complications

[a] Gram reactions are not applicable.

[b] IM occurs in both acute and chronic forms. Refer to the text chapter for additional details.

V. CHAPTER SELF-TEST

Continue with this section only after you have read Chapters 9-13, 17-22, and 27. A score of 80 percent or better is good. If your score is less than 65 percent, reread the chapter.

Correct answers to all questions can be found at the end of this section. Write your responses in the appropriate space provided.

Completion Questions

Provide the correct term or phrase for the following. Spelling counts.

1. The inner surface of the heart is lined with endothelial tissue, which forms the ______________________.
2. The heart contains valves that allow blood to flow in ______________________ direction(s) when open.
3. The valves located between the atria and ventricles are called the ______________________ valves.
4. Blood vessels that collect blood from capillaries and return it to the heart are called ______________________.
5. The sexual cycle of the malarial parasite is known as ______________________.
6. The ameboid, single-nucleus feeding form of *Plasmodium* species found in red blood cells is called a(an) ______________________.
7. The causative agent if Burkitt's lymphoma is ______________________.
8. The full name of the AIDS virus is ______________________.

Matching

Select the answer from the right-hand side that corresponds to the term or phrase on the left-hand side of the question sheet. An answer may be used more than once. In some cases, more than one answer may be required.

Topic: Disease Transmission

______ 9. Typhoid fever	a.	flea bite
______ 10. Endemic typhus fever	b.	mites
______ 11. Rickettsialpox	c.	ticks
______ 12. Scrub typhus	d.	ingestion of contaminated food or water
______ 13. Plague	e.	handling infected rabbits
______ 14. Kala azar	f.	fomites
______ 15. Rocky Mountain spotted fever	g.	sexually transmitted
______ 16. Relapsing fever	h.	none of these
______ 17. Malaria		
______ 18. AIDS		
______ 19. Infectious mononucleosis		
______ 20. ARC		

Topic: Properties of Disease Agents

______ 21. *Yersinia pestis*

______ 22. *Salmonella typhi*

______ 23. *Rickettsia akari*

______ 24. *Francisella tularensis*

______ 25. *Borrelia* species

______ 26. *Plasmodium* species

a. gram-negative rod

b. gram-positive coccus

c. acid-fast rod

d. spirochete

e. none of these

True or False

Indicate all true statements with a "T" and all false statements with an "F".

______ 27. Established pathogens in a main site of infection generally cannot spread to other regions of the body by means of the circulatory system.

______ 28. If the heart were not enclosed in a pericardial sac, friction forces would occur that could produce lesions on other organs.

______ 29. Arteries have thicker, stronger walls than do capillaries.

______ 30. Enlargement of the heart is one of the important signs of heart disease.

______ 31. Fibrosis of the tricuspid valve usually is the most common result of rheumatic heart disease.

______ 32. The heart can suffer from infection just as any other part of the body can.

______ 33. Rheumatic heart disease is caused by abnormal growth of connective tissue on heart valves.

______ 34. Buboes are characteristically found with cases of trench fever.

______ 35. The causative agent of tularemia is *Francisella tularensis.*

______ 36. Diagnosis of malaria can be made by means of blood smears.

______ 37. The asexual development cycle of malarial parasites is known as sporogony.

______ 38. ARC results from HIV infection.

Essay Questions

Answer the following questions on a separate sheet of paper.

39. Distinguish between sporogony and schizogony as they occur in malaria.

40. How is infectious mononucleosis transmitted?

41. What are the major means of HIV transmission?

Answers

1. endocardium
2. one
3. atrioventricular
4. veins
5. sporogony
6. trophozoite
7. Epstein Barr virus
8. human immuno-deficiency virus
9. d
10. a
11. b
12. b
13. a
14. f
15. c
16. e
17. f
18. g
19. h
20. g
21. a
22. a
23. a
24. a
25. d
26. e
27. F
28. T
29. T
30. T
31. F
32. T
33. T
34. F
35. T
36. T
37. F
38. T

39. A typical life cycle of the malaria parasite begins with the sexual phase, *sporogony,* developing within the invertebrate host (anopheline mosquito). *Schizogony,* the asexual phase, begins with the invasion of the host red blood cells.

40. Infectious mononucleosis is transmitted chiefly by kissing (direct oral means) or by drinking from a shared bottle or glass (indirect oral means) with the exchange of saliva and perhaps leukocytes. For this Epstein-Barr virus (EBV) infection to occur, an adequate quantity of virus must be introduced into the mouth of an antibody-negative host.

41. In adults, the HIV virus is transmitted primarily through sexual contact (homosexual or heterosexual) and through inoculation or transfusion with contaminated blood or commercial blood-product preparations such as those used for the treatment of hemophiliacs. The only other instances of adult transmission reported have involved artificial insemination or organ transplants from infected donors. In pediatric cases, the virus can be transmitted by infected women to their fetuses or offspring during pregnancy, during labor and delivery, and possibly shortly after birth.

VI. DISEASE CHALLENGE

The situation described has been taken from an actual case history. It has been designed to show how clinical, laboratory, microbial, and related information is used in disease diagnosis. A review of treatment and epidemiological aspects is given at the end of the presentation. Answers to questions, laboratory findings, and interpretations are given immediately following a specific question. Test your skill and take the *Disease Challenge.*

Case

A two-month-old boy was admitted with an 18-hour history of fever as high as 38.7° C and vomiting (emesis). Physical examination showed a slightly enlarged spleen and liver. The child's parents, a Cambodian refugee couple, who had entered the United States two months before their son's birth, indicated abnormal drowsiness, and a decreasing intake of milk was also evident. A further history revealed that the pregnancy and delivery (at another hospital) had been without complications. However, the mother had had an episode of shaking chills the day before delivery. In addition, the baby had developed jaundice shortly after birth, and had been kept in the hospital nursery for 14 days before discharge.

At this point, what disease(s) could be suspected? Bacterial infections, including syphilis and salmonellosis, the protozoan diseases toxoplasmosis or malaria, and the virus infections caused by herpes simplex virus, cytomegalovirus, and enteroviruses. (Refer to this chapter and Chapters 23, 26, 28, and 29 for descriptions of most of these diseases.)

Should treatment be started? If so, what form should it take? Since a bacterial infection is suspected, empiric treatment with antibiotics such as ampicillin and gentamicin should be started.

At this point, what type(s) of laboratory specimens should be taken? Blood for smears and culture, and urine and spinal fluid for culture.

Laboratory Results

All specimens for culture were negative. Examination of the baby's blood smear showed that approximately 5% of red blood cells contained ring or trophozoite stages of *Plasmodium vivax.*

Probable Diagnosis

A congenital form of malaria.

Epidemiology

When *P. vivax* was found in the blood smear of the baby, blood specimens were obtained from the mother. Various blood stages of the protozoon were found. Treatment was initiated shortly after the laboratory furnished these results.

Treatment

The antibiotics given to the baby had virtually no effect. The baby still had a high fever for three days. Shortly after the laboratory results were known, the infant was treated with chloroquine phosphate. The response to this therapy was dramatic. Within three days, the baby's temperature was normal, and his blood smears were negative for *P. vivax*. Recovery continued normally without complications.

VII. ENRICHMENT

Coxiella burnetii, a Typical Rickettsiae?

In 1935, an outbreak of a fever of unknown etiology occurred among field workers in Brisbane, Australia. E. H. Derrick, director of the Laboratory of Microbiology and Pathology of the Queensland Health Department at Brisbane, was assigned the task of investigating the nature of the disease. During his investigation, Derrick coined the name "Q" (for "Query") fever for the disease and accurately described most of the clinical features of the infection.

Q fever showed certain clinical similarities to typhus fever and to typhoid and paratyphoid fevers, but the absence of a rash and a negative Weil-Felix reaction ruled out these infections. Although Derrick was unable to identify or isolate the causative agent, he succeeded in transmitting Q fever to guinea pigs with the blood or urine of infected patients. Derrick later sent a sample of infected guinea pig liver to M. Burnet, who used it to transmit Q fever to mice and guinea pigs. On autopsy of the animals, Burnet identified typical rickettsialike organisms. Using his findings and closely reasoned arguments, Burnet concluded that the agent of Q fever was a rickettsial agent not unlike the typhus, scrub typhus, and spotted fever rickettsiae. The causative agent was subsequently named *Coxiella burnetii.*

More than four decades after the first reports of the disease, Q fever has been found to be distributed worldwide and remains a potential hazard in almost any habitable region of the earth.

The principal modes of entry of *C. burnetii* into humans are via inhalation of contaminated dust particles and aerosols generated in the milieu of stockyards and dairies and the ingestion and handling of infected meat and milk. The causative agent originates from the major reservoirs of *C. burnetii*—dairy cows, sheep, and goats. Infected cows shed enormous numbers of rickettsias in their milk, even though the animals appear healthy. Q fever in domestic animals is of major importance and continues to be the principal source of the infection in humans. Although cattle, sheep, and goats are the main reservoirs, the agent has been detected as well in dogs, camels, buffaloes, geese, and other fowl.

The typical picture of Q fever in humans includes an incubation period of 15 days or less, depending on the route of exposure, dosage of the infecting agent, and the age of the victim. This period is usually followed by the onset of a fever that reaches a plateau of 40°C within two to four days. The febrile period is accompanied by general discomfort, loss of appetite, muscular weakness, pain, and a severe generalized headache. Liver damage followed by an enlarged liver and injury to the kidneys and heart valves also occur. Rickettsias are present in the early stages of fever and may be found in urine at this time.

A long-lasting immunity generally results from the infection, although chronic and latent infections have been reported. Q fever is managed by tetracycline and chloramphenicol therapy, but *C. burnetii* is less sensitive to these antibiotics than are typical rickettsioses. Erythromycin has been found to be especially effective. At the present time, no commercial vaccines are available for human use.

28

Microbial Diseases of the Reproductive and Urinary Systems

I. INTRODUCTION

The genitourinary system consists of the kidneys, ureters, urinary bladder, urethra, and gonads. This chapter considers the general arrangement of these organs and the variety of microbial diseases that affect them. General symptoms, approaches to diagnosis, and treatment and control of major infections and complications of the urinary and genital organs, including the sexually transmitted diseases, are presented. Diseases of the genital organs and urinary system are considered together because of the intimate anatomic relationships that exist between certain structures. Two Disease Challenges are included at the end of the chapter.

II. PREPARATION

Chapters 11-13, 17-22, and 28 should be read before continuing. The following terms are important for you to know. Refer to the glossary and the appropriate chapters of the text if you are uncertain of any of them. A pronunciation guide for selected terms is provided as a learning aid.

1. acute (ak-KYOOT)
2. AIDS
3. anaerobe
4. antigen
5. catheterization (kath-eh-terr-ee-ZAY-shun)
6. chancre (SHANK-ker)
7. chancroid (SHANK-kroid)
8. chemotherapy
9. chronic
10. coagulase
11. clue (kloo) cells
12. coliforms
13. complement fixation
14. congenital (kon-JEN-ee-tal)
15. conjunctivitis (kon-kunk-tee-VYE-tis)
16. contagious (kon-TAY-jiss)
17. dark-field microscopy
18. dipstick device
19. electrolyte
20. epididymitis (eh-pea-did-ee-MY-tis)
21. exotoxin
22. fomite

23. fluorescent antibody
24. genital herpes
25. glomerulonephritis (glow-mer-you-low-neh-FRY-tis)
26. glomerulus
27. gonorrhea (gohn-or-REE-ah)
28. granuloma inguinale
29. gumma (GOO-mah)
30. hepatitis B virus
31. hemolysis (hee-mol-EE-sis)
32. immunoperoxidase
33. inflammation
34. kidney cortex
35. kidney medulla
36. lymphatic system
37. mannitol fermentation
38. neonate (knee-OH-nate)
39. nephron (NEFF-ron)
40. pelvic inflammatory disease
41. penicillinase
42. phagocytosis (fag-oh-sigh-TOE-sis)
43. prophylaxis (pro-fee-LACK-sis)
44. prostate gland
45. prostatitis (pros-ta-TIE-tis)
46. pyelonephritis (pie-low-nef-FRY-tis)
47. pyogenic (pie-oh-GEN-ick)
48. sexually transmitted diseases
49. spirochete (SPY-row-keet)
50. syphilis (SIFF-ee-liss)
51. TORCH complex
52. toxic shock syndrome
53. treponema (trep-oh-KNEE-mah)
54. urease
55. urethritis (you-ree-THRY-tis)
56. urogenous
57. vaginitis (vaj-in-EYE-tis)
58. venereal disease

III. PRETEST

Correct answers to all questions can be found at the end of this section. Write your responses in the appropriate space provided.

Completion Questions

Provide the correct term or phrase for the following. Spelling counts.

1. The microorganism most frequently present in cortical infections of the kidney is ____________________ .
2. Three major pathways by which microorganisms can gain access to the kidney and cause infections are the (a) ____________________, (b) ____________________, and (c) ____________________.
3. Inflammation of the urinary bladder is called ____________________.
4. The external genitalia of the female collectively are called the ____________________.
5. Two major products of the seminiferous tubules are (a) ____________________ and (b) ____________________ .
6. The causative agent of herpes genitalis is ____________________.
7. The soft, gummy tumors associated with tertiary syphilis are called ____________________.
8. The first lesion of syphilis that develops at the site of entry is the ____________________.

9. The acronym TORCH stands for the following complex of disease agents: T (a) ____________, O (b) ____________, R (c) ____________, C (d) ____________, H (e) ____________.

Matching

Select the answer from the right-hand side that corresponds to the term or phrase on the left-hand side of the question sheet. An answer may be used more than once. In some cases, more than one answer may be required.

Topic: Genitourinary Tract Diseases

______	10. *Caylmmatobacterium granulomatis*	a. gram-negative rod
______	11. *Treponema pallidum*	b. virus
______	12. Ophthalmia neonatorum	c. gram-negative coccus
______	13. *Staphylococcus aureus*	d. chlamydia
______	14. *Trichomonas vaginalis*	e. protozoan pathogen
______	15. *Neisseria gonorrhoeae*	f. anaerobic spore former
______	16. *Escherichia coli*	g. yeast
______	17. *Clostridium* species	h. spirochete
______	18. Herpes simplex virus (type 2)	i. toxic shock syndrome agent
______	19. Causative agent of lymphogranuloma venereum	j. none of these
______	20. *Gardnerella vaginalis*	
______	21. *Candida* species	
______	22. *Chlamydia trachomatis*	

True or False

Indicate all true statements with a "T" and all false statements with an "F".

______ 23. The frequent flushing of the urethral surface by urine helps to decrease microbial numbers in regions near the urinary bladder.

______ 24. Urine normally is sterile.

______ 25. Urinary tract infections are more frequent in males than in females.

______ 26. The toxemia of pregnancy can predispose individuals to urinary tract infections.

______ 27. The infecting organism in large part determines the form and intensity of kidney infections.

______ 28. Sources of bladder infections in the female include the gastrointestinal tract and the cervix.

______ 29. *Trichomonas vaginalis* may cause infection in males.

______ 30. Clue cells are characteristically found in most vaginitis cases caused by *Candida albicans.*

______ 31. Toxic shock syndrome is associated with a skin-destroying toxin produced by *Staphylococcus aureus.*

_____ 32. In industrialized countries, *Chlamydia trachomatis* is believed to be the most common cause of sexually transmitted diseases.

_____ 33. Clinical examination usually is sufficient to detect HSV infection.

Answers

1. *Staphylococcus aureus*
2a. bloodstream
2b. lymphatics
2c. ureter
3. cystitis
4. vulva
5a. spermatozoa
5b. testosterone
6. herpes simplex virus (type 2)
7. gummas
8. primary or Hunterian chancre
9a. *Toxoplasma*
9b. other disease agents involving the newborn or fetus
9c. rubella virus
9d. cytomegalovirus
9e. herpesviruses
10. a
11. h
12. d
13. i
14. f
15. c
16. a
17. f
18. b
19. d
20. a
21. g
22. *a* and *d*
23. T
24. T
25. F
26. T
27. T
28. T
29. T
30. F
31. T
32. T
33. F

IV. CONCEPTS AND TERMINOLOGY

The many functions of the urinary system include eliminating waste products, regulating the chemical composition and acid-base balance of the body tissues, and maintaining a constant water content in the body.

The Basic Anatomy of the Urinary System

The human urinary system includes two kidneys, two ureters, the urethra, and the urinary bladder. The general features of the urinary system components are listed in Table 28-1.

TABLE 28-1 Urinary System Components

Component	*General Organization and Composition*	*Function(s)*
Kidneys	Consist of an outer cortex and inner medulla. The basic functional unit of the kidney is the *nephron.*	Establish and help maintain the concentration of electrolytes (e.g., chloride, potassium, sodium, and bicarbonate) and excrete poisonous substances.
Ureters	Muscular tubes.	Transport urine to urinary bladder from kidney.
Urinary bladder	Storage structure.	Stores urine for elimination.
Urethra	Tubelike channel.	Serves as passageway for urine elimination. The female urethra functions only in urination; the male counterpart serves for the passage of both urine and semen.
Nephron	Composed of a *glomerulus* (blood filter) and a long tubule (concentrator and urine collector). The glomerulus, a tuft or collection of capillaries, is surrounded by a spherically expanded portion of the nephron, the *Bowman's capsule.*	Blood is filtrated through the glomeruli and is concentrated by the tubules, which empty into the ureters. The renal capsule filters dissolved substances and some water out of the blood plasma; the renal tubule reabsorbs water, electrolytes, and other substances to help maintain the body's normal internal environment.

The Flora of the Normal Urinary Tract

The normally sterile kidney is highly resistant to bacteria. Many well-known urinary tract pathogens are unable to invade the kidney and produce infection there. The few exceptions to this rule include certain enterococci strains, *Escherichia coli, Proteus* spp., and staphylococci.

In humans, bacteria are commonly present in the lower urethra. However, their numbers decrease in regions near the bladder. This is apparently caused by some antibacterial effect exerted by the urethral lining and the frequent flushing of the epithelial surface of the urethra by urine, which is normally sterile.

Diseases of the Urinary Tract

Predisposing conditions for urinary tract infections include diabetes mellitus, neurologic diseases, and lesions that interfere with urine flow. The major routes of infection are by blood, lymphatics, and from the ureter. Common urinary infections are caused by *coliforms,* intestinal organisms. Table 28-2 lists representative diseases.

TABLE 28-2 Representative Diseases of the Urinary System

Disease	*Causative Agent*	*Gram Reaction*	*Morphology*
Cystitis (inflammation of the urinary bladder)	*Escherichia coli, Proteus vulgaris, Pseudomonas aeruginosa*	—	Rods
Glomerulonephritis (inflammation of kidney glomeruli)	*Streptococcus pyogenes*	+	Coccus
Kidney cortex infections	*Staphylococcus aureus*	+	Coccus
Nonspecific urethritis (inflammation of the urethra)	*Candida albicans*	+	Oval yeast
	Mycoplasma hominis and *Ureaplasma urealyticum*	—	Pleomorphic
	Staphylococci and streptococci	+	Cocci
	Trichomonas vaginalis	Not useful	Protozoan
Prostatitis (inflammation of the prostate gland)	*E. coli, P. aeruginosa*	—	Rods
	Neisseria gonorrhoeae	—	Coccus
	Staphylococci and streptococci	+	Cocci
	Mixed infections of the above organisms		
Puerperal sepsis (child-bed fever)	Anaerobic streptococci	+	Cocci
	Clostridia	+	Rods
	E. coli, Proteus spp.	—	Rods
	Neisseria gonorrhoeae	—	Coccus
Pyelonephritis (inflammation of the kidney and pelvis)	*Bacteroides* spp., *Enterobacter aerogenes, E. coli, Proteus* spp., *Pseudomonas aeruginosa*	—	Rods
	Staphylococci and streptococci	+	Cocci
Pyonephrosis (inflammation of the kidney with pus formation)	Anaerobic streptococci	+	Cocci
	Bacteroides spp.	—	Rods

The Functional Anatomy of the Reproductive System

The female reproductive system consists of two button-shaped organs called the *ovaries,* two *oviducts (Fallopian tubes),* the *uterus (womb),* and the *vagina.* The ovaries are situated near the kidneys on either side of the uterus. The male system includes two oval glandular *testes,* a system of *ducts,* auxiliary glands, and the *penis.*

Table 28-3 describes the general properties of the human reproductive system.

Diseases of the Reproductive System

The reproductive organs and associated structures are subject to a wide variety of inflammatory conditions. Several of the nonvenereal diseases are listed in Table 28-4.

The Sexually Transmitted or Venereal Diseases

Sexually transmitted or venereal diseases are those diseases that are transmitted by sexual means (see Table 28-5). The infectious agents that cause these diseases are important both because they cause discomfort and disability and because some may result in sterility or life-threatening situations. Most venereal diseases occur among adults and adolescents. However, current findings show that children as young as 10 to 12 years of age also acquire such infections through sexual contact with children or adults. Infants may be infected *in utero* or during the birth process. The signs and symptoms of common sexually transmitted or venereal diseases are given in Table 28-6.

TABLE 28-3 Human Reproductive System Components

Component	*General Organization and Composition*	*Function(s)*
Female		
Ovaries	Each ovary consists of a medula (interior) and a cortex (outer region).	Site for the development and production of ova (eggs) and important hormones such as estradiol and progesterone.
Oviducts	Tubelike channels leading from the ovaries to the uterus.	Transport eggs to the uterus.
Vagina	Chamber lined with mucous-producing cells.	Serves as both a birth canal and copulatory canal.
Vulva	Includes the *clitoris, labia majora, labia minora,* and vestibule. The clitoris, a small erectile organ homologous to the male penis, is located toward the front of the vulva. The labia majora and the labia minora are folds of skin that line the vaginal opening. The labia minora, which are situated between the labia majora, contain the *glands of Bartholin.*	The glands of Bartholin are responsible for the secretion of an alkaline fluid that functions as a lubricant during copulation.
Male		
Testes	Located outside the abdominal cavity suspended by the spermatic cords in a saclike structure. The testes are divided into lobes that obtain several seminiferous tubules.	Production of *spermatozoa* and *testosterone* (male sex hormone) and mature sex cells (sperm).
Vas deferens (ductus deferens)	Tubes.	Carries the sperm to the urethra.
Auxiliary glands	The *seminal vesicles, prostate gland,* and *bulbourethral glands* of Cowper.	Involved with the formation of *seminal fluid* or *semen.* The secretions of the seminal vesicle empty into the vas deferens, while those of the prostate and bulbourethral glands empty into the urethra.

TABLE 28-4 Nonvenereal Diseases of the Reproductive Tract

Disease	*Causative Agent(s)*	*Means of Transmission*	*General Description*
Anaerobic infections	*Bacteroides* spp., *Clostridium* spp., *Streptococcus* spp.	Unsterile instruments or other materials used in the performance of an abortion.	Most result from complications during the confinement period following labor, the *puerperium,* or from complications of septic abortion.
Puerperal (childbirth) sepsis	Anaerobic streptococci, clostridia, *Escherichia coli, Neisseria gonorrhoeae,* and *Proteus* spp.	Direct contact with persons harboring pathogens in their upper respiratory passages, the patient herself, and droplet nuclei and fomites.	An infection that develops from the invasion of open wounds that occur normally or as the result of surgical procedures in the genital tracts of women who have just given birth. Postpartum (after-birth) hemorrhage, premature rupture of membranes, and prolonged labor predispose mothers to puerperal sepsis.
Acute prostatitis	*Neisseria gonorrhoeae, E. coli, P. aeruginosa,* staphylococci, and streptococci	Bacterial pathogens reach the prostate by routes similar to those listed for kidney infections.	Inflammations of the prostate gland.
Toxic shock syndrome (TSS)	Certain toxin-producing strains of *Staphylococcus aureus*	Usage of tampons; sexual activity also has been implicated.	Typical symptoms include a rash followed by skin shedding, low fever, and involvement of muscles, kidneys, liver, and gastrointestinal tract. Males also can exhibit the condition.
Vaginitis	A variety of microbes including anaerobic bacteria, *Candida albicans, Gardnerella vaginalis,* and *Trichomonas vaginalis*	Direct contact with infected persons, fomites.	Infections may result from a combined (synergistic) effect by anaerobic bacteria and *G. vaginalis.* Obvious signs of *G. vaginalis* infection include a gray-white homogenous discharge, fishy odor, and microscopically visible clue cells.

TABLE 28-5 Sexually Transmissable Diseases

1. Chancroid (B)[a]
2. Chlamydial infections (B)
3. Condyloma acuminata (genital warts) (V)
4. *Entamoeba histolytica* (P)
5. *Giardia intestinalis* (P)
6. Gonorrhea (B)
7. Granuloma inguinale (B)
8. Hepatitis A infection (V)
9. Hepatitis B infection (V)
10. Herpes genitalis (V)
11. Human immunodeficiency virus infection (V)
12. Lymphogranuloma venereum (B)
13. Molluscum contagiosum (V)
14. Monilial vaginitis (Y)
15. Nongonococcal urethritis (B)
16. Pediculosis pubis infestation (L)
17. Reiter's syndrome (arthritis and inflammations of the eyelid and urethra) (B)
18. Scabies (L)
19. *Shigella* infection (B)
20. Syphilis (B)
21. *Trichomonas* infection (P)

[a] B = bacterial; L = lice; P = protozoan; V = virus; and Y = yeast.

TABLE 28-6 Features of the Venereal Diseases

Disease	*Causative Agent*	*Gram Reaction*	*Morphology*	*General Features of Disease*
Chancroid	*Haemophilus ducreyi*	—	Small rod (coccobacillus)	Symptoms include small elevated lesions that form irregular soft-edged sores, soft chancres, and some pain.
Chlamydia infections	*Chlamydia trachomatis*	—	Small rod	Inflammation of the urethra and cervix. *C. trachomatis* also causes eye infections and pneumonia in newborns.
Gonorrhea	*Neisseria gonorrhoeae*	—	Coccus	Symptoms in the male include uncomfortable sensation along the course of the urethra and painful and frequent urination; the majority of females do not show definite symptoms. Disseminated gonorrhea occurs where the bloodstream is invaded, spreading skin lesions, and involvement of tendons and joints are common symptoms.
Granuloma inguinale	*Calymmatobacterium granulomatis*	—	Small rod (coccobacillus)	The appearance of a moist pimple (papule) on or in vicinity of external genitalia.
Genital herpes	Herpes simplex virus type 2	a	Virus	Initial infections may be without symptoms. The more common situations include painful vesicles (blisters) that develop into ulcers and enlarged lymph nodes in the groin. In a fetus or newborn, severe disease with high morbidity and serious effects are possible.
Lymphogranuloma venereum (LGV)	*Chlamydia (lymphogranulomatis) trachomatis*	— or variable (not useful)	Small rod (coccobacillus)	Swollen lymph nodes, which become filled with pus and eventually rupture and drain; these and other lesions occur in the vulvovaginal and/or rectal areas of females; leathery patches also occur.
Monilial vaginitis[b] (Candidiasis)	*Candida albicans*	+	Yeast cell	Severe itching in and around involved areas.
Syphilis	*Treponema pallidum*	Not useful	Spirochete	*Primary:* Formation of a hard sore (Hunterian chancre at infection site); healing occurs without scarring. *Secondary:* Occurs about two months after chancre; symptoms due to generalized spreading of the disease include swollen lymph nodes, general discomfort, fever, headache, and skin rash; large numbers of organisms may be found in mucous patches in various mouth parts. *Tertiary:* Soft, gummy, swollen areas or tumors (*gummas*) may form in any tissue. *Congenital:* Involvement of skin, teeth, and mucous membranes; anemia, enlarged liver and spleen, deafness, and blindness.
Trichomoniasis	*Trichomonas vaginalis*	Not done	Protozoan	Profuse discharge and irritation.

[a] Virus infection; the Gram stain is not done.

[b] Monilial vaginitis is common in pregnant women and has been associated with the use of oral contraceptives.

Diagnoses of venereal diseases are generally based on clinical signs and symptoms and specific immunological tests. Treatment involves the use of appropriate antibiotics in sufficient dosages. Preventive measures for venereal diseases are directed toward detecting and treating the sexual contacts who transmitted the disease to the patients and educating the public on the means of transmission, availability of treatment, and other pertinent information.

Sexually Transmitted Enteric Disease Agents

Increases in the sexual transmission of pathogens associated with the gastrointestinal tract are a major health problem. The general mechanism of transmission is through the fecaloral passage. Examples of involved pathogens include the bacteria: *Campylobacter fetus* and species of *Salmonella, Shigella,* and *Streptococcus*; the protozoa: *Entamoeba histolytica, Balantidium coli,* and *Giardia lamblia;* the viruses: cytomegaloviruses, hepatitis A and B viruses, and herpes simplex viruses 1 and 2; and the worms: *Enterobius vermicularis* (the cause of pin worm).

The TORCH Complex

Infection of a fetus or a newborn during delivery with *Toxoplasma* (a protozoan), rubella virus, cytomegalovirus, and the herpesviruses—the TORCH complex members—may yield an asymptomatic disease state. Even when the infections are clinically apparent, the associated signs and symptoms may be indistinguishable so that diagnosis and treatment are difficult. Furthermore, all the agents can produce long-term ill effects in the infected fetus or newborn, so that prognosis must be guarded.

Pelvic Inflammatory Disease

Pelvic inflammatory disease (PID) is a sexually acquired disease of young women. The most frequently found pathogens include *C. trachomatis, M. hominis,* and *N. gonorrhoeae.*

V. CHAPTER SELF-TEST

Continue with this section only after you have read Chapters 11-13, 17-22, and 28. A score of 80 percent or better is good. If your score is less than 65 percent, reread the chapter.

Correct answers to all questions can be found at the end of this section. Write your responses in the appropriate space provided.

Completion Questions

Provide the correct term or phrase for the following. Spelling counts.

1. Diseases transmitted through sexual activity are called ______________________.
2. The basic functional unit of a kidney is the ______________________.
3. Two major functions of the kidneys are to (a) ________________ and to (b) ________________.
4. The insertion of a surgical tubular device into body organs is known as ______________________.
5. Three major pathways by which bacteria can gain access to the kidney are the (a) ________________, (b) ________________, and (c) ________________.
6. The presence of ________________ bacteria per milliliter of urine indicates an infection.

Matching

Select the answer from the right-hand side that corresponds to the term or phrase on the left-hand side of the question sheet. An answer may be used more than once. In some cases, more than one answer may be required.

Topic: Genitourinary Infection

______ 7.	Chancroid	a.	*Escherichia coli*
______ 8.	Ophthalmia neonatorum	b.	*Staphylococcus aureus*
______ 9.	Kidney cortex infections	c.	*Treponema pallidum*
______ 10.	Acute cystitis	d.	*Candida albicans*
______ 11.	Monilial vaginitis	e.	*Neisseria gonorrhoeae*
______ 12.	Acute pyelonephritis	f.	*Haemophilus decreyi*
______ 13.	Pediculosis	g.	mite infestation
______ 14.	Herpes genitalis	h.	louse infestation
______ 15.	Gonorrhea	i.	virus infection
______ 16.	Scabies	j.	*Treponema* species
______ 17.	Syphilis	k.	none of these
______ 18.	Pinta		
______ 19.	AIDS		
______ 20.	PID		
______ 21.	Clue cell formation		

True or False

Indicate all true statements with a "T" and all false statements with an "F".

______ 22. Normally functioning kidneys help to maintain the concentration of poisonous substances in the body.

______ 23. Diseases of the kidney generally are called venereal.

______ 24. Coliforms are commonly associated with urinary tract infections.

______ 25. Diabetes mellitus can be a predisposing factor in urinary tract infections.

______ 26. Chronic pyelonephritis is probably the most common kidney disease.

______ 27. The urinary bladder is considerably resistant to infection.

______ 28. *Trichomonas vaginalis* can cause cystitis in females.

______ 29. Glomerulonephritis can result from the formation of antigen-antibody complexes trapped within the glomeruli.

______ 30. *Treponema pallidum* crosses the placenta.

______ 31. Moon's molar is a characteristic feature of congenitally acquired syphilis.

______ 32. *Trichomonas hominis* infections can be acquired by newborn babies by way of an infected birth canal.

______ 33. *Treponema pallidum* outside of the body is extremely resistant to the effects of both chemical and physical agents.

Essay Questions

Answer the following questions on a separate sheet of paper.

34. Why are genital herpes infections of major concern among public health experts?

35. What is the TORCH complex? Should there be any additions to the group?

Answers

1. venereal
2. nephron
3a. maintain the concentration of electrolytes (electrolyte balance)
3b. excrete poisonous substances
4. catheterization
5a. blood
5b. lymphatic system
5c. ureter
6. 100,000
7. f
8. e
9. b
10. a
11. d
12. a
13. h
14. i
15. e
16. g
17. c
18. j
19. i
20. e
21. k
22. F
23. F
24. T
25. T
26. F
27. T
28. T
29. T
30. T
31. T
32. T
33. F

34. Genital herpes infections are not only of major concern to public health experts, but also the general public. This concern stems from the increasing number of cases of the disease and the ability of herpes virus type 2 to remain latent in the host and periodically cause recurrent infections and is the result of the lack of effective treatment to eradicate recurrent herpes.

35. The acronym TORCH was devised to focus attention on the group of microbial agents (T, *Toxoplasma;* R, rubella virus; C, cytomegalovirus; H, herpes viruses; and O, others) that can cause long-term ill effects in an infected fetus or newborn baby. Another microbial agent that should be added to the group is *Treponema pallidum,* the causative agent of congenital syphilis.

VI. DISEASE CHALLENGE 1

The situations described have been taken from actual case histories. They have been designed to show how clinical, laboratory, microbial, and related information is used in disease diagnosis. A review of treatment and epidemiological aspects are given at the end of each presentation, if appropriate. Answers to questions, laboratory findings, and interpretations are given immediately following a specific question. Test your skill and take the *Disease Challenges.*

Case

A 39-year-old male was admitted to a major medical center for the evaluation of painful vesicles in his penis, and dysuria (painful urination). On admission, the patient complained of extreme pain in his penis and groin. His temperature was 37.5° C, and his pulse rate was slightly above normal.

On physical examination, the mouth, lips, and throat were found to be free of lesions. The chest, heart, and nervous system also were normal. The penis was swollen and inflamed. In addition, the following were observed: a blackened mass of dead tissue (*eschar*) surrounded the meatus, an irregular crusty ulcer with a pus-filled discharge extended over the glans, and 1- to 5-mm vesicles on a reddened surface were present at the base of the penis and on the left inguinal area. The rectal examination was normal.

At this point, what presumptive diagnosis could be made? Syphilis, gonococcal urethritis, herpes simplex infection.

What type(s) of specimen(s) should be taken and which diagnostic procedure should be performed with them? Blood specimens should be used for bacterial isolations and standard serological tests for syphilis. Urethral, ulcer, and vesicular specimens should be used for Gram stains and bacterial and viral isolations.

Laboratory Results

Blood and urethral cultures were negative. Gram stain results showed the presence of a *Micrococcus* species. The standard tests for syphilis (specifically the rapid plasma reagin test) were negative. The urethral and vesicular specimens yielded herpes simplex virus type 1.

What test(s) could be performed to distinguish herpes simplex type 1 from type 2? Immunofluorescence test utilizing monoclonal antibodies would be quite specific in this case.

Treatment

The patient was treated symtomatically. Four days after admission, no new vesicles developed. However, he had great difficulty in urination. A corrective surgical procedure was performed successfully to relieve this problem. By the eighth day, all lesions were healing satisfactorily. The patient was discharged on the tenth day. When seen for a follow-up examination one year later, the patient was found to be in excellent health.

DISEASE CHALLENGE 2

Case

Part 1: A 19-year-old female lifeguard was admitted to a local hospital by her physician for evaluation of the abrupt appearance of diarrhea, vomiting, severe headache, a sunburnlike rash, and a temperature of 39.6° C. This patient was also in the second day of her menses.

At this point, what type(s) of laboratory specimen(s) should be taken? Blood, urine, spinal fluid, and stool specimens for bacteriological culture and viral study.

Laboratory Results

The study of the specimens produced nothing of diagnostic significance.

Part 2: The patient made a spontaneous recovery and was discharged after four days. Approximately three weeks later, the same symptoms abruptly recurred and at the same point during her menses. Upon readmission to the hospital, the patient was found to have a standard slender tampon in place. Her clinical picture included an elevated pulse and respiratory rates; inflamed lips, tongue, and throat; a sunburnlike rash covering her arms and legs; swollen palms; joint tenderness; an inflamed vaginal lining; and a menstrual discharge with pus.

At this point, what type(s) of laboratory specimen(s) should be taken? Blood, spinal fluid, urine, and vaginal specimens. Appropriate serological tests are necessary to rule out measles, systemic staphylococcal or streptococcal infection, and Rocky Mountain spotted fever. Urine and vaginal specimens should be used for bacteriological cultures.

Laboratory Results

All serological test results were negative. However, both urine and vaginal cultures were positive for *Staphylococcus aureus.*

Probable Diagnosis

Toxic shock syndrome.

Treatment

The patient was treated symptomatically and given appropriate daily doses of nafcillin (a beta-lactamase-resistant penicillin). She made a complete and uneventful recovery within four days and was discharged. The patient was advised to avoid the use of tampons or similar inserts during her menses.

VII. ENRICHMENT

Urinary Infections in the Elderly

Infection and death from infection have always been recognized as common in the elderly. Several years ago, many physicians tended to consider such disease as inevitable in this patient population. This view is no longer tolerated since it is recognized that infections can be prevented if adequate attention is given to the elderly patient.

When infection does occur, it usually is treatable. Unfortunately, many of the aged die prematurely of unrecognized or untreated bacterial or viral disease. An autopsy survey conducted by the Internal Medicine Geriatric Program at the University of Iowa showed that septicemia (the presence of bacteria in the blood) and pneumonia are the third and fourth leading causes of death among those over 70, claiming the lives of 28% of that population.

The elderly are more vulnerable to infections. With age, the immune system is weakened and the ability to respond to stress is greatly reduced. Underlying long-lasting or chronic disorders can make an infection more likely, increase its severity, and contribute to its lethality. Of the elderly patients who die of infection, only half succumb to direct effects. The others die because of the added abnormal function that infection superimposes on an underlying defect in the body. In the elderly patient, the weakest organ or organ system is the most vulnerable to infection.

One of the systems readily associated with infections in the elderly is the urinary tract. The incidence of urinary tract infections rises with age, particularly in those over 65, for any of a number of reason: immobility, residual urine in the bladder, senile vaginitis, obstructive uropathy caused by prostatic disease, and diminished bactericidal activity of the urine and of prostatic secretions. Although these factors are known to influence the establishment of infection, the consequences of bacteriuria itself are less clear. A majority of elderly patients with urinary tract infections do not exhibit symptoms.

In women, *E. coli* is the primary pathogen, causing about 50 to 70% of urinary infections, whereas *Proteus vulgaris* causes another 25%. In men, more infections are caused by *Proteus* than by *E. coli. Pseudomonas, Staphylococcus,* and *Streptococcus faecalis* also can be involved.

Vaginitis occurs in the elderly usually as a result of vaginal atrophy, due to lowered hormonal stimulation of the epithelium following menopause. The most common pathogen in these cases is *Candida albicans,* which should be treated with local applications of an antifungal agent, such as nystatin or miconazole. Vaginal infections can be caused also by Trichomonas or by *Gardnerella vaginalis,* both of which are treated with metronidazole.

A major cause of recurring urinary tract infection in men is prostatitis, which can be diagnosed by its chronic character and an increase in bacteriuria following prostate massage. Up to 50% of prostatic infections are cured after three to four months of treatment with co-trimoxazole, trimethoprim sulfamethoxazole, or metronidazole. Certain infections may require aminoglycosides, such as gentamicin. Oral carbenicillin should be used only for *Pseudomonas* or *Proteus* infections.

In younger individuals, there is a well-established link between pyelonephritis and hypertension (high blood pressure) as well as a strongly suspected link between bacteriuria and hypertension.

Drugs for the treatment of urinary infections in elderly patients must be carefully chosen. This is especially true since certain medications may interfere with the patient's metabolism, or excretion, or the effectiveness of other drugs or treatments the patient may be receiving.

29

Microbial Diseases of the Central Nervous System and the Eye

I. INTRODUCTION

The central nervous system consists of the brain and spinal cord, their coverings, the meninges, and the cerebrospinal fluid. This chapter discusses the general organization of the central nervous system and a variety of microbial infections that affect it. Microbial infections of the eye are included as well. The general symptoms and approahes to diagnosis, treatment, and control are also considered. A Disease Challenge is included at the end of the chapter.

II. PREPARATION

Chapters 9-13, 17-22, and 29 should be read before continuing. The following terms are important for you to know. Refer to the glossary and the appropriate chapters of the text if you are uncertain of any of them. A pronunciation guide for selected terms is provided as a learning aid.

1. abscess (AB-sess)
2. aerosol (AIR-oh-sol)
3. African sleeping sickness
4. AIDS
5. alphavirus
6. anaerobe
7. arbovirus
8. aseptic (a-SEP-tik) meningitis
9. cerebrospinal (ser-ee-bro-SPY-nal) fluid
10. complement (COM-plee-ment) fixation
11. conjunctiva (kon-junk-TIE-vah)
12. conjunctivitis (kon-junk-ti-VYE-tis)
13. cornea (KOR-nee-ah)
14. diarrhea (die-ah-REE-ah)
15. droplet
16. ELISA
17. encephalitis (EN-sef-ah-lie-tis)
18. flavirus
19. fomite
20. hemorrhage
21. hydrophobia
22. inflammation
23. lacrimal gland
24. listeriosis (lis-ter-i-OH-sis)

25. meninges (meh-NIN-jeez)
26. meningitis (men-in-JYE-tis)
27. myocarditis (my-oh-kar-DIE-tis)
28. olfactory
29. ophthalmia neonatorum (of-THAL-me-ah nee-oh-NAY-tor-um)
30. otitis media
31. paralysis (pah-RAL-ee-sis)
32. pink eye
33. poliomyelitis (poh-lee-oh-my-ee-LIE-tis)
34. primary amebic meningoencephalitis
35. prion (PREE-on)
36. pus
37. rabies (RAY-beez)
38. sinusitis
39. slow virus
40. spinal column
41. toxofactor
42. toxoplasma (tok-so-PLAZ-mah)
43. trachoma (trah-KOH-mah)
44. trypanosome (tri-pan-oh-SOM)
45. tumor (TOO-mor)
46. vaccine (VAK-seen)

III. PRETEST

Correct answers to all questions can be found at the end of this section. Write your responses in the appropriate space provided.

Completion Questions

Provide the correct term or phrase for the following. Spelling counts.

1. The human central nervous system consists of the (a) ________________ and (b) ________________ .
2. The relay component between the brain and the spinal cord is the ________________________ .
3. The coverings of the brain and spinal cord are known as the ________________________ .
4. To date, ________________________ antigenic types of poliomyelitis have been identified.
5. A contaminated inanimate object is known as a(an) ________________________ .
6. The vector of African sleeping sickness is the ________________________ .
7. The causative agent of Chagas' disease is ________________________ .
8. The mucous membrane lining the inner part of the eyelids is the ________________________ .
9. Prevention of ophthalmia neonatorum can be achieved by the use of ________________________ .

Matching

Select the answer from the right-hand side that corresponds to the term or phrase on the left-hand side of the question sheet. An answer may be used more than once. In some cases, more than one answer may be required.

Topic: Central Nervous System Diseases

______ 10. Brain abscess	a. arbovirus infection

______ 11. Gonococcal conjunctivitis	b. rabies

______ 12.	Western equine encephalitis	c.	*Staphylococcus aureus*
______ 13.	Yellow fever	d.	*Neisseria meningitidis*
______ 14.	Meningococcal encephalitis	e.	*Veillonella* species
______ 15.	St. Louis encephalitis	f.	coxsackie viruses
______ 16.	Hydrophobia	g.	bunyaviruses
______ 17.	Epidemic pleurodynia	h.	*Trypanosoma* species
______ 18.	Slow virus disease	i.	prion
______ 19.	Toxoplasmosis	j.	*Toxoplasma gondii*
______ 20.	African sleeping sickness	k.	*Listeria monocytogenes*
______ 21.	Listeriosis	l.	none of these
______ 22.	Scrapie		

True or False

Indicate all true statements with a "T" and all false statements with an "F".

______ 23. Infections of the central nervous system can result from the direct extension of a disease process of another organ.

______ 24. The central nervous system contributes to the maintenance of the human's internal environment.

______ 25. The cerebrospinal fluid provides nutrients to the nervous system and serves as a shock absorber for the brain and spinal cord.

______ 26. Uncontrolled body movements can result from an injury to the nervous system.

______ 27. *Escherichia coli* is a cause of meningitis.

______ 28. A major source of *Toxoplasma gondii* is the fecal matter of infected domesticated cats.

______ 29. Fluorescent antibody tests are of diagnostic value for toxoplasmosis.

______ 30. *Naegleria fowleri* causes Chagas' disease.

Answers

1a. brain
1b. spinal cord
2. brain stem
3. meninges
4. three
5. fomite
6. tsetse fly
7. *Trypanosoma cruzi*
8. conjunctiva
9. erythromycin
10. c
11. h
12. a
13. a
14. d
15. a
16. b
17. f
18. i
19. j
20. h
21. k
22. i
23. T
24. T
25. T
26. T
27. T
28. T
29. T
30. F

IV. CONCEPTS AND TERMINOLOGY

Organization of the Central Nervous System

The survival of a complex organism such as the human body requires the coordination of vital processes taking place within it. The nervous system and the hormonal systems share this important function. Table 29-1 lists several of the components of the human central nervous system.

TABLE 29-1 Components of the Central Nervous System

Component	*Description*	*Function(s)*
Brain and spinal cord	The brain is encased within the bones of the skull. It is continuous with the spinal cord, which is surrounded by the segments of the vertebral column. The main regions of the brain are the *cerebrum, cerebellum,* and brain stem. The brain stem serves as the relay between the brain and the spinal cord. Both the brain and spinal cord are hollow and contain cerebrospinal fluid.	Control heart action, breathing; sense and respond to changes that take place in the external environment; regulate organ systems; maintain the organism's internal equilibrium in response to changes occurring within the organism
Meninges (membranes)	Three of these coverings of the brain and spinal cord are recognized: the *dura mater* (outer), *arachnoid* (middle), and *pia mater* (inner sheaths).	Protects the central nervous system components
Cerebrospinal fluid	Fills the subarachnoid space between the pia mater and arachnoid	Provides nutrients and serves as a shock absorber for the brain and spinal cord

Diseases of the Central Nervous System

Injuries and disorders involving the parts of the nervous system produce varied but definite effects. Symptoms of such conditions are particularly distressing. Fever, general weakness, headache, and a stiff neck are typical signs of a CNS infection. Among the most severe symptoms are paralysis and loss of control over body functions and movement. Table 29-2 lists representative microbial diseases of the nervous system.

Preventative and control measures against CNS infections are limited; they include immunization for diseases such as meningococcal meningitis and rabies, the control of arthropod vectors in arbovirus infections and African sleeping sickness, and general education programs.

The Eye

The eye consists of the eyeball and accessory structures, including eyebrows, eyelids, inner lining of the eyelids or conjunctiva, and the *lacrimal* apparatus, which produces tears.

Several factors play prominent roles in protecting the human eye from infection, including the washing and bactericidal action of tears and the mechanical barrier of an intact mucous membrane.

Table 29-3 lists bacterial and viral diseases of the eye.

Slow Virus and Prion Diseases

Slow, chronic, and latent virus diseases are associated with conventional viruses and less defined transmissable agents called *prions*. Examples of the virus infections include lymphocytic choriomeningitis, rabies, progressive multifocal choriomeningitis, and subacute sclerosing parencephalitis. Prion-associated conditions include the human diseases Kuru and Creutzfeldt-Jakob disease, as well as lower animal diseases—a brain disease of mink and scrapie of sheep.

TABLE 29-2 Representative Microbial Diseases of the Central Nervous System

Disease	*Causative Agent*	*Gram Reaction*	*Morphology*	*Transmitters and Sources of Causative Agents*	*General Features*
Aseptic meningitis	Enteroviruses, mumps virus	a		Direct or indirect contact with contaminated articles (fomites) and aerosols.	Fever, irritation, stiffness of the neck, and general fatigue.
Brain abscess	*Escherichia coli, Proteus* spp., *Haemophilus influenza* type B, *Staphylococcus aureus, Streptococcus pneumoniae, Streptococcus* spp.	— +	Rods Cocci	Complications that develop from chronic pus-producing sites in other portions of the body, such as the lungs, middle ear, paranasal sinuses, pelvis, and pleura. Dissemination of organisms from the foci of infection can occur (1) by direct extension through bones, (2) via the covering of the olfactory nerves, and (3) by way of the venous system.	Fever, headache, nausea, pus formation and possible interference with vision, breathing, hearing, and movement.
Listeriosis (menigitis)	*Listeria monocytogenes*	+	Rods	Contaminated foods; infected animals; and placental transmission.	High fever, drowsiness, headache, confusion, seizures, tremors, uncoord-inated movements, and coma.
Meningococcal meningitis[b]	*Neisseria meningitidis*	—	Coccus	Direct contact with droplets or fomites.	Fever, headache, pus formation, and infections of the bones in the skull or ear.
Rabies (hydrophobia)	Rabies virus	a		The principal source is believed to be wild animals. Humans and domesticated animals acquire this disease accidentally. Saliva con-taining the rabies virus introduced into humans by the bite of a rabid animal is the principal means of transmission. Infection through minor scratches contaminated by infectious feces has also been reported.	Fever, general discomfort, headache, visual difficulties, painful throat spasms,[c] convulsions, delirium, respiratory paralysis, and death.
Viral encephalitis	Arboviruses including Eastern equine, St. Louis, and Western equine encephalitis virus	a		Blood-sucking arthropods.	Fever, chills, nausea, general fatigue, drowsiness, pain and stiffness of neck, and general disorientation; blindness, deafness, and paralysis may develop as consequence of infection.
Toxoplasmosis	*Toxoplasma gondii*	a		*T. gondii* can be transmitted through the placenta, direct contact with aerosols from infectious cat feces, and ingestion of contam-inated raw meat.	When symptoms occur in adults, they include chills, fever, headache, extreme discomfort, and muscle pain; symptoms resemble infectious mononucleosis; symptoms of newborns who contract the disease *in utero* include fever, con-vulsions, enlarged spleen, and serious central nervous system defects causing blindness and mental retardation.

TABLE 29-2 Representative Microbial Diseases of the Central Nervous System *(continued)*

Disease	*Causative Agent*	*Gram Reaction*	*Morphology*	*Transmitters and Sources of Causative Agents*	*General Features*
Trypano-somiasis (African sleeping sickness)	*Trypanosoma brucei* variety *gambiense*, *T.* v.b. *rhodesiense*	a		Tsetse flies.	Disease may continue for years; symptoms include enlargement of lymph nodes, spleen, and liver; fever, chills, disturbed vision, general weakness, headache, loss of appetite, occasional rash, nausea, vomiting, and serious defects of the central nervous system, ending in death.

[a] Not applicable.

[b] Several other microorganisms can cause meningitis as a complication or secondary effect of disease states. These include the bacteria *Escherichia coli, Haemophilus influenzae, Staphylococcus aureus, Streptococcus pneumoniae,* species of *Proteus* and *Pseudomonas,* and anaerobes such as *Bacteroides, Clostridium,* and *Streptococcus.* The yeast *Cryptococcus neoformans* also is associated with meningitis.

[c] This fear of painful swallowing of fluids, which can cause attacks of convulsive choking, is responsible for naming this disease *hydrophobia.*

TABLE 29-3 Microbial Diseases of the Eye

Disease	*Causative Agent*	*Gram Reaction*	*Morphology*	*Transmitters and Sources of Causative Agents*	*General Features*
Conjunctivitis	*Neisseria gonorrhoeae*	—	Coccus	Direct contact with droplets and fomites.	Irritation around eyelid, tearing, swelling, redness, and discharge from the involved eye.
	Staphylococcus aureus	+	Coccus		
	Streptococcus pneumoniae	+	Coccus		
	Streptococcus spp.	+	Coccus		
	Newcastle disease virus, molluscum contagiosum virus, measles virus	a		Direct contact with droplets and fomites.	Irritation around eyelid, tearing, swelling, redness, and discharge from the involved eye.
Corneal ulcer	*Streptococcus pneumoniae*	+	Coccus	Injury, fomites, aerosols.	Destruction of cornea, leading to scaring and cataract (formation of opaque area).
Inclusion conjunctivitis (newborn)	*Chlamydia trachomatis*		Rod	Flies, contaminated fingers, fomites.	Reddening of eyelid and pussy discharge; no scarring.

TABLE 29-3 Microbial Diseases of the Eye *(continued)*

Disease	*Causative Agent*	*Gram Reaction*	*Morphology*	*Transmitters and Sources of Causative Agents*	*General Features*
Keratitis (inflammation of the cornea)	*Moraxella lacunata*	—	Rod	Injury, fomites, aerosols.	Fever, headache, and swelling of tissues around eyes; pus formation and scarring can occur as complication; condition can appear in combination with conjunctivitis.
	N. meningitidis	—	Coccus		
	S. aureus	+	Coccus		
	Streptococcus mitis	+	Coccus		
	S. pneumoniae	+	Coccus		
	Herpesvirus hominis (herpes simplex virus type 1)	a			
	Adenovirus, types 7 and 8[b]	a			
Ophthalmia neonatorum (conjunctivitis of the newborn)	*Neisseria gonorrhoeae*	—	Coccus	Placental transmission, fomites.	Discharge consisting of blood and pus, swollen eyelids.
	Staphylococcus aureus	+	Coccus		
Pink eye (acute mucopurulent conjunctivitis)	*Haemophilus aegypticus*	—	Rod	Injury, fomites, contaminated fingers.	Abundant discharge, redness, and extreme swelling of the eyelids; bleeding within the conjunctiva can occur.
Trachoma	*Chlamydia trachomatis*	—	Rod	Flies, fomites, contaminated fingers.	Typical effects include accumulation of blood vessels on the surface and penetration of cornea, tumor formation, scarring, and blindness.

[a]Not applicable.

[b] Causative agents of epidemic keratoconjunctivitis, a condition resulting from eye injuries contaminated with these viruses.

V. CHAPTER SELF-TEST

Continue with this section only after you have read Chapters 9-13, 17-22, and 29. A score of 80 percent or better is good. If your score is less than 65 percent, reread the chapter.

Correct answers to all questions can be found at the end of this section. Write your responses in the appropriate space provided.

Completion Questions

Provide the correct term or phrase for the following. Spelling counts.

1. An inflammation of the membrane coverings of the brain is called ______________________________.
2. The dura mater is an example of a(an) ______________________________.

3. The outermost covering of the human eye is the ______________________.

4. Tears contain an enzyme called ______________________, which is bactericidal to various microorganisms.

Matching

Select the answer from the right-hand side that corresponds to the term or phrase on the left-hand side of the question sheet. An answer may be used more than once. In some cases, more than one answer may be required.

Topic: Means of Disease Transmission

______ 5. Brain abscess	a.	warm-blooded animal bites
______ 6. Toxoplasmosis	b.	direct extension through bones
______ 7. Slow virus infections	c.	mosquitoes
______ 8. Ophthalmia neonatorum	d.	passage through infected birth canal
______ 9. Rabies	e.	ticks
______ 10. Chagas' disease	f.	infectious cat feces
______ 11. Yellow fever	g.	unknown
______ 12. African sleeping sickness	h.	kissing bugs
______ 13. Trachoma	i.	tsetse flies
______ 14. Listeriosis	j.	contaminated food or water
______ 15. PAM	k.	none of these
______ 16. Scrapie		
______ 17. Creutzfeldt-Jakob disease		

Topic: Disease Agents of the Central Nervous System and Related Structures

______ 18. Mumps virus	a.	aseptic meningitis
______ 19. *Toxoplasma gondii*	b.	myocarditis
______ 20. *Streptococcus* spp.	c.	African sleeping sickness
______ 21. *Trypanosoma brucei* variety *gambiense*	d.	Chagas' disease
______ 22. Group A coxsackie viruses	e.	pink eye
______ 23. *Trypanosoma cruzi*	f.	keratitis
______ 24. *Neisseria gonorrhoeae*	g.	none of these
______ 25. *Haemophilus aegypticus*		
______ 26. *Chlamydia trachomatis*		
______ 27. *Listeria monocytogenes*		
______ 28. *Naegleria fowleri*		
______ 29. *Acanthamoeba* species		

True or False

Indicate all true statements with a "T" and all false statements with an "F".

_______ 30. Brain abscesses are more commonly associated with the spinal cord than the brain.

_______ 31. A large number of brain abscesses are associated with middle ear infections.

_______ 32. Inflammation of the dura mater can occur as a complication of infected skull fractures.

_______ 33. Various bacterial species cause aspetic meningitis.

_______ 34. The principal reservoir of rabies virus is the human.

_______ 35. Rabies infection can be acquired through contamination of minor scratches with virus-containing materials.

_______ 36. Fruit-eating bats can transmit rabies.

_______ 37. Coxsackie viruses are examples of arboviruses.

Essay Questions

Answer the following questions on a separate sheet of paper.

38. What is meningitis? Explain the mechanisms by which the disease state can develop.

39. What principal measures are available to control African sleeping sickness?

Answers

1. meningitis
2. meninges
3. cornea
4. lysozyme
5. b
6. f
7. g
8. d
9. a
10. h
11. c
12. i
13. j
14. j
15. j
16. k
17. k
18. a
19. g
20. g
21. c
22. a
23. d
24. g
25. e
26. g
27. g
28. g
29. g
30. T
31. T
32. T
33. F
34. F
35. T
36. T
37. F

38. *Meningitis* is an inflammation of the membranes around the brain and spinal cord. It can result from one of several mechanisms. These include the introduction of microorganisms: (a) through penetrating injuries or primary infections involving the skull and spinal volumn, (b) by the direct extension of a disease process from primary foci of infection located in other parts of the body through bone via vascular channels or along the covering of the olfactory nerves, and (c) by means of the bloodstream (hematogenous route) during the course of a septicemia.

39. Practical control measures for reducing the incidence of African sleeping sickness include: (1) destruction of breeding sites of the tsetse flies, (2) diagnosis and treatment of the disease in patients, (3) quarantine of infected individuals, (4) wearing protective clothing against the tsetse flies, and (5) prophylactic drug administration, especially in areas where the risk of infection is great.

VI. DISEASE CHALLENGE

The situation described has been taken from an actual case history. A review of treatment and epidemiological aspects is provided at the end of the presentation. Answers to questions, laboratory findings, and interpretations are given immediately following a specific question. Test your skills and take the *Disease Challenge.*

Case

Part 1: A 10-month-old male baby was admitted at 11 P.M. to the emergency department with a fever of 40° C, slight nasal discharge, and soft cough. According to the mother of the child, the signs of illness began to appear some four hours earlier. Two older brothers of the baby were well and showed no signs or symptoms of illness.

During the physical examination, the young baby was initially quiet, but soon became very irritable and difficult to control. Respiration, pulse rates, and white blood count were elevated.

At this point, what microbial disease(s) would you suspect? Influenza, common cold, strep throat.

What laboratory specimen(s) should be taken? A blood specimen for routine laboratory tests and bacterial cultures as well as a throat swab for bacteriological isolation and identification.

Laboratory Results

All routine laboratory tests were normal. However, it was too early to obtain the bacteriological culture results.

Part 2: On the basis of the normal laboratory findings, the child was sent home at approximately 2 A.M. The mother was asked to observe the baby closely and to call if any worsening of the condition occurred. At 10 A.M. of the same morning, the mother called to report that the fever was as high and that the child was unresponsive and appeared to be in a dazed state. The physician in charge told the mother to return to the hospital as quickly as possible.

Upon examination, the baby was extremely irritable and sluggish.

At this point, what microbial disease(s) would you suspect? Meningitis.

What laboratory specimen(s) should be taken? Cerebrospinal fluid for bacteriological culture and identification and for routine laboratory tests.

Laboratory Results

Gram stains of the cerebrospinal fluid (CBF) revealed many gram-negative coccobacilli. Blood cultures taken earlier became positive and showed the presence of ampicillin-sensitive *Haemophilus influenzae* type b. CBF cultures also 24 hours later were found to contain the same organism.

What treatment would be appropriate?

Treatment

Intravenous ampicillin and chloramphenicol were administered for two days. After this period, ampicillin was continued for an additional ten days. The patient's temperature returned to normal. Recovery continued without complications.

As a precaution, all members of the patient's family—father, mother, and brothers—were treated with rifampin for five days. The use of rifampin eliminates *H. influenzae* from the nasopharynx and reduces the possible spreading of the organism to susceptible individuals.

VII. ENRICHMENT

Japanese B Encephalitis

Japanese B encephalitis virus is an RNA flavivirus that exhibits a marked natural liking for the central nervous system. The disease state caused by this virus was originally designated "B" to distinguish it from von Economos' disease or type A encephalitis and has been recognized in epidemic form in Japan probably since 1871. Epidemics also have occurred in China, Guam, Korea, Okinawa, the Philippines, and Taiwan and in military personnel during World War II and the Korean and Vietnam wars. Epidemics usually appear in late summer, but sporadic cases occur throughout the year.

Japanese B encephalitis virus is maintained enzootically, involving a range of animals, particularly pigs and birds such as egrets and herons. Transmission between hosts is by mosquitoes of the *Culex* species, including *C. gelidus, C. pseudovishnui,* and *C. tritaeniorhynchus.* These mosquitoes are able to maintain the virus during their winter hibernation. In addition, the *Culex* mosquitoes breed extensively in rice fields, thus establishing large mosquito populations capable of effectively transmitting the virus to susceptible animals. Pigs that are found in close association with humans act as contributing or amplifier hosts, transmitting the Japanese B encephalitis by means of *Culex* mosquitoes to humans. Human-to-human direct transmission of the viral agent has not been shown, although a case of transplacental transfer has been reported.

The incubation period for Japanese B encephalitis varies from 6 to 16 days. Before the fullblown effects of the disease appear, individuals experience fever, headache, general discomfort (malaise), loss of appetite, and vomiting. In the acute stage, which is usually the first presentation of Japanese B encephalitis, the symptoms appearing include a high fever that reaches a peak in two to four days and then subsides, a severe frontal or generalized headache, confusion, delirium, disorientation, mild diarrhea, muscular rigidity, tremors involving the fingers, tongue, and closed eyelids, and coma. Approximately one third of the individuals recovering from Japanese B encephalitis exhibit neurological and psychological complications. The most common complications are confusion, emotional instability, memory loss, and personality changes.

Treatment of Japanese B encephalitis patients is limited. Symptomatic and supportive management are the main approaches to treatment. Control of the disease also is quite limited. The mass eradication of the mosquito vectors is at present impractical, although local control is possible. The effectiveness of a live attenuated vaccine is under study.

30

Microbiology and Cancer

I. INTRODUCTION

Cancer, an often fatal disease, is a major or leading cause of death in several areas of the world. The term *cancer,* however, is not one disease state, but actually includes over 100 histologically distinct, malignant (harmful) growths or neoplasms. Each condition must be considered separately because of differences in behavior, treatment, and possibly their causes.

Roughly 90% of all malignancies are caused by cancer-causing agents (carcinogens) in the environment. This chapter describes the general features of cancers, then deals with the possible roles of microorganisms and their products in producing cancers. Consideration is given to current hypotheses for viral causes of cancer, immunological approaches to diagnosis and treatment, and the use of microorganisms for the detection of cancer-causing (carcinogenic) agents and for the treatment of cancer.

II. PREPARATION

Chapters 10, 17, 19, 20, 22, and 30 should be read before continuing. The following terms are important for you to know. Refer to the glossary and the appropriate chapters of the text if you are uncertain of any of them. A pronunciation guide for selected terms is provided as a learning aid.

1. acquired immune deficiency syndrome
2. acute (A-kyoot)
3. adoptive immunotherapy
4. aflatoxin (af-LAH-tock-sin)
5. Ames test
6. anaplasia (an-nah-PLAY-see-ah)
7. antibiotic
8. Bacillus of Calmette (CAL-met) and Guérin (GER-in)
9. benign (BEE-nine)
10. Burkitt's lymphoma
11. cancer (CAN-sir)
12. carcino-embryonic antigen
13. carcinogenic (car-SIN-oh-jen-ik)
14. carcinoma (car-SIN-oh-mah)
15. contact inhibition
16. DNA
17. extrinsic (eks-TRIN-sick)
18. fibroblast
19. focus assay
20. hemorrhage (HEM-oh-ridge)
21. hepatoma
22. hyperplasia (high-per-PLAY-zee-ah)
23. hypothesis
24. immunoperoxidase
25. intercellular
26. interleukin-2

27. intracellular
28. intrinsic (in-TRIN-sick)
29. leukemia (lou-KEY-me-ah)
30. lymphokine activated killer cells
31. lymphoma (lim-FOH-mah)
32. macrophage
33. malignant (mah-LIG-nant)
34. melanoma
35. metastasis (meh-TAS-tah-sis)
36. monoclonal antibodies
37. mycotoxin (my-koh-TOK-sin)
38. mutagen
39. neoplasm (knee-oh-PLAZ-em)
40. oncogenic (on-co-GEN-ick)
41. papilloma
42. perforin
43. retrovirus
44. reverse transcriptase
45. RNA
46. *sarc* gene
47. sarcoma (sar-CO-mah)
48. T cell lymphotropic viruses I and II
49. tissue culture
50. topoinhibition (toe-poh-in-hib-EE-shun)
51. transformation
52. tumor (TOO-mor)

III. PRETEST

Correct answers to all questions can be found at the end of this section. Write your responses in the appropriate space provided.

Completion Questions

Provide the correct term or phrase for the following. Spelling counts.

1. Three kinds of environmental factors considered to be major causes of cancers are (a) ______________ , (b) ______________ , and (c) ______________ .
2. Normal cell growth is inhibited by ______________ .
3. The intracellular controls that stop further cellular growth are associated with the property called ______ .
4. Uncontrolled cellular growths are known as ______________ .
5. Tumors localized in formation and generally harmless are referred to as being ______________ .
6. The separation of malignant cells from a major portion of a tumor and spreading to other body regions is known as ______________ .
7. Various factors capable of causing tumor formation are referred to as being (a) ______________ , while those that produce specific malignancies are referred to as being (b) ______________ .
8. Tumors develop as a result of the actions of (a) ______________ factors within cells and external or (b) ______________ factors.
9. Solid tumors derived from various tissues including skin, nerves, and the linings of the gastrointestinal and genital systems are called ______________ .
10. The uncontrolled growth and multiplication of white blood cells is a cancerous state known as ______ .

11. The most common causes of death with untreated leukemias are (a) ______________________ and (b) ______________________.
12. The form of cancer in which abnormal numbers of leukocytes are produced by the spleen and lymph nodes is called ______________________.
13. Cancerous tumors growing from bone, cartilage, connective tissue, fat, and muscle are called __________.
14. The process by which normal cells are altered by viral infection to become malignant is ______________.
15. The most important oncogenic viruses that infect a large number of animal species are ______________.
16. RNA tumor viruses can be distinguished on the basis of three properties: (a) ______________________, (b) ______________________, and (c) ______________________.
17. Viruses that are either possible or actual causes of human cancers include (a) ______________________, (b) ______________________, and (c) ______________________.
18. Routine screening tests for cancer detection include (a) ______________, (b) ______________, (c) ______________________, and (d) ______________________.
19. A specific test used to detect carcinogenic and mutagenic compounds is known as the ______________.
20. The attenuated strain of *Mycobacterium tuberculosis* used to stimulate a host's immune response to destroy neoplastic cells is ______________________.

True or False

Indicate all true statements with a "T" and all false statements with an "F".

______ 21. Uncontrolled cellular growths are called tumors.

______ 22. Benign tumors generally invade and destroy normal tissues.

______ 23. Cancerous cells are noted for their ability to spread from one body region to another.

______ 24. Only one type of cancer is recognized today.

______ 25. A carcinoma is a solid tumor derived from various tissues, including skin and nerves.

______ 26. The structural abnormality in which cells do not mature is called metastasis.

______ 27. According to the previous hypothesis of cancer formation, the genetic information for cancer is present in every cell.

______ 28. The protovirus hypothesis holds that cancer viruses arise from segments of RNA.

29. Aflatoxins can be carcinogenic.

______ 30. Carcinoembryonic antigen signals a specific type of leukemia.

______ 31. Tumor-specific antigens are absent in normal cells.

Answers

1a. chemical
1b. radiation
1c. certain microbial products
2. crowding
3. contact inhibition or topoinhibition
4. tumors or neoplasms
5. benign
6. metastasis
7a. oncogenic
7b. carcinogenic
8a. intrinsic
8b. extrinsic
9. carcinomas
10. leukemia
11a. hemorrhaging
11b. uncontrolled infection
12. lymphoma
13. sarcomas
14. viral transformation
15. type C RNA viruses
16a. morphology
16b. immunological differences
16c. modes of maturation
17a. adenoviruses
17b. papovaviruses
17c. Epstein-Barr virus
18a. pap smear
18b. physical examinations
18c. x-rays
18d. proctoscopic examination of gastrointestinal system
19. Ames test
20. Bacillus of Calmette and Guérin
21. T
22. F
23. T
24. F
25. T
26. F
27. F
28. F
29. T
30. F
31. T

IV. CONCEPTS AND TERMINOLOGY

Although some animal cancers are known to be caused by viruses, environmental factors generally are assumed to be the major causes.

General Characteristics of Cancerous States

Normal cell growth is inhibited by a crowding effect produced when a sufficient number of cells have occupied a given area. Such *contact inhibition* or *topoinhibition* does not occur when cellular reproduction becomes uncontrolled, as in the case of cell growths or swellings, known as *tumors* or *neoplasms. Benign tumors* form in localized areas, do not spread, and are generally harmless. Malignant or cancerous tumors spread to other areas, and invade and destroy normal tissue. The movement or spreading of such growths to new sites is called *metastasis.*

Various chemical, physical, and microbial agents have been shown to possess *oncogenic* (tumor-causing) and/or *carcinogenic* (cancer-producing) effects. Microbial carcinogens are produced by various bacteria and fungi.

A given tumor develops from normal cells as a result of changes influenced by *intrinsic* (internal) factors, for example, age, heredity, sex, hormones, and a natural predisposition to tissue overgrowth, and by *extrinsic* (external) factors, for example, chemicals, irradiation, and possibly some viruses.

The Forms of Human Cancer

As noted in the introduction, more than 100 clinically distinct types of cancer are recognized. The general properties of the four major categories are listed in Table 30-1.

Cancerous states can exhibit the following three characteristics:

1. *Anaplasia:* cells not maturing or unable to perform normally
2. *Hyperplasia:* uncontrolled cellular reproduction
3. *Metastasis:* detachment from main tumor and spreading to other areas

Viral Transformation

Viral transformation is the process by which normal cells are changed by viral infection to become malignant. Transformed cells may show changes in metabolism, antigenic properties, and morphology. Before such changes occur, it is believed that the invading virus must be integrated into the host cell's genetic apparatus (*genome*).

TABLE 30-1 Categories of Malignancies

Category	*Description*	*Cells or Tissues Involved*
Carcinoma	Solid tumor	Epithelial tissues (e.g., breast glands, skin, nerves, and linings of gastrointestinal, genital, respiratory, and urinary systems).
Leukemias	Uncontrolled multiplication and accumulation of leukocytes (white blood cells). Most of these cells are unable to carry out normal functions.	Granulocytes and agranulocytes (e.g., lymphocytes).
Lymphomas	Production of abnormally high numbers of lymphocytes and immature nonfunctional cells by the spleen and lymph nodes.	Lymphoid tissues (e.g., spleen).
Sarcomas	Tumors growing from various tissues.	Bones, cartilage, connective tissues, fat, and muscle.

The most obvious signs of neoplastic transformation are changes in cellular shape and the loss of surface attachment or place (topoinhibition). Transforming viruses added to unspecialized connective tissue (fibroblast) cultures cause infected cells to form small piles, which resemble miniature tumors, on normal cells. Each little tumorlike mass is called a focus of transformed cells and is the basis of the procedure called the *focus assay* for tumor viruses.

In cases of tissue culture systems infected by cancer virus, the presence of such agents may not always be apparent.

Discoveries Relating Viruses and Cancer

Several verified experiments have demonstrated virus-induced malignancies in rabbits, frogs, and mice. Both RNA- and DNA-containing viruses have been shown to cause cancers in lower animals. Leukemia viruses are found in a large number of lower animals including cats, chickens, dairy cattle, and the gibbon ape. Transmission of these viruses includes genetic inheritance and direct contact.

Characteristics of Oncogenic RNA Viruses

Oncogenic RNA viruses belong to two of the three designated classes: A, B, or C. The properties of agents in these classes are summarized in Table 30-2. Oncogenic RNA viruses have more hereditary information in their genomes than do other animal RNA viruses.

Cancer Virus Hypothesis

In 1970, the enzyme *reverse transcriptase* (RNA-directed DNA-polymerase) was discovered. This enzyme catalyzes and reverses the usual flow of genetic information from DNA to RNA. The discovery was important because it explained the genetic transmission and transformations of oncogenic viruses such as the retroviruses. It seems possible that some human cancers are caused by RNA viruses. Thus far, three current RNA virus cancer hypotheses have been developed to account for the way in which genetic information with a cancerous process is expressed in cells. These hypotheses are compared and described in Table 30-3.

The Role of DNA Viruses in Carcinogenesis

Several viruses have been considered as possible causes of human cancer. These include adenoviruses, papovaviruses, certain poxviruses, and several herpesviruses. A herpesvirus, the Epstein-Barr virus, has been shown to cause Burkitt's lymphoma, a lymphoid tumor condition involving the jaw.

TABLE 30-2 Oncogenic RNA Viruses

Class	*General Morphology*[a]	*Oncogenic Activity*
Type A	An RNA core surrounded by a protein covering or shell.	Not infectious or oncogenic
Type B	Eccentric nucleocapsid surrounded by a glycoprotein layer with projections or spikes.	Carcinomas of the breast
Type C	Consists of a spherical RNA-protein core surrounded by a lipid layer. Particle gives a targetlike appearance.	Leukemia, lymphomas, and sarcomas

[a] See Chapter 10 for descriptions of viral morphology.

TABLE 30-3 Cancer Virus Hypotheses

Hypothesis	*Originator(s)*	*Description*
Provirus	Temin (early 1960s)	After infection by an RNA virus, the host cell's DNA makes a *DNA copy* of the *viral RNA*, which is then incorporated into the host cell's DNA.
Oncogene (transforming)	Huebner and Todaro (1969)	Every cell contains an oncogene (a cancer-causing segment of DNA) that normally is prevented from functioning (repressed). When activated (depressed) by carcinogens or viruses, it produces a transforming protein, which changes a normal cell into a malignant one.
Protovirus	Temin (1970)	Random segments of DNA are brought together by cellular and genetic events to form the protovirus, which then brings about the formation of a cancer-transforming protein similar to that proposed in the oncogene hypothesis.

Cancer Detection

The basic approach used to find cancer includes physical examinations and various laboratory tests for the detection of tumors and several antigens associated with cancer, for example, fetal antigen and tumor-specific antigen.

The Ames/*Salmonella* microsome mutagenicity test is used to determine the carcinogenic potential of various chemicals. The test detects compounds that can induce permanent genetic changes in specially developed bacterial strains of *Salmonella typhimurium* and *Escherichia coli*. Carcinogenic compounds cause mutations at levels greater than do those occurring at normal (spontaneous) levels.

The Use of Microbial Antitumor Activities for Treatment

The major forms of cancer treatment used today are surgery, radiation, chemotherapy, or a combination of the three. New approaches to treatment involve the use of microorganisms to stimulate a host's immune system to kill neoplastic cells. Microorganisms used for such purposes include Bacillus of Calmette and Guérin, *Corynebacterium parvum*, and vaccinia virus.

Monoclonal Antibodies

The possible roles for monoclonal antibodies in tumor therapy include the delivery of high concentrations of cell-killing radioactive chemicals, drugs, or toxins to cancerous tissues. These antibodies also may be of value in locating and attacking tumor cells.

A Future Outlook

If human oncogenic viruses can be isolated and obtained in sufficient quantities in pure form, the possibility exists for the development of effective vaccines or other approaches with which to halt cancerous processes. Treatment of cancers with immune cells having antitumor reactivity is among new approaches having great promise for cancer victims.

V. CHAPTER SELF-TEST

Continue with this section only after you have read Chapters 10, 17, 19, 20, 22, and 30. A score of 80 percent or better is good. If your score is less than 65 percent, reread the chapter.

Correct answers to all questions can be found at the end of this section. Write your responses in the appropriate space provided.

Matching

Select the answer from the right-hand side that corresponds to the term or phrase on the left-hand side of the question sheet. An answer may be used more than once. In some cases, more than one answer may be required.

Topic: Oncogenesis

_____ 1. Tumors growing from bone

_____ 2. Heredity

_____ 3. Uncontrolled cellular growth

_____ 4. Destructive tumors

_____ 5. The process of spreading from one area to another

_____ 6. Harmless tumor

_____ 7. Cancer-causing

_____ 8. Irradiation

_____ 9. Uncontrolled growth and multiplication of different white blood cells

_____ 10. Reverses the flow of information from RNA to DNA

a. tumor
b. benign
c. malignant
d. metastasis
e. carcinogenic
f. intrinsic tumor-causing factor
g. lymphoma
h. none of these

Completion Questions

Provide the correct term or phrase for the following. Spelling counts.

11. The *in vitro* laboratory technique used to detect transforming viruses is the _______________ .

12. A structural abnormality of cancerous cells resulting in the inhibition of cell development and maturation is _______________ .

13. Sarc genes may transform cells of similar _______________ origin.

14. In lower animals, the transmission of leukemia viruses can occur through (a) ________________ and (b) ________________.
15. Three DNA viruses implicated in the case of human cancers are (a) ________________, (b) ________________, and (c) ________________.
16. The enzyme that controls the formation of a DNA copy of viral cancer RNA in the cytoplasm of affected cells is called ________________.
17. Two possible therapeutic functions for monoclonal antibodies are (a) ________________ and (b) ________________.

True or False

Indicate all true statements with a "T" and all false statements with an "F".

______ 18. Intracellular controls that stop further cellular growth involve a property known as metastasis.

______ 19. Malignant tumors usually are localized.

______ 20. Hormonal influences represent one type of intrinsic tumor-causing factor.

______ 21. Cancer is in reality over 100 clinically different conditions.

______ 22. Anaplasia is an uncontrolled reproduction of cells.

______ 23. Viral transformation converts malignant cells into normal ones.

______ 24. Epstein-Barr virus is a known cause of a human cancerous condition.

______ 25. Transcriptase reverses the flow of genetic information from RNA to DNA.

______ 26. Bacteria can synthesize carcinogens.

______ 27. Cancer is contagious.

______ 28. Antitumor agents can be obtained from bacteria such as *Salmonella enteritidis* and *Serratia marcescens*.

______ 29. Recombinant interleukin-2 is used to activate erythrocytes in adoptive immunotherapy.

Essay Questions

Answer the following questions on a separate sheet of paper.

30. What is viral transformation?
31. How does the enzyme RNA-directed DNA-polymerase (reverse transcriptase) function?

Answers

1. i
2. f
3. a
4. c
5. d
6. b
7. c
8. i
9. g
10. i
11. focus assay
12. anaplasia
13. embryonic
14a. genetic mechanisms
14b. direct contact
15a. herpes simplex viruses I and II
15b. Epstein-Barr virus
15c. cytomegalovirus
16. RNA-directed DNA-polymerase
17a. specific antitumor agents
17b. delivery of cell-killing radioactive drugs or other anticancer cell chemicals
18. F
19. F
20. T
21. T
22. F
23. F
24. T
25. F
26. T
27. F
28. T
29. F

30. Viral transformation is the process by which normal cells are altered by viral infection to become malignant. Transformed cells often undergo many changes in morphology, metabolic functions, and antigenicity.

31. The enzyme RNA-directed DNA-polymerase (or reverse transcriptase) found in RNA cancer viruses promotes the incorporation of viral genetic material into a susceptible host cell by transcribing complementary DNA (cDNA) from the viral RNA host cell. The cDNA can integrate into the host cell chromosomal DNA. Such integrated viral information (proviral DNA) is transmitted to future generations in the same way as are other genes.

VI. ENRICHMENT

Monoclonal Therapy and Bone Marrow Transplants

Cancer patients suffering from leukemia or lymphoma frequently experience damage to their bone marrow as a result of chemotherapy. Replacement of such damaged marrow by healthy new marrow could allow for more intensive chemotherapy and thus provide for better control of the cancerous states. A new procedure for transplanting bone marrow offers substantial promise. This procedure incorporates the use of monoclonal antibodies and does not require the marrow of a matched donor. The technique, which was developed by researchers at Harvard Medical School, is based on the concept that mature T cells are responsible for the graft/host reaction. If these particular T cells could be removed from the total population of bone marrow cells, the remaining donated cells might take hold without being rejected.

The new technique utilizes monoclonal antibodies produced against the mature T cells. The donor's bone marrow is treated with the monoclonal antibodies, causing the mature T cells to be identified and removed from the marrow. This treatment eliminates the need for tissue-matching procedures involving donor and recipient. Furthermore, the remaining new, immature T cells in the donated marrow learn to recognize the recipient as "self" and subsequently mature to become fully immunocompetent.

Bone marrow transplants previously were limited because of the difficulty in finding donor marrow that was an exact immunologic match to the patient's marrow. This monoclonal procedure opens new possibilities of treatment not only for certain forms of cancer, but also for congenital bone marrow disorders, such as sickle cell anemia.

31

Helminths and Diseases

I. INTRODUCTION

Helminths or worms that are of medical and agricultural importance include the roundworms (nematodes) and the flatworms (e.g., tapeworms and flukes). Helminth infestations and infections are extremely common worldwide. A disease such as schistosomiasis, caused by flukes of the genus *Schistosoma,* is one of the most significant in tropical regions because it produces extensive liver and bladder damage. Trichinosis, or pork roundworm infection, is an example of a disease seen with increasing frequency in other parts of the world. The disease agent here involves the muscle tissue and may cause death in cases of heavy infections. While many helminths cause serious public health problems and are of major economic importance, most of these multicelled forms of life are free living.

This chapter will present the characteristic properties of helminths, including their structures and life cycles, and representative helminthic human and plant diseases. Approaches to identification, prevention, control, and treatment also are described. An illustrated Disease Challenge is included at the end of the chapter.

II. PREPARATION

Chapters 17, 19, 21, 22, and 31 should be read before continuing. The following terms are important for you to know. Refer to the glossary and the appropriate chapters of the text if you are uncertain of any of them. A pronunciation guide is provided as a learning aid.

1. autoinfection
2. arthropod
3. cercaria (sir-KAH-ree-ah)
4. cestode (SESS-toad)
5. complement fixation
6. computerized tomography
7. coprozoic (cop-roh-ZOH-ik)
8. cutting plate
9. definitive host
10. ectoparasite
11. ELISA
12. endoparasite
13. eosinophilia (ee-oh-sin-oh-PHIL-ee-ah)
14. facultative (FAK-ul-tay-tiv)
15. filariform (fill-AIR-ee-form)
16. fluke (flook)
17. free living
18. gall
19. gravid (GRAV-id)
20. helminth
21. hepatitis (hep-ah-TIE-tis)
22. hermaphroditic (her-MAFF-roh-dit-ik)
23. hookworm
24. host

25. immunogen
26. infestation (in-fess-TAY-shun)
27. intermediate host
28. larvae (LARR-vee)
29. life cycle
30. metacercaria (met-ah-sir-KAH-ree-ah)
31. microfilaria (my-kroh-fee-lah-REE-ah)
32. miracidium (mee-rah-SID-ee-um)
33. nematode (NEM-ah-toad)
34. obligate parasite
35. ovum (OH-vum)
36. parasite (PAIR-ah-site)
37. proglottid (pro-GLOT-id)
38. prognosis (prog-NO-sis)
39. reservoir
40. retroinfection
41. rhabditiform (RAB-dit-ee-form)
42. schistosome (SHIS-toh-som)
43. schistosomule (SHIS-toh-soh-mule)
44. scolex (SKOH-leks)
45. sporocyst (SPORE-oh-sist)
46. strobila
47. stylet
48. toxin (TOK-sin)
49. trematode (TREM-ah-toad)

III. PRETEST

Correct answers to all questions can be found at the end of this section. Write your responses in the appropriate space provided.

Completion Questions

Provide the correct term or phrase for the following. Spelling counts.

1. The study of worms is called ______________________.
2. Hermaphroditism among worms is found in a large number of (a) ______________________ and (b) ______________________.
3. Attachment devices of parasitic worms include (a) ______________________, (b) ______________________, (c) ______________________, and (d) ______________________.
4. The condition in which parasites attach to the skin or temporarily invade the skin's top layers is called ______________________.
5. A host that harbors the adult or sexually mature parasite is known as the ______________________ host.
6. The pressure of a large number of eosinophils is called ______________________.
7. The basic stages of parasitic nematodes involve (a) ______________________, (b) ______________________, and (c) ______________________.
8. The head of a tapeworm is called a(an) ______________________.
9. The body segment of a tapeworm is called a(an) ______________________.
10. A proglottid is the major site of ______________________ production.

Matching

Select the answer from the right-hand side that corresponds to the term or phrase on the left-hand side of the question sheet. An answer may be used more than once. In some cases, more than one answer may be required.

Topic: Disease Transmission

_______ 11. Elephantitis
_______ 12. Trichinosis
_______ 13. Pinworm
_______ 14. Gall formation
_______ 15. Hydatid disease
_______ 16. Fish tapeworm
_______ 17. Beef tapeworm
_______ 18. Hookworm
_______ 19. Schistosomiasis

a. consumption of raw or improperly cooked contaminated pork
b. arthropod bites
c. swimming in contaminated water
d. physical contact with contaminated soil
e. none of these

Topic: Specimens of Choice for Diagnosis

_______ 20. Trichinosis
_______ 21. Hookworm
_______ 22. Elephantitis
_______ 23. Fish tapeworm
_______ 24. Beef tapeworm
_______ 25. Lung fluke
_______ 26. Schistosomiasis

a. stool
b. blood
c. sputum
d. muscle biopsy
e. none of these

True or False

Indicate all true statements with a "T" and all false statements with an "F".

_______ 27. Several nematode species are known to be damaging to plants.
_______ 28. Many intestinal worms have suckers to enable them to resist digestion by their hosts.
_______ 29. Some worms are capable of either a parasitic or free-living existence.
_______ 30. Incidental parasites invade a host intermittently during their life cycles to obtain nutrients.
_______ 31. Helminths have a higher rate of direct multiplication within their hosts than do most bacterial pathogens.
_______ 32. Some helminths cause injury to their hosts through the production of toxins.
_______ 33. Facultative helminths are capable of either parasitic or a free-living existence.
_______ 34. Metabolic or excretory/secretory (ES) antigens appear to be highly effective immunogens.
_______ 35. A major source of *Anisakis* infection is improperly cooked pork or pork products.

Answers

1. helminthology
2a. cestodes (tapeworms)
2b. trematodes (flukes)
3a. hooks
3b. spines
3c. cutting plates
3d. suckers
4. infestation
5. definitive or final
6. eosinophilia
7a. adult worm
7b. egg
7c. larvae
8. scolex
9. proglottid
10. egg (ovum)
11. b
12. a
13. e
14. e
15. e
16. e
17. e
18. d
19. c
20. d
21. a
22. b
23. a
24. a
25. c
26. a
27. T
28. F
29. T
30. F
31. F
32. T
33. T
34. T
35. F

IV. CONCEPTS AND TERMINOLOGY

Most worms are free living and do not cause serious problems. However, some are parasitic and cause serious diseases in humans, domestic and wild animals, and plants. Parasitic worms live in varying environments and must adapt to conditions to survive a host's body defenses and chemical peculiarities. These multicellular forms differ from pathogenic microorganisms by having much longer generation times and the absence or lower rate of direct multiplication within hosts.

One of the major subdivisions of parasitology is the study of helminths. This term, derived from the Greek word *helmins,* meaning "worm," designates both parasitic and free-living species of worms.

The Equipment of Parasitic Worms

A parasitic existence is in large part determined by the development of certain structural and metabolic modifications. Many intestinal worms (and certain others) have an especially hard outer covering (integument), enabling them to resist digestion by the host. Other modifications include the possession of hooks, spines, cutting plates, suckers, various enzyme secretions, and additional weapons for purposes of attachment or penetration, and the development of elaborate reproductive systems. Cestodes (tapeworms) and a large number of trematodes (flukes) are hermaphroditic; that is, they possess the reproductive organs of both sexes.

Several helminths produce toxins that can poison their hosts. Examples include the release of toxins by larvae of *Wuchereria bancrofti* (the cause of elephantitis), which cause tissue swelling and thickening; the effects of tapeworm waste products causing *verminous intoxication;* and the dropping of the rectum (prolapse) in whipworm infection, caused by a toxin affecting the nervous control of intestinal muscles.

Life Cycles of Parasitic Helminths

Worms and other organisms that are capable of either a parasitic or a free-living existence are referred to as *facultative* parasites. The *obligate* parasites, by contrast, cannot complete their life cycles without an appropriate host or hosts.

Among the several different types of host are the *definitive* or *final* host, which harbors the adult or sexually mature parasite, and the *intermediate* host, which provides the environment for some or all of the immature or larval stages.

The life cycles of parasites sometimes involve several hosts. Apparently, the more complicated life cycle greatly diminishes the parasite's chances for survival. However, some parasites with complex life cycles have compensating adaptations, including parthenogenesis (reproduction without a male) and "overdeveloped" reproductive organs.

Various types of parasites also are recognized:

1. *Endo-:* animal species found internally
2. *Ecto-:* parasites that attach to the skin or temporarily invade the superficial layers of the skin
3. Temporary: parasites that invade a host intermittently during their life cycles only to obtain nutrients
4. Incidental: parasites that may establish an infection in a host that ordinarily is not parasitized
5. Spurious: foreign organisms that pass through the intestinal tract without causing an infection
6. *Pseudo-:* particles misdiagnosed as parasites

Economic and social conditions are also important in the distribution of parasites. A low standard of living, inadequate sanitation, and ignorance of the means to control parasitic diseases favor the establishment of helminthic infections.

Symptoms of Infection

The symptoms of helminth infections depend on the number and location of parasites within the host and the host's general health and resistance. Symptoms, if they occur, include anemia, excess numbers of eosinophils, fever, muscle pains, and respiratory difficulties.

Diagnosis and Immunological Control

Laboratory tests used most often for the identification and diagnosis of helminth infections involve microscopic examination of host specimens. Certain immunologic tests also can be used for demonstration of past infection or diagnosis.

Immunological means are under study to control helminth infections effectively. Three types of immunogens are involved: irradiated-attenuated preparations of worms, live helminths, and metabolic or excretory/secretory (ES) substances produced by *in vitro* helminth cultures.

Transmission

The sources of parasitic diseases and their means of transmission include: (1) domestic or wild animals in which parasites can live (domestic and sylvatic reservoirs, respectively); (2) bloodsucking insects (e.g., mosquitoes); (3) foods containing immature infective parasites; (4) contaminated soil or water; and (5) humans and any portion of their environment that has been contaminated. Even the individual harboring a parasite can cause his or her own reexposure with the same species of parasite, a process known as *autoinfection.*

The finding and identification of ova (eggs), larvae, or adult worms are sufficient in most cases for the diagnosis of specific helminth infections.

The Nematodes, Cestodes, and Trematodes

Several properties of three major groups of helminths that are of major economic and public health importance are described in Table 31-1.

Parasitic Nematodes of Plants

Roundworm plant parasites cause enormous damage to cultivated plants and are responsible for major agricultural losses. Both internal and external plant parts are subject to attack. Examples of the most common effects include tumors (galls), rotting of plant tissue, and interference with normal plant growth and development. Several species of soil fungi are natural enemies of nematodes.

TABLE 31-1 Nematodes, Cestodes, and Trematodes

Helminth Category	*General Properties of Adult Worms*	*Fundamental Stages in Life Cycle*	*Representative Infections and Means of Transmission*	*Preventative Measures*
Nematodes (roundworms)	Unsegmented, cylindrical and long, tapered at both ends, bilaterally symmetrical; contain digestive, nervous, and reproductive systems. Sexes are in separate worms. Attachment devices include lips, cutting plates, or teeth. They are capable of attacking practically every tissue of the body.	Egg → larvae → adult worm.	1. Aniskakiasis—ingestion of larval forms in raw, salted, or pickled fish 2. Filariasis—infected mosquitoes introduce motile larvae (microfilaria) through bites 3. Hookworm—larval forms penetrate through skin, most often the soles of feet 4. Pinworm—ingestion or inhalation of ova containing infective larvae 5. Trichinosis—ingestion of larvae containing raw or improperly cooked pork or pork products	Better sanitation and hygiene; elimination of mosquitoes and their breeding sites; ample cooking of foods that may contain infective larvae; education; treatment of infective individual
Cestodes (tapeworms)	Contain heads (scolices with suckers, or sucker and hooks) and body segments (proglottids); are hermaphroditic and have excessive reproductive capacity; lack digestive, circulatory, respiratory, and skeletal systems.	Egg → larvae → adult worm. Most life cycles involve at least two hosts.	Fish, beef, pork, and sheep tapeworm ingestion of foods containing infective larvae	Better sanitation; ample cooking of foods containing infective larvae; education; treatment of infective individuals
Trematodes	Flat, covered by an external cuticular layer; have suckers, and excretory, digestive, and reproductive systems. With one exception (the schistosomes or blood flukes), they they are hermaphroditic.	Eggs → larval stages (miracidium, sporocyst, redia, and cercariae) → adult. Certain flukes have an additional infective stage known as metacercariae. Schistosomes have only three larval stages: miracidium, sporocyst, and cercariae. Snails, water fleas, crayfish serve as intermediate hosts for flukes.	1. Clonorchiasis—ingestion of larvae-containing raw or improperly cooked fish 2. Paragonimiasis—ingestion of larvae-containing crayfish, crabs, and related seafoods 3. Schistosomiasis—penetration of skin by infective larvae in water environments	Interruption of parasite's life cycle; destruction of adult flukes and ova

V. CHAPTER SELF-TEST

Continue with this section only after you have read Chapters 17, 19, 21, 22, and 31. A score of 80 percent or better is good. If your score is less than 65 percent, reread the chapter.

Correct answers to all questions can be found at the end of this section. Write your responses in the appropriate space provided.

Completion Questions

Provide the correct term or phrase for the following. Spelling counts.

1. Hermaphroditism is a characteristic property of medically important worms such as the (a) ________ and most (b) ________.
2. Organisms capable of either a parasitic or free-living existence are referred to as being ________.
3. A parasite found inside a host can be called a(an) ________.
4. Foreign organisms that pass through the intestinal tract without causing an infection are called ________.
5. Five specific sources of parasitic diseases and their means of transmission are (a) ________, (b) ________, (c) ________, (d) ________, and (e) ________.
6. Most diagnoses of specific helminth infections are based on the finding of (a) ________, (b) ________, or (c) ________.
7. A simple method for the identification and diagnosis of pinworm is the ________.
8. The head of a tapeworm is called a(an) ________.
9. The invasive larval form of schistosomes is called a(an) ________.
10. The active larval stage of trematodes that is released from an ovum is a(an) ________.

Matching

Select the answer from the right-hand side that corresponds to the term or phrase on the left-hand side of the question sheet. An answer may be used more than once. In some cases, more than one answer may be required.

Topic: General Means of Disease Transmission

______ 11. Ingestion of food containing larvae	a.	*Loa loa*
______ 12. Contact with contaminated soil	b.	*Ancylostoma duodenale*
______ 13. Infected arthropod bites	c.	*Trichinella spiralis*
______ 14. Contaminated fingers	d.	*Wuchereria bancrofti*
______ 15. Swimming in contaminated water	e.	*Ascaris* species
	f.	*Schistosoma* species
	g.	*Taenia solium*
	h.	filariasis
	i.	*Enterobius vermicularis*
	j.	none of these

True or False

Indicate all true statements with a "T" and all false statements with an "F".

______ 16. In certain situations, an individual harboring a parasite can cause his or her own reexposure with the same parasite species.

______ 17. Symptoms can be absent in cases of parasitic helminthic infections.

______ 18. Most nematodes possess the reproductive organs of both sexes.

______ 19. Nematodes do not attack the human nervous or respiratory systems.

______ 20. Most roundworms are the same size.

______ 21. Filariform larvae represent a major stage in the development of cestodes.

______ 22. Skin tests are the major means used for the diagnosis of specific helminth infections.

______ 23. Practically every tissue of the human body is vulnerable to attack by certain nematode species.

______ 24. Nematodes have numerous appendages.

______ 25. Some nematode species are microscopic, while others are several centimeters in length.

______ 26. Tapeworms are hermaphroditic.

______ 27. Schistosomiasis results from the penetration of skin by oncospheres.

______ 28. The use of footwear is considered to be an effective control measure against endemic hookworm infection.

______ 29. All proglottids of a tapeworm are sexually mature.

______ 30. Fish tapeworm can cause a vitamin B_{12} deficiency.

______ 31. Helminths have shorter generation times in their hosts than do most bacterial pathogens.

______ 32. Toxins produced in whipworm infections cause swelling and thickening in the various body areas.

______ 33. Verminous intoxication is a characteristic feature in several tapeworm infections.

______ 34. Metabolic or excretory substances produced by *in vitro* helminth cultures have no protective value in controlling helminth infections.

______ 35. Anisakis infections are associated with contaminated fish and fish products.

Essay Questions

Answer the following questions on a separate sheet of paper.

36. How do helminths differ from other types of infectious disease agents?

37. What types of immunogens are under consideration for vaccines against helminths?

Answers

1a. cestodes
1b. trematodes
2. facultative
3. endoparasite
4. spurious
5a. domestic or wild animals
5b. bloodsucking arthropods
5c. foods containing immature infective parasites
5d. contaminated soil or water
5e. humans or any portion of their contaminated environment
6a. ova (eggs)
6b. larvae
6c. adult worms
7. Scotch tape swab procedure
8. scolex
9. cercaria
10. miracidium
11. *c* and *g*
12. b
13. a, d, h
14. i
15. f
16. T
17. T
18. F
19. F
20. F
21. F
22. F
23. T
24. F
25. T
26. T
27. F
28. T
29. F
30. T
31. F
32. F
33. T
34. F
35. T

36. Helminths are multicellular forms of life. They also differ from other infectious disease agents by having much longer generation times and by either not multiplying directly in their hosts or multiplying at low levels.

37. Three types of immunogens are under consideration as vaccine materials: irradiated-attenuated live helminths, extracts of helminth bodies, and metabolic or excretory/secretory (ES) substances produced by the *in vitro* culture of helminths. Of these immunizing materials, secretory antigens appear on the whole to be highly effective and induce protection against reinfection with cestodes and nematodes in lower animals.

VI. DISEASE CHALLENGE

The situation described has been taken from an actual case history. It has been designed to show how clinical, laboratory, microbial, and related information is used in disease diagnosis. A review of treatment and epidemiological aspects is given at the end of the presentation. Answers to questions, laboratory findings, and interpretations are given immediately following a specific question. Test your skills and take the *Disease Challenge.*

Case

Part 1: A 19-year-old male was admitted to a major medical center with a fever of 39.1°C and complaints of abdominal swelling and pain, constipation, severe headaches, myalgia (muscle pain), periorbital edema (puffiness around the eyes), and extreme fatigue. These symptoms began to develop approximately six months earlier.

On questioning the patient, it was learned that he was a sheep herder from Utah, who also raised sheep dogs. In addition, two of his prize dogs became quite ill recently, and were taken to the local veterinarian for examination and treatment.

At this point, what specimen(s) should be taken from the patient? Blood for standard laboratory tests such as red and white blood cell counts and differential blood cell determinations. Stool specimens also should be taken for microscopic examination and bacteriological cultures.

Laboratory Results

Laboratory blood test results showed a highly elevated eosinophil count. Both the microscopic analysis and cultures of the stool specimens were found to be negative for worm ova and pathogenic bacterial species, respectively.

Probable Diagnosis

A possible allergic condition and/or a worm infection.

Part 2: Considering the possibility that the patient's condition was related to his sick dogs, a telephone call was made to the attending veterinarian. Suspecting the possibility of a helminth infection in the dogs, the veterinarian obtained the appropriate stool specimens. Upon examination, forms such as the ones shown in Figure 31-1 were found on microscopic examination of the specimens.

Figure 31-1
Microscopic View of Stool Specimen Material

Probable Diagnosis in the Human Patient

Tapeworm. Actually, sheep tapeworm (*Echinococcus* species) was found to be the basis of the problems with the sheep dogs.

What additional test(s) should be taken with the sheep herder? X-ray examination and serological tests for *Echinococcus.*

Laboratory Results

The x-ray examination revealed three masses on the walls of the small intestine. In addition, the serological tests were positive for *Echinococcus.*

Part 3: Based on the x-ray findings, the masses were surgically removed. The patient was placed on a chemotherapy regimen of mebendazole and advised to return in six months for additional x-rays.

The masses were sent to the pathology laboratory for examination. The results of a microscopic examination revealed the presence of numerous immature scolices, thereby confirming the diagnosis.

VII. ENRICHMENT

Schistosomiasis Control

Schistosomiasis has been recognized since the time of the pharaohs; yet it was not until the 1920s that the life cycles of its causative agent was satisfactorily understood so that realistic approaches for control could be organized. In 1965, the World Health Organization Expert Committee estimated that between 180 and 200 million people in the world were infected. This estimate has been increased significantly in recent years. Increases in the infection rate are caused by several factors. The combination of high rates of population growth, increased numbers of water resource development projects, and widespread population movements have led to greater numbers of persons at risk and, in some instances, the introduction of infection to areas previously free of it. Unfortunately, many control projects have had a minimal impact on the global problem of schistosomiasis.

Five species of the parasite commonly infect humans: *Schistosoma haematobium, S. mansoni, S. japonicum, S. intercalatum,* and *S. mekongi.* The first occurs in Africa and the Middle East; the second in Latin America, the Caribbean, Africa, and the Middle East; the third in Asia; the fourth in Central Africa; and the fifth in Laos and Cambodia.

The life cycle of schistosomes is complex. Adult worms live in the definitive host, where the female lays up to 300 eggs per day (*S. japonicum,* up to 3,500). Eggs are passed in the urine (*S. haematobium*) or stool (*S. mansoni, S. japonicum, S. intercalatum*), and in water, they hatch to release a water-phase, free-living miracidium that penetrates an appropriate snail intermediate host. In the snail, the parasite reproduces asexually to develop into the next stage, a second free-living aquatic organism, released from the snail. From a single miracidium, as many as 100,000 cercariae may develop and be shed by the snail. The cercariae are capable of penetrating the unbroken skin of the human host and then developing into an adult worm. No multiplication occurs in the definitive host.

Infection with schistosomiasis may be initially mild; with high levels of exposure to infected water, disease symptoms are likely to develop over time. The disease caused by schistosomes is primarily related to the accumulation of eggs in the tissues and the body's reactions to them. With *S. haematobium,* there are often serious complications in the urinary tract; with *S. mansoni* and *S. japonicum,* complications involve the liver and spleen; and with *S. japonicum,* complications may also involve the central nervous system. Severe infections can be seriously debilitating and, in some cases, can lead to death, although this is not frequent. Infection with schistosomiasis is likely to be a contributing factor to reduced life expectancies in the developing countries.

Control of transmission is both an important and a challenging problem. Specific items that must be taken into account include a wide range of ecological and biological factors affecting the snail and parasite and socioeconomic and environmental factors that affect the behavioral patterns of the human population.

There are two main approaches to control: (1) reduce or prevent contamination of surface waters with schistosome eggs by extensive use of chemotherapy or by construction of adequate lavatory facilities (latrines) or (2) prevent infection of the human population through snail control or by reducing exposure to contaminated water by provision of adequate safe alternative water sources for domestic and recreational use. In the case of chemotherapy, there is the possibility of a two-pronged approach. Specifically, these are a short-term intensive attack phase with chemotherapy to rapidly reduce the existence and intensity of infection (egg-load) followed by a less costly concentration or maintenance phase to prevent a rapid reappearance of transmission. Chemotherapy should play an increasingly important role in future control programs.

PART VIII
MICROBIOLOGY TRIVIA PURSUIT
(Chapters 21 through 31)

This is your last opportunity to test your attention to detail, and your fact-gathering ability. Since Part VIII is the largest component of the *Study Guide,* there will be *six* challenging Parts for you to tackle. Remember that a certain number of questions must be answered in each Part before you proceed up the MICROBIOLOGY TRIVIA PURSUIT trail. The answers and directions for continuing are given at the end of each Part. Good luck!

PART 1

Completion Questions

Provide the correct term or phrase for the following. Spelling counts.

1. The enzymes produced by *Haemophilus influenzae* and streptococci known to decrease IgA molecules are called ______________________.
2. Complete the following formula used to experimentally establish an infectious disease in a particular host. ______________________

$$\text{Infectious Disease} = \frac{\text{Virulence x} \quad (?)}{\text{Host Resistance}}$$

3. Human T cell lymphotropic virus I (HTLV-I) causes ______________________.
4. A cancer developing in skin-pigment producing cells is called a(an) ______________________.
5. The detachment, movement, and subsequent relocation of a malignant cell is called __________.

Directions

Check your responses with the *Answers* section, and add up your score. Enter the number of correct answers in the space provided.

Total Correct Answers for Part 1: __________.

If your score was 4 or higher, proceed to Part 2.

Answers

1. IgA proteases; 2. dosage of pathogens; 3. adult T cell leukemia; 4. melanoma; 5. metastasis.

PART 2

Very good! Congratulations on the successful completion of Part 1. Now try your hand with Part 2.

Completion Questions

Provide the correct term or phrase for the following. Spelling counts.

6 and 7. The phylum of Platyhelminthes contains two major types of worms: (6) ______ and (7) ______ .

8. ______ is the only effective means of controlling tetanus.

9. Athlete's foot also is known as ______ .

10. The Schultz-Charlton test is an aid in the diagnosis of ______ .

Directions

Check your responses with the *Answers* section, and add up your score. Enter the number of correct answers in the space provided.

Total Correct Answers for Part 2: ______ .

If your score was 4 or higher, proceed to Part 3.

Answers

6. tapeworms; 7. flukes; 8. immunization; 9. tinea pedis; 10. scarlet fever.

PART 3

Excellent! You are doing very well. Here is a challenging Part 3.

Completion Questions

Provide the correct term or phrase for the following. Spelling counts.

11 and 12. The gummy accumulation that covers teeth and is called dental plaque consists of (11) ______ and (12) ______ .

13. One of the earliest colonizers of an infant's oral cavity is ______ .

14. ______ is the predominant organism found in acute cases of epiglottitis.

15. The embryo of a cestode is referred to as the ______ .

Directions

Check your responses with the *Answers* section, and add up your score. Enter the number of correct answers in the space provided.

Total Correct Answers for Part 3: ______ .

If your score was 4 or higher, proceed to Part 4.

Answers

11. salivary mucin; 12. bacteria; 13. *Streptococcus salivarius*;
14. *Haemophilus influenzae*, type b; 15. oncosphere.

PART 4

Great! You are doing very well. Here is a challenging Part 4.

Completion Questions

Provide the correct term or phrase for the following. Spelling counts.

16. The causative agent of pseudomembranous gastroenteritis is ______________________.

17. A halophilic, anaerobic vibrio known for its causing gastroenteritis is ______________________.

18, 19, and 20. Most patients with typhoid fever develop antibodies to three specific antigens of *Salmonella typhi*: (18) ________________, (19) ________________, and (20) ________________.

Directions

Check your responses with the *Answers* section, and add up your score. Enter the number of correct answers in the space provided.

Total Correct Answers for Part 4: ____________.

If your score was 4 or higher, proceed to Part 5.

Answers
16. *Clostridium difficile*; 17. *Vibrio parahaemolyticus*; 18. *O*; 19. *H*; 20. *Vi*.

PART 5

Very good! Congratulations on the successful completion of Parts 1 through 4. Now try your hand with Part 5.

Completion Questions

Provide the correct term or phrase for the following. Spelling counts.

21a and 21b. Valve damage in rheumatic heart disease is caused by (a) ____________ and (b) ____________.

22 and 23. In the IMViC set of reactions, the *I* represents ______________________ and the *C* represents ______________________.

24. Puerperal sepsis also is known as ______________________.

25. Clue cells are characteristic of an infection with the bacterium ______________________.

Directions

Check your responses with the *Answers* section, and add up your score. Enter the number of correct answers in the space provided.

Total Correct Answers for Part 5: ____________.

If your score was 4 or higher, proceed to Part 6.

Answers

21a. stenosis; 21b. valvular insufficiency; 22. indole; 23. citrate;
24. childbed fever; 25. *Gardnerella*.

PART 6

Great going! This is the last opportunity to test your attention to detail, and your fact-gathering ability. Good luck!

Completion Questions

Provide the correct term or phrase for the following. Spelling counts.

26. Gummas are characteristic of the sexually transmitted disease ______________________.
27. Most human listeriosis cases are acquired from the ______________________.
28. The ______________________ has been the major source of *Toxoplasma gondii* for humans.
29. The full name of one causative agent of African sleeping sickness is ______________________.
30. Brazilian purpuric fever is caused by the bacterium ______________________.

Directions

Check your responses with the *Answers* section, and add up your score. Enter the number of correct answers in the space provided. Then find your ranking for the entire *Microbiology Trivia Pursuit.*

Total Correct Answers for Part 6: ______________.

Total Correct Answers for Parts 1 through 6: ______________.

Answers

26. syphilis; 27. contaminated food; 28. cat;
29. *Trypanosoma brucei* variety *gambiense; 30. Haemophilus aegyptius.*

PERFORMANCE SCORE SCALE

Number Correct	*Ranking*
22	You should have done better
24	Better
26	Good
28	Excellent
30	Outstanding